AF696988

CELL CULTURE ENGINEERING VI

Cell Culture Engineering VI

Edited by

MICHAEL J. BETENBAUGH
The Johns Hopkins University, Baltimore, Maryland, USA

JEFFREY J. CHALMERS
The Ohio State University, Columbus, Ohio, USA

ROB ARATHOON
Genentech Inc., South San Francisco, California, USA

FRANK W. R. CHAPLEN
Oregon State University, Corvallis, Oregon, USA

and

ALISON J. MASTRANGELO
The Johns Hopkins University, Baltimore, Maryland, USA

Reprinted from Cytotechnology, volume 28, 1998.

Kluwer Academic Publishers

Dordrecht / Boston / London

A C.I.P. Catalogue record for this book is available from the Library of Congress.

ISBN 0-7923-5575-X

Published by Kluwer Academic Publishers
P.O. Box 17, 3300 AA Dordrecht, The Netherlands

Kluwer Academic Publishers incorporates
the publishing programmes of
D. Reidel, Martinus Nijhoff, Dr W. Junk and MTP Press.

Sold and distributed in North, Central and South America
by Kluwer Academic Publishers
101 Philip Drive, Norwell, MA 02061, U.S.A.

In all other countries, sold and distributed
by Kluwer Academic Publishers
P.O. Box 322, 3300 AH Dordrecht, The Netherlands

Printed on acid-free paper

All rights reserved
©1998 Kluwer Academic Publishers
No part of the material protected by this copyright notice may be reproduced or
utilized in any form or by any means, electronic or mechanical,
including photocopying, recording or by any information storage and
retrieval system, without written permission from the copyright owner.

Printed in the Netherlands

CONTENTS

Cytotechnology **28:** 1–8, 1998.
© 1998 *Kluwer Academic Publishers. Printed in the Netherlands.*

Cell therapy in kidney failure

H. David Humes*, Angela J. Funke & Deborah A. Buffington
Department of Internal Medicine, 3101 Taubman Center, University of Michigan Health System, 1500 East Medical Center Drive, Ann Arbor, MI 48109-0368, U.S.A.

Received 26 August 1998; accepted 26 August 1998

Key words: bioartificial organs, renal failure, stem cells, tubule cells

Abstract

Current therapy for acute renal failure continues to have an exceedingly high mortality rate, exceeding 50% even with dialytic or hemofiltrative support. Current renal replacement therapy in ARF only substitutes for filtration function of the kidney but not its cellular metabolic functions. Replacing these metabolic functions may optimize current therapy for this devastating disease process. In this regard, a renal tubule assist device (RAD) has been developed to be placed in an extracorporeal continuous hemoperfusion circuit in series with a hemofilter. The RAD consists of porcine renal proximal tubule cells grown as confluent monolayers in a multifiber bioreactor with a membrane surface area from 0.4 to 1.6 m^2. The cells along the inner surface of the hollow fibers are immunoprotected from the patient's blood by the hollow fiber membrane. *In vitro* experiments demonstrate that this device possesses differentiated renal transport, metabolic and endocrinologic properties. These properties, in fact, are responsive to normal physiological regulatory parameters. In preliminary experiments in uremic dogs, this device has also been shown to tolerate a uremic environment while providing reabsorptive, metabolic, and endocrinologic activity. Pilot human trials of the RAD are anticipated within the next year to improve current renal replacement therapy in this devastating disease process.

Introduction

The improved understanding of the cellular and molecular basis of organ function and disease has been translated during the last two decades into new diagnostic and therapeutic approaches to a wide range of disease processes including renal failure. A whole new biotechnology industry has evolved to translate this knowledge in basic biological sciences into more effective therapeutic and diagnostic modalities. Some notable successes have ensued. The most successful applications of biotechnology to date have been to apply recombinant genetic engineering to produce new pharmacological agents. This method consists of identifying a disease due to a lack of a single protein made by cells, isolating the gene for this protein, and finally using recombinant molecular biologic techniques to introduce the gene into an expression system to produce large amounts of the gene product. Both erythropoietin and human insulin have been produced in this manner. Although this approach has been proven successful, the opportunities to use such a strategy are limited, as many physiological responses are due to a complex interaction of a series of cell products rather than to a lack of one component.

In such situations, in which the desired effect is dependent on an array of cell products, a possible solution is the development of a more complex approach using 'cell therapy'. Cell therapy is based on the concept that specific cells can be cultured *in vitro* to perform differentiated biological tasks. The simplest manifestation of cell therapy is to implant living cells to produce a natural hormone or protein in short supply as a result of disease, usually with a regulatable element to a biological signal (Humes, 1997). Oxygen regulation of erythropoietin production and glucose control over insulin secretion are notable examples.

The use of cells as gene-product delivery vehicles is another application of cell therapy. The ability to

* Author for all correspondence.

grow and expand cells *in vitro* allows for efficient transfection or transduction of a targeted gene into a population of cells grown in tissue culture. The introduction of these transduced cells into a recipient allows for the production of a gene product that the patient lacks or that is present in an abnormal form. The development of immunocapsulation techniques will allow the use of heterologous cells (Tai, 1993). The genes for coagulation factors VIII and IX are two examples of the potential treatment of the genetic diseases of hemophilia A and B.

A final technology has been termed 'tissue engineering', a developing field in which techniques from the biological and engineering sciences are combined to create structures that mimic the functions of human organs (Langer, 1993). The majority of applications currently envisioned for tissue engineering involve placing modified animal or human cells within an artificial construct. Current efforts are focused on producing bioartificial organs using cells seeded on hollow-fiber bioreactors perfused in an extracorporeal circuit using modified dialysis equipment. Pilot human studies have already taken place, with an extracorporeal liver-assist device designed to replace hepatic function while waiting for the patient's own liver to regenerate after an acute insult (Sussman et al., 1994). Given the successes of renal replacement therapy in the last four decades, a natural application of tissue engineering is in the treatment of acute and chronic renal failure.

Acute renal failure

Acute renal failure (ARF) is a result of toxic or ischemic insults to the kidney. It is a common disorder affecting nearly 200,000 patients per year in the United States (Lake, 1994; Thadhani et al., 1996). It presents as a devastating clinical disorder with whole organ failure occurring within days of the initiating injurious event. The patients with this condition are gravely ill, requiring intensive care unit care.

Current therapy for ischemic or toxic acute renal failure, or acute tubular necrosis (ATN), is predominantly supportive in nature. The therapeutic goals are the maintenance of fluid and electrolyte balance, adequate nutrition, and, when present, treatment of infection and uremia. Uremia is treated with either intermittent hemodialysis or continuous hemofiltration. Although this approach has had substantial impact on this disease process over the past 40 years, patients with ATN still have an exceedingly high mortality rate of greater than 50%, even with dialytic or filtrative support. The precise reason for this mortality rate in the face of normal electrolyte and fluid balances, and a nonuremic condition, is unclear.

Perhaps an explanation for this high mortality rate resides in the recognition that hemodialysis or hemofiltration only substitutes for the filtration function of the kidney but does not replace the homeostatic, regulatory, metabolic, and endocrine functions of the kidney. Review of the causes of death in patients suffering from ATN demonstrates that the single factor most responsible for death was development of disseminated bacterial infection due to impairment of host defense (Lordon, 1972; Whelton, 1969). This impairment is likely the result of the loss of cellular metabolic function of the kidney rather than lost filtrative function. Accordingly, the development of cell therapy modalities replacing these reabsorptive, synthetic, metabolic, and endocrinologic functions of the kidney may add significant value to the current suboptimal supportive options available to treat established ARF. An approach to this form of therapy is the development of a bioartificial tubule to replace these functions and to optimize current treatment modalities.

Bioartificial renal tubule

Critical to providing organ function replacement through cell therapy is the need for the isolation and growth *in vitro* of specific cells from adult tissue. These cells are those that possess stem cell-like characteristics with a high capacity for self-renewal and the ability to differentiate under defined conditions into specialized cells to develop correct structure and functional components of a physiologic organ system (Hall, 1989; Potten, 1990). Recent data by our laboratory have demonstrated methodology to isolate and grow renal proximal tubule progenitor cells from adult mammalian kidneys (Humes et al., 1996; Humes, 1992). These studies were promoted by the clinical and experimental observations suggesting that renal proximal tubule progenitor cells must exist, as tubule cells have the ability to regenerate after severe nephrotoxic or ischemic injury to form a fully functional and differentiated epithelium.

In this regard, the adult mammalian kidney tubule epithelium exists in a relatively dormant, slowly replicative state, but has a large potential for regenerative morphogenesis following severe ischemic or toxic injury. Under selective serum-free growth conditions, which included epidermal growth factor and retinoic acid, a subpopulation of renal proximal tubule cells isolated from adult rabbit kidney were grown in cell culture. This report defines conditions in which to selectively grow from adult mammalian kidney a subpopulation of cells with an ability to differentiate morphogenically and with a high capacity for replication. Under the growth conditions of these experiments, these cells were able to both differentiate individually into a renal proximal tubule cell phenotype with cell polarity, apical microvilli, and tight junctional complexes between cells along the luminal border and to pattern form collectively into cylindrical arrays of cell monolayers surrounding a centralized lumen. These cells also were shown to have a high capacity for self-renewal. Genetic marking of the cells with a recombinant retrovirus and dilution analysis demonstrated that *in vitro* tubulogenesis often arose from clonal expansion of a single genetically tagged progenitor cell. These tubules were derived from tubule cells grown in primary culture and serial passages (at least 4 replication cycles) before suspension into collagen gels. Because many tubules in the collagen gel contained as many as 150 cells (at least 7 replication cycles), these findings demonstrate that *in vitro* tubulogenesis arose from clonal expansion of a single cell with the ability to undergo at least 11 replication cycles. These results suggest that a population of proximal tubule progenitor cells exist within the adult kidney in a relatively dormant, slowly replicative state but with a rapid potential to proliferate, differentiate, and pattern form to regenerate the lining proximal tubule epithelium of the kidney following severe ischemic or toxic injury commonly seen in clinical situations.

The bioartificial renal tubule is clearly feasible when conceived as a combination of living cells supported on polymeric substrata. A bioartificial tubule uses epithelial progenitor cells cultured on water and soluble-permeable hollow fiber membranes seeded with various biomatrix materials, such that expression of differentiated vectorial transport, metabolic, and endocrine function is attained. With appropriate membranes and biomatrices, immunoprotection of cultured progenitor cells can be achieved concurrently with long-term functional performance as long as conditions support tubule cell viability (Humes, 1997; Cieslinski, 1994). The technical feasibility of an implantable epithelial cell system derived from cells grown as confluent monolayers along the luminal surface of polymeric hollow fibers has been recently achieved (McKay, 1998).

As a first step towards developing a tissue engineered renal tubule assist device, Madin-Darby Canine Kidney (MDCK) cells, a permanent renal epithelial cell line, were seeded into the lumen of single hollow fibers (McKay, 1998). Functional confluence of the cells was demonstrated by the recovery of intraluminally perfused ^{14}C-inulin, which averaged greater than 98.9%, versus less than 7.4% with the control non-cell hollow fibers under identical pressure and flow conditions. The baseline absolute fluid transport rate averaged 1.4 ± 0.4 μL 30 min^{-1}. To test the dependency of fluid flux with oncotic and osmotic pressure differences across the bioartificial tubule, albumin was added to the extracapillary space followed by addition of ouabain, an inhibitor of $Na^{+}K^{+}$ATPase, the enzyme responsible for active transport across the renal epithelium. Addition of albumin resulted in a significant increase in volume transport to 4.5 ± 1.0 μL 30 min^{-1}. Addition of ouabain inhibited transport back to baseline levels of 2.1 ± 0.4 μL 30 min^{-1}. These results were the first demonstration that renal epithelial cells could be successfully grown as a confluent monolayer along a hollow fiber and developed functional active transport capabilities.

Bioartificial renal tubule assist device: In vitro performance

The next step in the development of a bioartificial renal tubule assist device (RAD) is to scale up from this single hollow fiber renal tubule to a multifiber bioreactor with renal proximal tubule cells that maintain not only transport properties, but also differentiated metabolic and endocrine functions. To accomplish this next step, a reliable tissue source for renal progenitor cells is required. Although successful renal tubule progenitor cell expansion has been achieved with human adult kidneys, a nonhuman animal tissue source for tubule cells has been strongly considered. Because an expensive screening process for infectious agents must be accomplished to ensure the safety of a human donor source of tissue along with the lack of consistent access and procurement of human tissue, an animal tissue source for renal tubule cells for RAD construction has been elected to be developed. The short-term use of this device for acute therapy in the intensive care unit (ICU) setting allows a nonhuman tissue source as a preferred strategy. For economic and safety concerns, pigs can be used as a tissue source for this extracorporeal short-term RAD. Because of its anatomic and physiological similarities with humans and the relative simplicity with which it can be bred in large numbers, the pig is currently considered the best source of organs for both human xenotransplantation and immunoisolated cell therapy devices (Cozzi, 1995; Cooper et al., 1991; Calne, 1970). Kidneys are taken from 4–6 week old Yorkshire breed pigs. A full clinical profile of each donor pig for pathogens and blood and tissue pathology is accomplished to ensure the safety and noninfectivity of donor tissue. From these kidneys, renal proximal tubule segments are isolated and renal tubule progenitor cells are expanded with techniques previously described (Humes et al., 1996; Humes, 1992).

Further experiments are now under way to scale up to a clinically applicable device with the use of commercially available high-flux hollow fiber cartridges. Preliminary experiments have tested transport and metabolic functions of these cells grown intraluminally within these cartridges with membrane surface areas of from 97 cm^2 to 1.6 m^2. Starting with a high flux hemofilter cartridge, the intraluminal surface of the hollow fibers were coated with laminin. Renal tubule cells were then seeded at a density of 10^5 cells mL^{-1} into the intracapillary space with four cell infusions separated by 30 min and a 90° rotation of the cartridge. The seeded cartridge was connected to the bioreactor perfusion system, in which the extracapillary space was filled with culture media and the intracapillary space perfused with similar media at a rate of 4–5 mL hr^{-1}. Culture media, both intracapillary and extracapillary, were changed every 2–3 days to maintain adequate metabolic substrates for growth. After 7 to 14 days of growth, the unit was studied. Preliminary *in vitro* experiments utilizing porcine renal proximal tubule progenitor cells have clearly shown differentiated transport and metabolic function of the RAD unit as summarized in Table I.

Of importance, the transport properties were inhibitable by specific inhibitors–ouabain for active sodium transport, phlorizin for active glucose transport, probenecid for para-aminohippuric acid (PAH) secretion, and acivicin for glutathione transport and metabolism. The metabolic processes of the RAD also demonstrated sensitivity to normal physiological variables: ammoniagenesis was pH sensitive and vitamin D_3 activation was PTH and phosphate sensitive. The absolute values of these various functions and responsivity to inhibitors and physiologic modulators are detailed in the attached data summary.

Ex vivo performance of the RAD

While assessing the functionality of a RAD unit *in vitro* is an important component of bioreactor design, the true test of functionality and utility comes in testing the device *in vivo*, or *ex vivo*. Before clinical trials can be undertaken with such a device, extensive testing must be done in large animals, where the system can be evaluated under physiologic conditions and can be optimized for functionality and ease of use. The renal assist device will be used as a component in an extracorporeal circuit, where care must be taken to ensure proper operating conditions. Conditions under operation should mimic, as best as possible, those that the cells in the device are accustomed to seeing physiologically or *in vitro*. Important parameters to monitor and adjust are, for example, flow rates, pressures, and temperature.

The bioartificial kidney set-up consists of a filtration device (a conventional hemofilter) followed in series by the tubule unit. Specifically, blood is pumped out of a large animal using a peristaltic pump. The blood then enters the fibers of a hemofilter, where ultrafiltrate is formed and flows to the RAD downstream to the hemofilter. Processed ultrafiltrate exiting the RAD is collected and can be discarded as 'urine.' The filtered blood exiting the hemofilter enters the RAD through the extracapillary space (ECS) port and disperses among the fibers of the device. Upon exit of the RAD, the processed blood is delivered back to the animal. This additional pump is required to maintain appropriate hydraulic pressures within the RAD. In this regard, the pressure of the blood and ultrafiltrate just before entry into the RAD are monitored. Heparin is delivered continuously into the blood prior to entering the RAD to diminish clotting within the device. The RAD is oriented horizontally and placed into a temperature controlled environment. The temperature of the cell compartment of the RAD must be maintained at 37 °C throughout its operation to ensure optimal functionality of the cells. Maintenance of a physiologic temperature is a critical factor in the functionality of the RAD.

The tubule unit is able to maintain viability, because metabolic substrates and low-molecular weight growth factors are delivered to the tubule cells from the ultrafiltration unit and the blood in the ECS. Fur-

Table 1. In vitro functional characteristics of the renal assist device (RAD)

Transport	**Metabolic**
Fluid reabsorption	Ammoniagenesis
Ouabain inhibition	pH responsive
Sucrose inhibition	
	Gluconeogenesis
Bicarbonate reabsorption	
Acetazolamide inhibition	Glutathione synthesis
Glucose reabsorption	
Phlorizin inhibition	**Endocrinologic**
PAH secretion	1,25-Dihydroxyvitamin D_3 activation
Probenecid inhibition	PTH enhancement
	Phosphate inhibition

thermore, immunoprotection of the cells grown within the hollow fiber is achieved due to the impenetrance of immunoglobulins and immunologically competent cells through the hollow fiber if the encapsulating membrane has a pore size which excludes compounds with a molecular weight greater than 150,000 Da. Rejection of these cells will, therefore, not occur. This arrangement thereby allows the filtrate to enter the internal compartments of the hollow fiber network, lined with confluent monolayers of renal tubule cells for regulated transport and metabolic function.

The use of such a device in uremic and non-uremic large animals has shown that inulin leak rates increased by a few percentage points from less than 5–10% to slightly greater than 10% immediately after use, but returned to pre-study values after being maintained in culture for two weeks, meaning cells remain viable and can continue growing. The use of scaled-up cartridges can increase metabolic production. A surface area expansion from 0.4 to 0.8 m^2 results in an increase in the number of cells from 2×10^9 to 4×10^9 cells. Of note, RAD cartridges have been maintained in culture in excess of two months after use in large animal studies.

With the development of this extracorporeal circuit and design considerations, preliminary studies have been completed. Dogs weighing approximately 25 kg were placed on a controlled, low protein diet five days prior to being made uremic by performing bilateral nephrectomies. A double lumen catheter was placed into the internal jugular vein, extending into the heart, and sutured into place on the skin to prevent dislodgment. After 24 hr of post-operative recovery, the dogs were treated either with hemofiltration and the RAD or with hemofiltration and a control cartridge containing no cells. Dogs were treated daily for either 7 or 9 hr for 3 days or for 24 hr continuously. Preliminary *ex vivo* experiments in these uremic dogs have demonstrated that the RAD performs differentiated transport, metabolic and endocrinologic function characteristic of the proximal tubule *in vivo*.

The successful completion of proven functional performance of the RAD in uremic dogs prepares for the progression of the testing of this device in human subjects. Upon Food and Drug Administration (FDA) approval for an IND, a phase I/II trial is planned for safety and toxicity studies followed by a full controlled phase III efficacy clinical trial. Before proceeding to human clinical trials, the FDA, the Centers for Disease Control and Prevention (CDC) and the National Institutes of Health (NIH) are currently developing new requirement before allowing human studies to proceed. The need for newer requirements and safeguards were mandated with the discovery that pigs, the favored donor animal, contain porcine endogenous retroviruses (PERV) which can, under selected conditions, infect human cells (LeTissier et al., 1997). The technology described in this report, however, has much less risk than xenotransplantation (animal-to-human solid organ transplantation) due to the short exposure time, the immunoisolation of the cells, and the use of the RAD in non-immunosuppressed patients. The ability to proceed with these clinical studies will be dependent upon meeting the soon-to-be-established safeguards for animal-to-human cell therapy.

Added value of the RAD to current therapy

Current therapy for ischemic or toxic ARF is predominantly supportive in nature. The therapeutic goals are the maintenance of fluid and electrolyte balance, adequate nutrition, and treatment of infection and uremia when they are present. Uremia is treated with either intermittent hemodialysis or continuous hemofiltration. Although this approach over the last 40 yr has had substantial impact on this disease process, patients with ATN still have an exceedingly high mortality of greater than 50%, even with dialytic or hemofiltrative support (Thadhani, 1996; Humes, 1995). Perhaps an explanation for this high mortality is that hemodialysis or hemofiltration only substitutes for the filtration function of the kidney and does not replace the homeostatic, regulatory, metabolic, and endocrine functions of the kidney. Review of the causes of death in patients suffering from ATN demonstrated that the single factor most responsible for death was development of infection. These infectious complications, with resulting sepsis and septic shock, may develop in ATN due to the loss of the nonfiltrative and metabolic functions of the kidney.

The kidney not only is important as an excretory organ but provides important resorptive, homeostatic, metabolic, and endocrinological functions. There are multiple renal cellular functions that the RAD may provide to a patient with ATN which may be important in preventing the septic complications of this disorder. Three key examples can be provided to demonstrate this added value.

The loss of both glutathione synthetic function and production of key free-radical scavenging enzymes in renal failure undoubtedly diminishes host defense function (Dröge et al., 1994; Avissar et al., 1994) and contributes to the pathogenesis of septic shock (Kinscherf et al., 1994; Zimerman, 1995). Increasing evidence demonstrates that the redox state within immunocompetent cells critically regulates immunologically relevant genes so that the balance between glutathione (GSH) and glutathione disulfide (GSSG) regulates these processes. This evidence clearly supports the concept that GSH is a key limiting factor that determines the magnitude of immunologic functions both *in vitro* and *in vivo*, thereby playing an important role in host defense. GSH depletion may well be responsible for the induction of immunologic nonresponsiveness under conditions of antigenic stimulation (Dröge et al., 1994). Even a small depletion of intracellular glutathione pool alters dramatically the process of myeloblast transformation, proliferation, and cytotoxic T-cell generation (Kinscherf et al., 1994).

A variety of studies have clearly demonstrated that excessive free radical generation contributes to the vascular and tissue damage in sepsis. The sepsis syndrome triggers a complex series of coagulation, complement and cytokine cascades to defend against bacterial invasion (Zimerman, 1995). Overwhelming sepsis, however, results in excessive host defense responses, including neutrophil activation with excessive release of free radicals (Goode, 1993). Since the action of free radicals is normally limited by antioxidant defense systems, in which GSH/GSSG axis is central, it is not surprising that GSH was protective to oxidant injury and improved the hemodynamic and metabolic derangements in both experimental animal and clinical studies of sepsis (Goode, 1993; Lash et al., 1986).

In this regard, the kidney plays an integral role in glutathione metabolism of the body (Curthoys, 1983). The initial step in glutathione turnover is its release from the cell where it is transported to the kidney where it is almost completely extracted and degraded to its constituent amino acids, predominantly in the proximal tubules. Within these epithelial cells, glutathione is resynthesized and returned to the systemic circulation. This process is so dynamic that renal glutathione turnover rate has a half-life of only 30 min. The kidney, and specifically proximal tubule cells, are also the major source of synthesis of antioxidant glutathione-related enzymes (Mohandas et al., 1984; Avissar et al., 1994). The key role of the kidney in this GSH regulatory pathway is exemplified by the fact that renal failure patients have severely low plasma levels of GSH and glutathione peroxidases (Avissar et al., 1994; Ross et al., 1997) and therefore are at increased risk for oxidative stress, especially in bacterial infection.

The loss of key endocrinologic functions in ARF may also diminish host defense to infection in patients with this disorder. The 1-hydroxylation of 25-OH vitamin D_3 in the proximal tubule cells converts vitamin D_3 into its most active metabolite. Both acute and chronic renal failure results in declines in the circulating level of this active metabolite and leads to vitamin D deficiency. A number of studies have clearly demonstrated that 1,25-dihydroxyvitamin D_3 plays an important role in the regulation of the immune system (Koren et al., 1992). High affinity receptors are found in peripheral blood lymphocytes and thymocytes and vitamin D deficiency impairs cell mediated immunity (Yang et al., 1993). Neutrophils from patients with vitamin D deficiency have abnormal motility and phagocytic ability (Bhalla, 1989). Administration of vitamin D_3 to patients on hemodialysis restores mitogen stimulated T cell responses to normal (Bikle, 1992). A critical role of cytosolic calcium in the oxidative burst of granulocytes has been acknowledged (Sullivan et al., 1989). The immune system has, therefore, been clearly recognized to be an important target tissue of this important hormone.

The kidney may play another role in host defense by participating in the complex and dynamic network of pro- and anti-inflammatory cytokines. The proximal tubule cells derive embryonically from mesodermal progenitors closely related to bone marrow precursor cells and have retained many elements of immunologically competent cells, including the ability for antigen presentation and production of a variety of immunologic active cytokines (Ong, 1994). Little has been explored into the role of the kidney in the sepsis syndrome, even though this disorder is a defining factor in the high mortality rate associated with ATN. Septic shock is an acute syndrome that is characterized by hypotension, coagulopathy, and eventual multiorgan failure primarily to ischemic tissue injury. This disorder is associated with dramatic elevations in inflammatory cytokines, including tumor necrosis factor (TNF)-α, interleukin (IL)-1 and IL-6 (Galley, 1996). This reactive and uncontrolled inflammatory response results in the adverse hemodynamic and metabolic disturbances in septic shock. More detailed understanding of the pathophysiology of this process, however, has suggested that the problem arises not from the expression of pro-inflammatory cytokines but the inadequate modulation by anti-inflammatory cytokines (Galley, 1996). One central anti-inflammatory mediator in this cascade is IL-10. IL-10 is a very proximal inhibitor of the cell-mediated (Th1-type) immune response, thereby promoting Th2-type humoral immune responses (Galley, 1996; Standiford, 1997). This compound has potent anti-inflammatory properties, including deactivation of neutrophils and macrophages, and diminishes the production of TNF-α, interferon-γ and members of both the C-X-C and C-C chemokine families. In fact, endotoxemia results in elevated plasma levels of IL-10 in response to endotoxin challenge in mice with lipopolysaccharide (LPS) but the source of this IL-10 in endotoxemia is not clear (Howard et al., 1993). The administration of IL-10 to mice challenged with LPS protects from lethal endotoxemia and neutralization of IL-10 increases lethality after LPS challenge, thereby demonstrating the important role of IL-10 in endotoxin shock (Standiford et al., 1995; Howard et al., 1993). The manner in which these insights relate to the kidney in ATN is suggested by the recent data from our laboratory that LPS exposure to proximal tubule cells (PTC) *in vitro* results in substantial production of IL-10 by PTC. This finding leads to the hypothesis that ATN, which is characterized by predominant PTC damage, may diminish the systemic response of increased IL-10 plasma levels in response to septic shock. This blunted IL-10 anti-inflammatory response could then lead to worsening hemodynamic and metabolic abnormalities in the sepsis syndrome, incremental multiorgan failure and mortality.

For clinical usefulness, these devices must be prefabricated with a defined and carefully screened porcine tissue source and held in storage in cell incubators. Upon request to treat a patient suffering from acute renal failure, which develops over a three to five day time course, the bioartificial device must be transported similar to a transplanted organ to the distant site for patient use. The logistics required to successfully achieve this effort are significant but may well be worth developing if the RAD has a significant positive outcome on the morbidity and mortality of acute renal failure. With the successful technological achievements with recent *in vitro* and *ex vivo* experiments with this device, initial clinical studies are planned in patients within the year.

As discussed, prior studies suggest that the kidney and its proximal tubule cells play a critical role in host defense and the immunologic processes responsible to combat bacteriologic and viral infections. Cell replacement therapy with an extracorporeal bioartificial renal tubule assist device (RAD), by providing critical metabolic, endocrinologic and cytokine factors during ARF, will hopefully result in a decrease in the mortality rate of this devastating disorder.

References

Avissar N, Ornt DB, Yagil J, Horowitz S, Watkins RH, Kerl EA, Takahashi K, Palmer IS and Cohen HJ (1994) Human kidney proximal tubules are the main source of plasma glutathione peroxidase. Am J Physiol 266: C367-C375.

Bhalla AK (1989) Hormones and the immune response. Annals of the Rheumatic Diseases 48: 1–6.

Bikle DD (1992) Clinical counterpoint: Vitamin D: New actions, new analogs, new therapeutic potential. Endocrine 13: 765–784.

Calne RY (1970) Organ transplantation between widely disparate species. Transplantn Proc 2: 550–553.

Cieslinski DA and Humes HD (1994) Tissue engineering of a bioartificial kidney. Biotech Bioeng 43: 678–681.

Cooper DKC, Ye Y, Rolf JLL and Zuhdi N (1991) The pig as potential organ donor for man. In: Cooper DKC, Kemp E, Reemtsma K and White DJG (eds.) Xeno-Transplantation Springer, Berlin, pp. 481–500.

Cozzi E and White D (1995) The generation of transgenic pigs as potential organ donors for humans. Nature Medicine 1: 965–966.

Curthoys NP (1983) Role of γ-glutamyltranspeptidase in the renal metabolism of glutathione. Mineral Electrolyte Metab 9: 236–245.

Dröge W, Schulze-Osthoff K, Mihm S, Galter D, Schenk H, Eck HP, Roth S and Gmunder H (1994) Functions of glutathione and glutathione disulfide in immunology and immunopathology. FASEB J 8: 1131–1138.

Galley HF and Webster NR (1996) The immuno-inflammatory cascade. Brit J of Anesthesia 77: 11–16.

Goode HF and Webster NR (1993) Free radicals and antioxidants in sepsis. Crit Care Med 21: 1770–1776.

Hall PA and Watt FM (1989) Stem cells: The generation and maintenance of cellular diversity. Development 106: 619–633.

Howard M, Muchamuel T, Andrade S and Menon S (1993) Interleukin 10 protects mice from lethal endotoxemia. J Exp Med 177: 1205–1208.

Humes HD and Cieslinski DA (1992) Interaction between growth factors and retinoic acid in the induction of kidney tubulogenesis. Exp Cell Res 201: 8–15.

Humes HD (1995) Acute renal failure: Prevailing challenges and prospects for the future. Kidney Int 48: S26–S32.

Humes HD, Krauss JC, Cieslinski DA and Funke AJ (1996) Tubulogenesis from isolated single cells of adult mammalian kidney: Clonal analysis with a recombinant retrovirus. Am J Physiol 271(40): F42–F49.

Humes HD (1997) Application of cell and gene therapies in the tissue engineering of renal replacement devices. In: Lanza RP, Langer R and Chick WL (eds.) Principles of Tissue Engineering, Academic Press, San Diego, pp. 577–589.

Kinscherf R, Fischbach T, Mihm S, Roth S, Hohenhaus-Sievert E, Weiss C, Edler L, Bartsch P and Droge W (1994) Effect of glutathione depletion and oral N-acetyl-cysteine treatment on $CD4^+$ and $CD8^+$ cells. FASEB J 8: 448–451.

Koren R, Ravid A and Liberman UA (1992) Peipheral blood mononuclear cells: A model for the human vitamin D endocrine system in health and disease. Mol and Cellular Endocrinology 83: C9–C12.

Lake EW and Humes HD (1994) Acute renal failure: Directed therapy to enhance renal tubular regeneration. Semin Nephrol 14: 83–97.

Langer R and Vacanti JP (1993) Tissue engineering. Science 260: 920–926.

Lash LH, Hagen TM and Jones DP (1986) Exogenous glutathione protects intestinal epithelial cells from oxidative injury. Proc Natl Acad Sci USA 83: 4641–4645.

LeTissier P, Stoye JP, Takeuchi Y, Patience C and Weiss RA (1997) Two sets of human-tropic pig retrovirus. Nature 389: 681–682.

Lordon RE, and Burton JR (1972) Post-traumatic renal failure in military personnel in Southeast Asia. Am J of Med 53: 137–147.

McKay SM, Funke AJ, Buffington DA and Humes HD (1998) Tissue engineering of a bioartificial renal tubule. ASAIO J 44: 179–183.

Mohandas J, Marshall JJ, Duggin GG, Horvath JS and Tiller DJ (1984) Differential distribution of glutathione and glutathione-related enzymes in rabbit kidney: possible implications in analgesic nephropathy. Biochem Pharm 33(11): 1801–1807.

Ong ACM and Fine LG (1994) Tubular-derived growth factors and cytokines in the pathogenesis of tubulointerstitial fibrosis: Implications for human renal disease progression. Am J Kidney Dis 23: 205–209.

Potten CS and Loeffler M (1990) Stem cells: Lessons for and from the crypt. Development 110: 1001–1020.

Ross EA, Koo LC and Moberly JB (1997) Low whole blood and erythrocyte levels of glutathione in hemodialysis and peritoneal dialysis patients. Am J Kidney Diseases 30: 489–494.

Standiford TJ, Strieter RM, Lukacs NW and Kunkel SL (1995) Neutralization of IL-10 increases lethality in endotoxemia. Cooperative effects of macrophage inflammatory protein-2 and tumor necrosis factor. J of Immunol 155: 2222–2229.

Standiford TJ and Huffnagle GB (1997) Cytokines in host defense against pneumonia. J Inv Med 45: 335–345.

Sullivan R, Fredette JP, Griffin JD, Leavitt JL, Simons ER and Melnick DA (1989) An elevation in the concentration of free cytosolic calcium is sufficient to activate the oxidative burst of granulocytes primed with recombinant human granulocyte-macrophage colony-stimulating factor. J Biol Chem 264: 6302–6309.

Sussman NL, Gislason GT, Conlin CA and Kelly JH (1994) The Hepatix extracorporeal liver assist device: Initial clinical experience. Artificial Organs 18: 390–396.

Tai IT and Sun AM (1993) Microencapsulation of recombinant cells: A new delivery system for gene therapy. FASEB J 7: 1061.

Thadhani R, Pascual M and Bonventre JV (1996) Acute renal failure. N Engl J of Med 334: 1448–1460.

Whelton A and Donadio Jr JV (1969) Post-traumatic acute renal failure in Vietnam. A comparison with the Korean War experience. Johns Hopkins Med J 124: 94–105.

Yang S, Smith C, Prahl JM, Luo X and DeLuca HF (1993) Vitamin D deficiency suppresses cell-mediated immunity *in vivo*. Archives of Biochem and Biophysics 303: 98–106.

Zimerman JJ (1995) Defining the role of oxyradicals in the pathogenesis of sepsis. Crit Care Med 23: 616–617.

Cytotechnology **28:** 9–17, 1998.
© 1998 *Kluwer Academic Publishers. Printed in the Netherlands.*

Improved bicistronic mammalian expression vectors using expression augmenting sequence element (EASE)

Teri L. Aldrich, James N. Thomas & Arvia E. Morris*
Immunex Corporation, 51 University Street, Seattle, Washington 98101, U.S.A.
E-mail: morrisa@Immunex.com

Received 25 August 1998; accepted 25 August 1998

Key words: bicistronic, dihydrofolate reductase, expression vectors, gene amplification, recombinant protein

Abstract

We have recently reported on the isolation of a 5.7 kb segment of Chinese hamster ovary cell genomic DNA, Expression Augmenting Sequence Element (EASE), which when used in bicistronic expression vectors allows the development of stable Chinese hamster ovary cell pools in a five to seven week time period that express high levels of recombinant protein (6–25 μg 10^{-6} cells/day depending on the protein). In the present study, we have mapped the activity of the EASE to a 2.1 kb region using colony forming assays and developed bicistronic expression vectors with the smaller EASE or control lambda DNA. The recovery of pools expressing the hematopoietic growth factor, FLT3 Ligand, in methotrexate-containing media took 1 to 4 weeks less when using EASE expression vectors compared with control vectors. The cell pools developed with the EASE and control vectors had similar final protein expression levels. Southern blot analysis suggested the expression cassette from the EASE containing vectors integrated in tandem arrays arranged in either head to head or head to tail fashion. By contrast, control vectors appeared to integrate with multiple interruptions to the expression vector. Thus, the EASE, within a bicistronic expression vector, appeared to facilitate tandem vector integration and reduce the time required to develop cell pools for protein expression.

Introduction

Bicistronic expression vectors are effective tools for expressing recombinant protein in host cell lines (Kaufman et al., 1987). Expression of recombinant protein from these vectors is made more stable with the use of a viral internal ribosomal entry site sequence (Kaufman et al., 1991). Typically, a bicistronic expression vector encodes an amplifiable selectable marker on the same transcript as a gene of interest. When these vectors are transferred into host cells and the cells are grown step wise under increasing selective pressure, the level of expression of the recombinant protein usually follows the expression of the selectable marker. To generate stable cell lines, cloning steps can then be performed.

The Expression Augmenting Sequence Element (EASE) improves expression of recombinant proteins from bicistronic vectors in Chinese hamster ovary (CHO) cells by decreasing the number of selective steps needed to develop pools that express high levels of recombinant protein (Morris et al., 1997). In the present study, we have further characterized EASE activity by mapping it to a 2.1 kb region of CHO genomic DNA. Expression vectors using the minimal EASE were compared with control vectors. Stable pools expressing recombinant protein could be developed faster using the EASE vectors, and recombinant protein expression was comparable. In addition, we found that EASE may act by improving integration and amplification of plasmid DNA.

* Author for all correspondence.

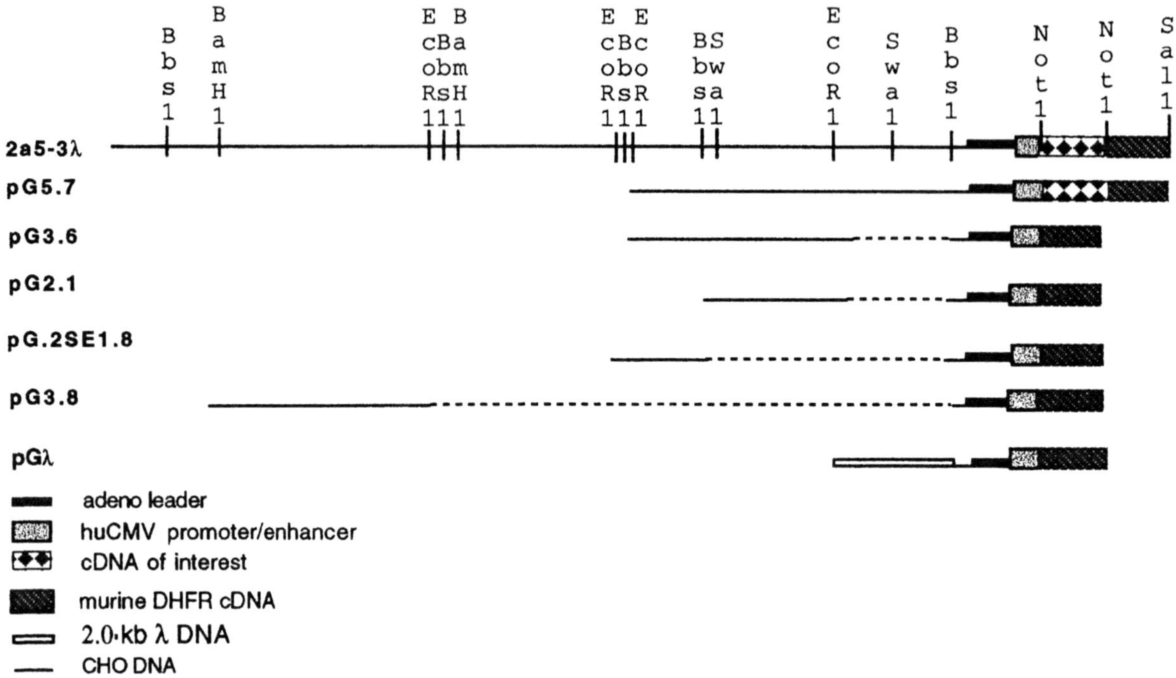

Figure 1. Constructs used for mapping study. Each vector contains the huCMV promoter/enhancer driving the expression of either a bicistronic mRNA encoding a gene of interest and murine DHFR or muDHFR alone. 5′ to the huCMV promoter/enhancer various lengths of CHO DNA containing EASE activity or control DNA were inserted as indicated. Elements used in each construct are indicated in key at the bottom left. Dashed line represents deleted regions of CHO DNA. Restriction enzyme sites are indicated.

Materials and methods

Construction of plasmids

The manufacturers' instructions were followed for all enzymes used in plasmid constructions. Enzymes were purchased from New England Biolabs, Beverly, MA, Boehringer-Mannheim Biochemicals, Indianapolis, IN or Stratagene, La Jolla, CA. Plasmid pG.2 was constructed from plasmid pG5.7 (Figure 1 and Morris et al., 1997) as follows. The TNFrFc cassette was removed by digestion with *Not*I and ligation. The resulting plasmid was cut with *EcoR*I and *Bbs*I, the ends filled with Klenow and ligated. To facilitate cloning of some of the constructs used for mapping studies, the *Bbs*I site of pG.2 was replaced with a *Bam*HI site. Plasmid pG.2 was digested with *Bbs*I, the ends filled using *Pfu* polymerase and then a *BamH*I linker (New England Biolabs) added. This plasmid was named pG.2B. The 3.6 kb *EcoR*I fragment and the 1.8 kb *EcoR*I-*Swa*I fragment from pG5.7 were each cloned into PG.2 to make plasmids pG3.6 and pG.2SE1.8, respectively. In both cases, the ends of the fragments were filled and ligated to *Bbs*I cut and filled pG.2. The 2.1 kb region of pG5.7 used to construct pG2.1 was amplified by PCR using the following oligonucleotides: 5′-GCGCGCGGATCCTACATTGTAATTCCTAAATTCTA-3′ and 5′-AAAGGATCCGTTCGTGGTGGACATCAC-3′. The PCR product was purified using the Wizard DNA Clean-Up System (Promega, Madison, WI) It was digested with *BamH*I and then cloned into the *BamH*I site of phagemid pBluescript II SK +/- (Stratagene.) Following dideoxy sequence analysis, the 2.1 kb *BamH*I fragment was removed from this intermediate construct and cloned into the *BamH*I site of pG.2B to create pG2.1. The 3.8 kb *BamH*I fragment from plasmid 2a5-3λ (Morris et al., 1997) was cloned into the *BamH*I site of pG.2B to make pG3.8. Plasmid pGλ was constructed by cloning the 2.0 kb *BamH*I-*Bcl*I fragment from bacteriophage lambda into the *BamH*I site of pG.2B. Plasmids used for protein expression studies were derivatives of pG2.1 and pGλ. A cDNA for FLT3L (Brasel et al., 1996), the encephalomyocarditis virus internal ribosomal entry site (nucleotides 260–824) (Jang and Wimmer, 1990) and the dihydrofolate reductase gene were cloned into the *Not*I and *Sal*I sites of pG2.1 and pGλ to create pG2.1FLT3L and pGλFLT3L, respectively (Figure 2).

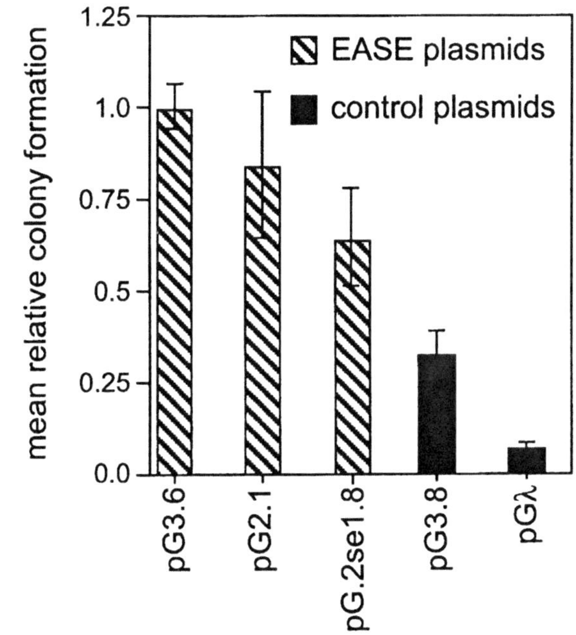

Figure 2. Mapping of EASE activity to a 2.1 kb region of CHO genomic DNA. Each data point represents results from 5 or 6 independent transfections for each plasmid. Colony forming assays were performed in 10 nM MTX. Colony formation is shown relative to results for pG3.6. Error bars represent standard error of the mean.

DXB-11 cell culture, transfection and colony forming assay

DXB-11 CHO cells were maintained in nonselective GHT medium (DMEM:F12 supplemented with 7.5% bovine calf serum [Hyclone, Logan, UT], 2 mM L-glutamine, 90 μM thymidine [T], 90 μM hypoxanthine [H] and 120 μM glycine [G]) in a 37 °C humidified incubator containing 5% CO_2. Transfections by electroporation were performed as follows. For each electroporation, 10 μg of plasmid DNA was linearized using an appropriate single cutting enzyme. Restriction reactions were dialyzed and the DNA quantitated again prior to transfection. Mid-log phase DXB-11 cells were harvested, washed once with cold electroporation buffer (20 mM HEPES, pH 7.4, 137 mM NaCl, 5 mM KCl, 0.7 mM Na_2HPO_4, 6 mM dextrose) and then resuspended in electroporation buffer at 10^7 cells per mL. Linearized DNA (5–10 μg) was added to 4×10^6 cells. After 5 min on ice, the cells were electroshocked using a BTX Electro Cell Manipulator 600 (San Diego, CA) at 500 V, 400 μF, 13 ohms and charging voltage of 174. The cells were returned to ice for a 10 min incubation before plating.

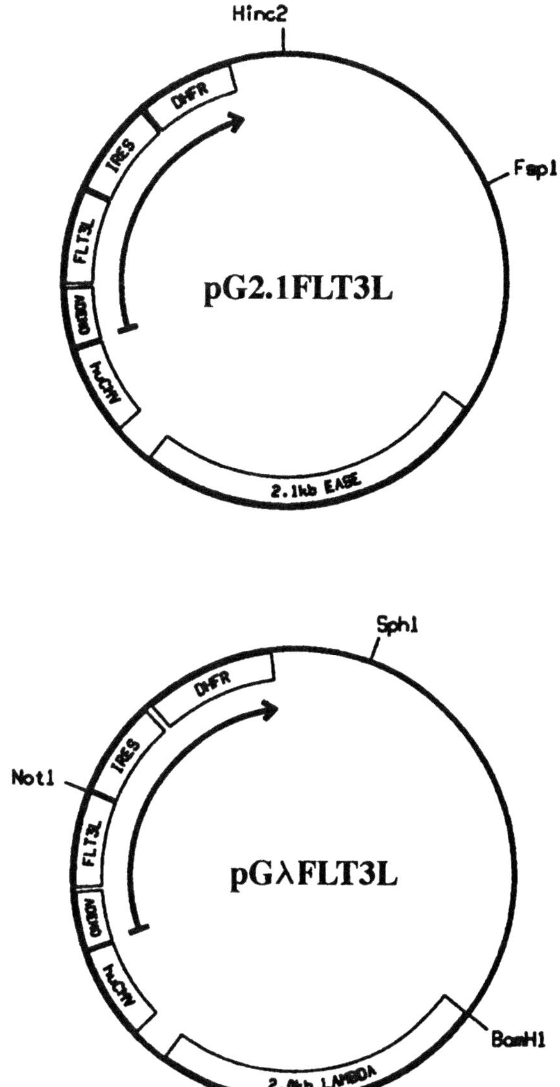

Figure 3. Plasmids used to test the activity of EASE in stable pools. Adeno = Adenovirus tripartite leader sequence, FLT3L = cDNA coding for FLT3L, IRES= internal ribosomal entry site, DHFR = cDNA coding for murine DHFR. Arrow indicates the direction of transcription and structure of the bicistronic mRNAs. Restriction sites are indicated.

For colony forming assays, transfected cells were plated in a 10 cm dish containing 10 mL of nonselective medium. Three days after transfection, 3×10^5 cells were plated per 10 cm dish in -GHT medium (DMEM:F12 supplemented with 7.5% dialyzed fetal bovine serum [Sigma, St. Louis, MO], 6 mM L-glutamine, 1 mM asparagine) containing 10 nM methotrexate (MTX) (Lederle Laboratories, Pearl River, NY). Nine to eleven days later, the cells

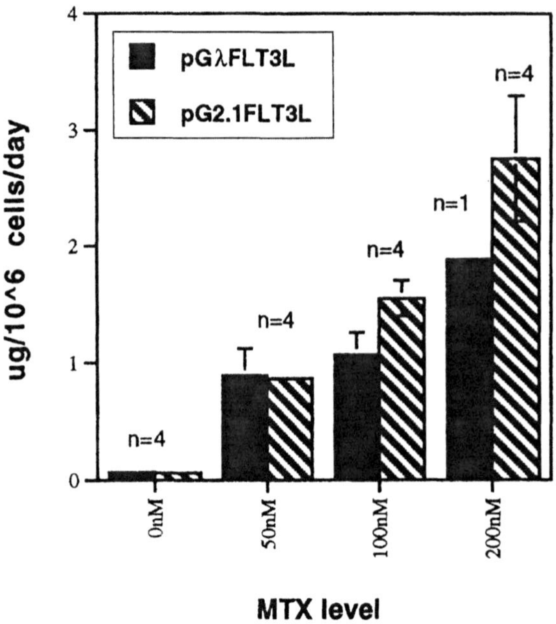

Figure 4. Recombinant protein production from pools made with EASE and non-EASE vectors. Black bars represent specific productivity of pools generated using the non-EASE containing vector pG λFLT3L. Diagonal striped bars represent specific productivity of cell pools generated with the EASE containing vector, pG2.1FLT3L. FLT3L was measured using ELISA assays. Only one pool using pG λFLT3L grew at 200 nm MTX compared with four using pG2.1FLT3L. Error bars represent standard error of the mean.

were fixed (2% formaldehyde [Sigma], 0.2% glutaraldehyde [Sigma] in phosphate buffered saline) and then stained with 2% methylene blue. Colonies were counted, and the counts were normalized to the amount of DHFR encoding sequence used for each electroporation.

Amplification, specific productivity and plasmid stability

For experiments involving recombinant protein expression and amplification of DHFR, cells were recovered from electroporation for 3 days in nonselective medium, and then transfected cells were seeded at 1×10^6 cells per T75 flask in -GHT medium. They were allowed to grow to confluency and then seeded again at 1×10^6cells per T75 flask in -GHT medium. After the second round of selection, they were seeded at 1×10^6 cells per T75 flask in -GHT medium supplemented with either 50 nM, 100 nM or 200 nM MTX and grown to confluency. They were grown a second time in 50, 100 or 200 nM MTX before specific production of recombinant protein was measured.

For production cultures and the stability study the amount of recombinant protein expressed was quantitated by determining the average microgram of protein produced per 10^6 cells per day (specific productivity). Protein titers were measured by ELISA (Engvall and Perlmann, 1972). Plasmid stability was investigated by comparing specific productivity of a recombinant protein expressed by cell pools cultured in -GHT media supplemented with 100 nM MTX. Specific productivity was measured every five passages. For the production cultures, cells were seeded at 2×10^6 cells mL^{-1} in T125 shake flasks. Cultures were then induced and protein production, cell viability and glucose consumption was monitored for 10 days at which time the cultures were terminated. All cultures were handled identically.

Southern hybridizations

High molecular weight genomic DNA was extracted using previously described methods (Mitchell et al., 1986). The DNA was quantitated with a fluorometer (Hoefer, San Francisco, CA), digested with appropriate enzymes and then fractionated on 1% tris-acetate buffered agarose gels as described (Sambrook et al., 1989). DNA fragments were separated using a continuous electrophoresis apparatus (Owl Scientific, Woburn, MA). The DNA was transferred to Zeta-Probe GT Genomic Tested Blotting Membrane (Bio-Rad, Hercules, CA) as recommended by the supplier. Prehybridization and hybridization to P^{32} (Amersham, Arlington Heights, IL) *flt3l* probes prepared using a Random Primed DNA Labeling Kit (Boehringer-Mannheim Biochemicals) were performed at 60 °C. The filters were exposed to Kodak XAR5 film for an appropriate time. The screens were scanned using a Personal Densitometer SI (Molecular Dynamics, Sunnyvale, CA) and analysis and quantitation were accomplished using ImageQuant software (Molecular Dynamics).

Statistical analysis

Analysis of variance of group means and linear regression was performed using the JMP Statistical Discovery Software (Cary, NC)

Results and discussion

Mapping of EASE activity with colony forming assays

In a previous study, it was shown that 2a5-3λ and a derivative of 2a5-3λ, pG5.7, (Figure 1 and Morris et al., 1997) could be used to develop stable cell pools expressing recombinant protein in a five- to seven-week time period. In the present study, a colony forming assay was used to map the activity in the 5.7 kb region of CHO DNA to a minimal element able to facilitate the production of CHO pools. A series of constructs were made containing various lengths of CHO DNA from 2a5λ or control λ DNA placed 5′ of the CMV promoter/enhancer of a plasmid expressing the amplifiable dominant selectable marker DHFR (Figure 1). Plasmid pG3.8 also served as a negative control plasmid as it contains a 3.8 kb segment of CHO DNA from 2a5λ which previous studies had indicated was not necessary for EASE activity (Morris et al., 1997). In preliminary experiments, pG3.6 was shown to have the same activity as pG5.7 in the colony forming assays; thus, for the present study pG3.6 was used as the positive control. The transfected cells were plated in 10 nM MTX, scored for colony formation, and colony formation relative to pG3.6 was calculated. Mean relative colony formation for each vector was analyzed using analysis of variance (ANOVA) and differences in means were found to be highly significant ($p<.0001$). pG2.1 had an equivalent amount of colony forming activity compared to pG3.6 (Figure 2), and both had significantly more colony forming activity compared with the negative controls pGλ and pG3.8. Plasmid pG.2SE1.8 had more colony forming activity than the negative controls but less colony forming activity than pG3.6 and pG2.1 (Figure 2). From this study it appeared that pG2.1 contained most if not all of the EASE activity in pG3.6, so it was used for additional experiments.

Enhanced colony formation could reflect improved expression due to several mechanisms: 1) increased number of cells transfected (higher transfection efficiency), 2) higher copy number of transfected DNA at each integration site per cell or 3) increased transcription at each integration site. The experiments described below suggest that the EASE element facilitated colony formation by increasing gene copy number at a faster rate than control plasmids.

Expression of recombinant protein using EASE and non-EASE containing bicistronic vectors

To determine if the 2.1 kb EASE identified in the colony forming assays could be used in expression vectors to facilitate recombinant protein production, this element and the 2.0λ control DNA were used in bicistronic vectors to express FLT3L, a soluble hematopoietic growth factor. A cDNA for FLT3L (Brasel et al., 1996) and the encephalomyocarditis virus internal ribosomal entry site (nucleotides 260-824) (Jang and Wimmer, 1990) were cloned into pG2.1 and pGλ to create pG2.1FLT3L and pGλFLT3L (Figure 3), respectively. The plasmids were linearized using *Fsp*I for pG2.1FLT3L and *Bam*HI for pGλFLT3L. Four independent electroporations into DXB11 CHO cells were performed for each construct. Electroporated cells were grown to confluency in DHFR selective media and subsequently grown in 50, 100 or 200 nM MTX. Cells electroporated with pG2.1FLT3L grew in 100 nM MTX selective media significantly faster than cells electroporated with pGλFLT3L (t-test with $p<0.05$) (Table I). Only one pGλFLT3L pool grew in the 200 nM MTX compared with four for pG2.1FLT3L, and it took about 5 weeks longer to reach harvest (Table I). There was no difference in the growth rate of cell pools in the 50 nM MTX supplemented media. The fact that the cell pools containing EASE vectors grew faster under higher selective pressure supports the hypothesis that the pG2.1FLT3L vector was able to amplify more rapidly compared with the control plasmid.

To determine if the production of recombinant protein in pools generated with the two plasmids was equivalent for each set of pools, specific production of FLT3L was measured for pools grown in different levels of MTX. At higher selective pressure, the pools from both vectors expressed increased amounts of protein indicating that both plasmids were able to increase expression due to MTX selection (Figure 4). Though the EASE containing pools had a higher average FLT3L expression at 100 and 200 nM MTX levels, upon statistical analysis there was no significant difference in the specific productivity between the pools generated with the different vectors at the same level of MTX.

The highest producing pools from selections done at 100 nM MTX using either pG2.1FLT3L (pG2.1FLT3L F5) or pGλFLT3L (pGλFLT3L B3) were tested for production of recombinant protein in a small scale shake flask production format. After cell

Table 1. Selection of cell pools in various levels of methotrexate requires less time with EASE vector

Vector	Days in		Total cells at confluency	
	100 nM MTX	200 nM MTX	100 nM MTX	200 nM MTX
pG2.1FLT3L	43 n=4	55±11 n=4	$1.49 \times 10^6 \pm 0.75$	$0.70 \times 10^6 \pm 0.42$
pGλFLT3L	54±7 n=4	93 n=1	$1.15 \times 10^6 \pm 0.74$	0.24×10^6

Only one pool using pGλFLT3L grew at 200 nm MTX compared with four using pG2.1FLT3L.

pools were adapted for growth in suspension for several passages, shake flasks were induced and cultured for 10 days. In a representative experiment, the EASE derived pG2.1FLT3L F5 pool maintained a higher specific productivity and titer throughout the course of the culture than the pGλFLT3LB3 pool (Figure 5). These data indicate that both vectors were able to generate pools that could be used to produce milligram amounts of soluble recombinant protein, but the pool generated with the EASE containing vector, pG2.1FLT3L F5, was available for production more quickly than the non-ease containing pool.

Prior to the production culture described above, the stability of recombinant protein expression was investigated for these two pools (Figure 6). Linear regression was used to determine if the protein production from the two pools was changing over the course of the study. The r^2 for the pG2.1FLT3L F5 and pGλFLT3LB3 were 0.19 and 0.45 respectively. The low r^2 values for both cell pools indicates that no statistically significant decrease in protein expression was detected. The higher r^2 value for pGλFLT3LB3 suggests there maybe a downward trend in protein expression for this pool. From a practical point of view, it is important to note that the production cultures (Figure 5) were performed approximately 10 passages subsequent to the stability study. This indicates that the stability of the pools was sufficient for generating milligram amounts of material under production conditions for an extended period.

Southern blot analysis of integration of plasmids in pools made with EASE and non-Ease vectors

To investigate the mechanism of action of the EASE containing vector at the molecular level, the integration pattern of the expression vectors for both sets of pools generated with 0 and 100 nM MTX was compared using Southern blot analysis. High molecular weight genomic DNA from pools derived from pG2.1FLT3L transfected cells was digested with *Hinc*II, an enzyme that linearized the plasmid (Figure 3). The pGλFLT3L derived pool DNA was digested with *Not*I and *Sph*I, enzymes that cut the plasmid once each to release a 6 kb fragment containing the FLT3L cDNA, the λ control DNA insert and the pGem1 plasmid backbone (Figure 3). A probe encoding FLT3L was used to detect integrated plasmid DNA. Pools created with pG2.1FLT3L showed two major bands at about 8 kb and >10 kb in pools selected at 0 and 100 nM MTX (Figure 7). The 8 kb band detected in the pG2.1FLT3L pools was the expected size for the unit length plasmid, suggesting that some of the EASE plasmid integrated and amplified in head to tail tandem arrays. The larger band could represent a head to head integration event which would produce a 13 kb band when cut with *Hinc*II. The relative intensity of this higher molecular weight band was increased relative to the 8 kb band in the 100 nM pools, indicating that this was the preferred amplification target. The pools generated with the EASE containing vector did show some bands whose sizes could not be explained by either head to head or head to tail integration events particularly after amplification, thus indicating the presence of some novel junctions. In contrast to the relatively simple patterns generated by pG2.1FLT3L transfected cells, the pools created with the pGλFLT3L plasmid showed complex banding patterns in pools at both MTX levels. Head to tail and head to head integration events would generate 6–7 kb bands which were not consistently detected at either MTX level. The amount of amplification for each set of pools was highly variable and no significant difference in the amount of amplification between pools was detected.

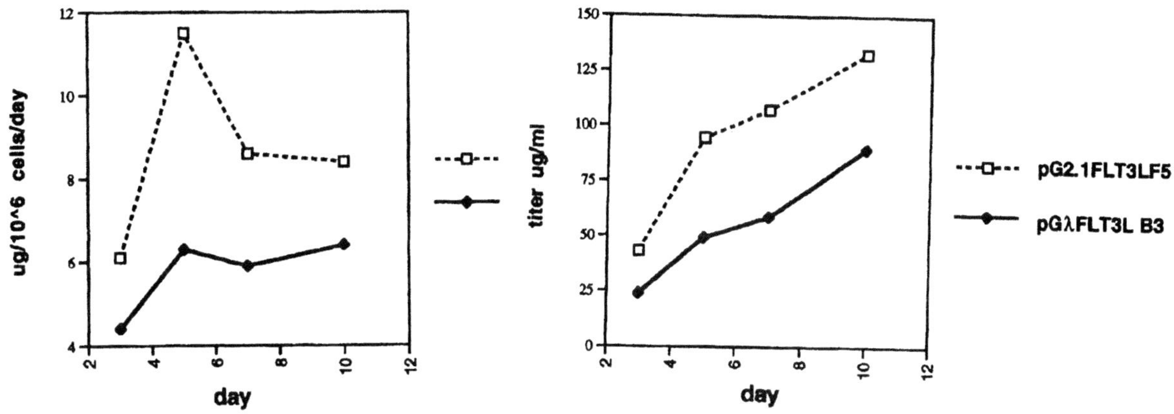

Figure 5. Production of FLT3L from pool pG2.1FLT3L F5 or pG λFLT3LB3 was compared. Each cell pool was suspension adapted and used in a small scale shake flask production experiment. FLT3L production was measured by ELISA. Each point represents the mean of duplicate samples.

Conclusions

In the current study it has been shown that a truncated 2.1 kb EASE element can be used in bicistronic vectors to create CHO cell pools expressing recombinant protein faster than a control vector which contained a 2.0 kb segment of λ DNA. The mechanism of action of the EASE element seems to be at the level of DNA integration and amplification. Southern blot analysis of DNA from cell pools generated with the EASE vector selected at low selective pressure (no MTX) had similar banding patterns. By contrast the pools selected with the control vector showed complex banding patterns with many interruptions to the expression plasmid. The amount of plasmid DNA initially integrated in the cell pools and amount of plasmid sequence in the DNA from the amplified pools were highly variable with no statistical difference between the sets of pools. In addition, there was no statistical difference in the protein expression level between pools. These data taken together indicate that the EASE element is causing an increased rate of amplification which would account for the shorter time line to get to an equivalent level of DNA copies and protein expression as the controls. By contrast, if the EASE were acting to enhance transcription, we would expect to see less integrated plasmid DNA in the EASE containing pools compared with the controls.

Other groups have described DNA sequences which promote gene amplification (Beitel et al., 1991; Holst et al., 1988; McArthur and Stanners, 1991), and their use in expression vectors (Hemann et al., 1994; Meyer et al., 1993; Weidle et al., 1988). The HSAG family of middle repetitive DNA sequences (Beitel et al., 1991; McArthur and Stanners, 1991) is similar to EASE in that they increased colony formation in colony forming assays. The EASE sequence described here is distinct from the HSAG sequences since it does not correspond to a previously described middle repetitive sequence family and it is only able to increase colony formation 3 to 5 fold compared to the HSAG sequences which were able to increase colony formation by two orders of magnitude. In addition the HSAG DNA did not seem to change the integration pattern of plasmid DNA compared with controls as was the case for EASE. This effect could be due to the methods used for introducing DNA into the cells; for the HSAG studies calcium phosphate precipitation was used, and electroporation was used for the EASE studies reported here.

Weidle et al. (1988) have successfully used a sequence from murine rRNA, muARS4 in expression constructs containing a crippled promoter driving the selectable marker. These authors report that stable expression in individual clones was dependent on culturing cell lines in the presence of selective pressure and that the expression plasmid DNA appears to be maintained in the transformed cells both episomally as well as integrated in the chromosome. Cell lines cloned from pools made with EASE containing plasmids are stable when cultured for at least 70 generations without selective pressure (data not shown), indicating that EASE containing plasmids are stably maintained integrated in the chromosome.

An additional murine ribosomal element muNTS1 has been used for recombinant protein expression (Hemann et al., 1994; Meyer et al., 1993) and does not

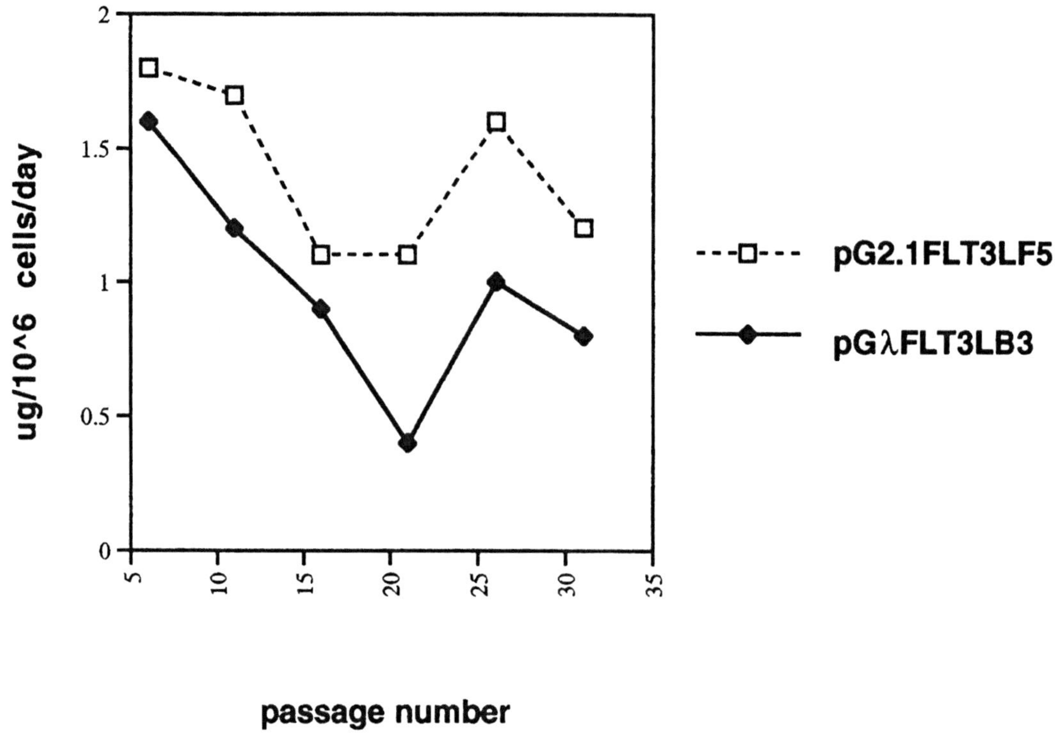

Figure 6. Stability study of pools pG2.1FLT3L F5 and pG λFLT3LB3 grown in selective media supplemented with 100 nM MTX for 32 passages. Specific productivity was measured using ELISA.

require selective pressure to maintain stable expression (Meyer et al., 1993). It is difficult to determine the similarities and differences between the mechanism of action of the EASE, muARS4 and muNTS1 elements as they have been developed in different experimental systems. It appears that the muARS4 and muNTS1 elements increase the total amount of amplification at a given selective pressure whereas the EASE decreases the amount of time necessary to reach a set amount of amplification.

In summary, the EASE is able to improve protein expression most likely by increasing the rate of plasmid DNA amplification when cells are grown under increasing selective pressure. The EASE appears to be unique since its ability to promote amplification is moderate compared to the HSAG middle repetitive sequence family. In this report we have shown that the cell pools expressing recombinant protein can be generated with the EASE vectors without any cloning steps making them apparently easier to use than the muNTS1 vector which require cloning to select high expressing cell lines (Hemann et al., 1994; Meyer et al., 1993). EASE containing bicistronic vectors have been used to generate milligram amounts of protein from serum free and serum containing cell pools for many proteins (Morris et al., 1997), using efficient two step selection protocols.

References

Beitel LK, McArthur JG and Stanners CP (1991) Sequence requirements for the stimulation of gene amplification by a mammalian genomic element. Gene 102: 149–156.

Brasel K, McKenna HJ, Morrissey PJ, Charrier K, Morris AE, Lee CC, Williams DE and Lyman SD (1996) Hematological effects of Flt3 ligand *in vivo* in mice. Blood 88: 2004–2012.

Engvall E and Perlmann P (1972) Enzyme-linked immunosorbent assay, Elisa. 3. Quantitation of specific antibodies by enzyme-labeled anti-immunoglobulin in antigen-coated tubes. J Immunol 109: 129–135.

Hemann C, Gartner E, Weidle UH and Grummt F (1994) High-copy expression vector based on amplification-promoting sequences. DNA Cell Biol 13: 437–445.

Holst A, Müller F, Zastrow G, Zentgraf H, Schwender S, Dinkl E and Grummt F (1988) Murine genomic DNA sequences replicating autonomously in mouse L cells. Cell 52: 355–365.

Jang SK and Wimmer E (1990) Cap-independent translation of encephalomyocarditis virus RNA: structural elements of the internal ribosomal entry site and involvement of a cellular 57-kD RNA-binding protein. Genes Dev 4: 1560–1572.

Kaufman RJ, Davies MV, Wasley LC and Michnick D (1991) Improved vectors for stable expression of foreign genes in mam-

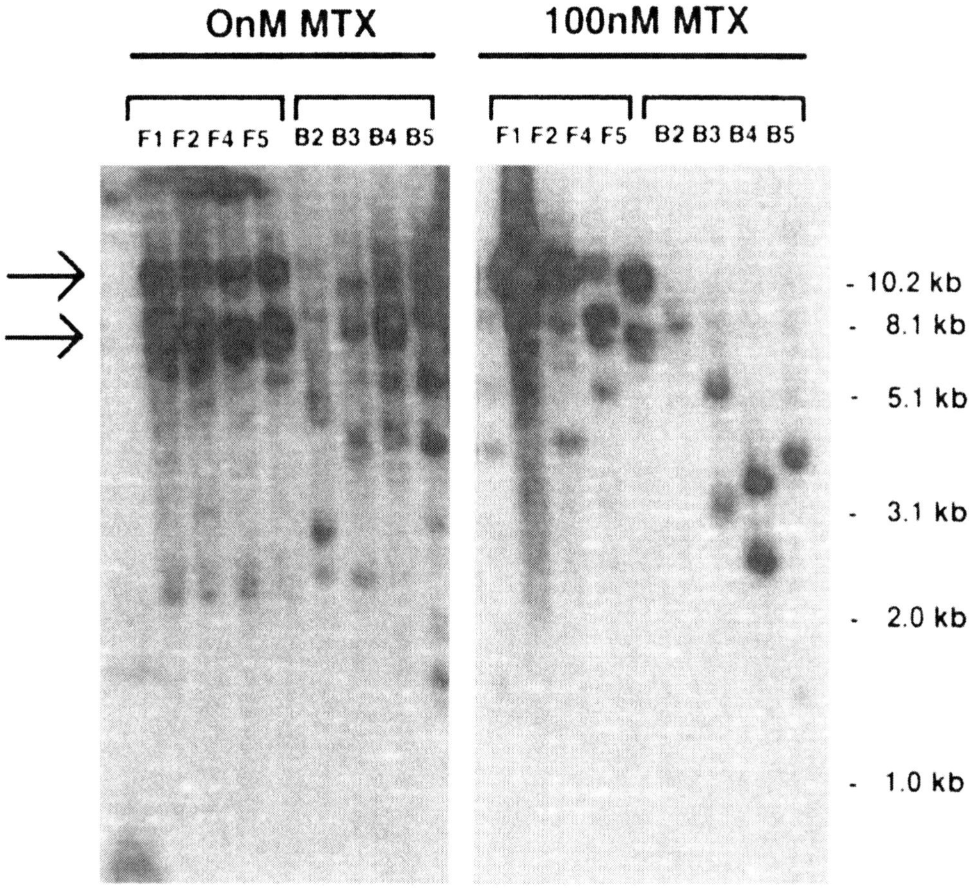

Figure 7. Southern blot analysis of DNA isolated from cell pools generated with EASE and non-EASE containing plasmids. The amount of MTX used to select each pool is indicated above the lanes. Pools F1, F2, F4, F5 were generated using plasmid pG2.1FLT3L. Pools B2, B3, B4, B5 were generated using plasmid pG λFLT3L. The DNA from pools generated with pG2.1FLT3L were digested with *Hinc*II. The DNA from pools generated with pG λFLT3L were digested with *Not*I and *Sph*I restriction enzymes. The locations of the molecular weight markers are indicated on the right. The panel on the right represents the autoradiograph image after 72 h. The panel on the left represents the autoradiograph image after 24 h.

malian cells by use of the untranslated leader sequence from EMC virus. Nucleic Acids Res 19: 4485–4490.

Kaufman RJ, Murtha P and Davies MV (1987) Translational efficiency of polycistronic mRNAs and their utilization to express heterologous genes in mammalian cells. EMBO J 6: 187–193.

McArthur JG and Stanners CP (1991) A genetic element that increases the frequency of gene amplification. J Biol Chem 266: 6000–6005.

Meyer J, Nick S, Stamminger T, Grummt F, Jahn G and Lipps HJ (1993) Inhibition of HIV-1 replication by a high-copy-number vector expressing antisense RNA for reverse transcriptase. Gene 129: 263–268.

Mitchell PJ, Carothers AM, Han JH, Harding JD, Kas E, Venolia L and Chasin LA (1986) Multiple transcription start sites, DNase I-hypersensitive sites, and an opposite-strand exon in the 5′ region of the CHO dhfr gene. Mol Cell Biol 6: 425–440.

Morris AE, Lee CC, Hodges K, Aldrich TL, Krantz C, Smidt PS and Thomas JN (1997) Expression augmenting sequence element (EASE) isolated from chinese hamster ovary cells. In: Carrondo MJT, Griffiths B and Moreira JLP (eds.) Animal Cell Technology, Kluwer Academic Publishers, The Netherlands, pp. 529–534.

Sambrook J, Fritsch E and Maniatis T (1989) Molecular Cloning: A Laboratory Manual, 2nd Edition. Cold Spring Harbor Press, Cold Spring Harbor.

Weidle UH, Buckel P and Grummt F (1988) A new expression system for mammalian cells based on putative replicator sequences of the mouse and a truncated thymidine kinase gene. Gene 73: 427–437.

Cytotechnology **28:** 19–29, 1998.
© 1998 Kluwer Academic Publishers. Printed in the Netherlands.

Effects on growth behavior in continuous hybridoma cell cultures: The role of viral contamination

Andrea Hawerkamp*, Dirk Lütkemeyer, Frank Gudermann, Anna Falkenhain, Heino Büntemeyer & Jürgen Lehmann
Institute of Cell Culture Technology, University of Bielefeld, P.O. Box 100131, 33501 Bielefeld, Germany
E-mail: aha@zellkult.techfak.uni-bielefeld.de

Received 26 August 1998; accepted 26 August 1998

Key words: continuous hybridoma cell cultivation, growth inhibitory effect, retrovirus

Abstract

This article describes the retrovirus expression with optimal nutrient supply and its potential growth inhibition effects in continuous hybridoma cell cultivation. A special reactor setup with total cell retention was developed to examine growth inhibition effects. Using this fermentation strategy we observed a decrease of viability cell rate which occurred at a defined state of the process despite sufficient nutrient supply. Therefore we assume that inhibitory substances are responsible for these effects. The molecular weight range of the inhibitory substances and the possible retrovirus cooperation of these growth inhibition effects were examined. To determine the molecular weight range we used the following methods: ultrafiltration, gelfiltration, ultracentrifugation and gel electrophoresis. Furthermore, RT-PCR and western-/immunoblot are used to detect retrovirus particles in the supernatant and to show a retrovirus participation on growth inhibition effects. The possible growth modulation was tested in a biological assay (MTT-assay).

Abbreviations: BSA: Bovine Serum Albumine; E-MuLV: Ecotrope Murine Leukemia Virus; H_2O_2: Hydrogen Peroxide; IAP: Intracisternal A-type particle; K_S: Monod constant [mol L^{-1}] NaCl: Sodium Chloride; PBS: Phosphat Buffered Saline; TAE: Tris-Acetate-EDTA; TBS: Tris Buffered Saline; Tris: [Tris(hydroxymethyl)aminomethane]; Tris-HCl: [Tris(hydroxymethyl)aminomethane hydrochloride]; TTBS: Tris Buffered Saline + Tween 20.

Introduction

Hybridoma cells are commercially used for the production of monoclonal antibodies for diagnostic and therapeutic purposes. Typical cell densities in continuous perfusion fermentation processes are about 1×10^7 cells mL^{-1}. During this type of fermentation strategy a decrease of the viability can often be observed. Some plausible reasons for this are either the consumption of essential medium components or the production of some kind of inhibitory substances. Rønning et al. (1991) described two potential inhibitors with molecular weights of 67 and <5 kDa.

Another kind of potential inhibitory substances are IAP's (Intracisternal A-type Particles). Johnson et al. (1997) examined the relationship between IAP production, cell growth in culture and antibody production in murine hybridoma cultivation. They suggested that high levels of retrovirus particle production might inhibit cell growth and antibody yield significantly. Most of the commercially used murine hybridoma cell lines are known to contain retrovirus-like particles and in many cases infectious c-type murine leukemia viruses (MLV) which can be either ecotropic (infectious for cell lines of murine origin or closely related species) or xenotropic (infectious for species other than murine (Froud et al., 1997)). These particles are derived from MuLV-related sequences inserted into the mouse genome. Sequences belonging

* Author for all correspondence.

to the ecotropic and xenotropic group can produce replication-competent viral particles (Irving et al., 1993). The level of those murine retroviruses particles during batch fermentation (Froud et al., 1997) followed the viable cell density. It appeared that maximum titres of virus particles were found in the exponential growth phase of the cells at optimum nutrient supply. Reduction of nutrient supply caused a decrease of virus titres.

In accordance with these results we examined the virus expression and a potential growth inhibition effect of retrovirus-like particles in a continuous perfusion fermentation process. Therefore a special reactor setup (Figure 1) was developed which guaranteed a sufficient nutrient supply and the continuous removal of waste products.

Materials and methods

Cell cultures

The cell line used in the present study was the rat-mouse hybriboma 187.1 (ATCC, HB 58) cultured in a serum free medium (Jäger et al., 1988) based on DMEM/F12 (1:1) with human transferrin (5 mg L^{-1}) and bovine insulin (5 mg L^{-1}). The amino acids were supplemented to necessary concentrations (see also Figure 3).

Bioreactor

The cultivation was performed in a 2 litre reactor (Figure 1), Biostat BF 2 (B. Braun Biotech International) with bubble free membrane aeration and internal perfusion membrane (Lehmann et al., 1988). This perfusion membrane enables the medium feed and to harvest the cell free supernatant. The bioreactor was equipped with separate cooled reservoires for glutamine and basal medium. Glutamine (100 mmol L^{-1}) was added according to its consumption by the cells and kept on a concentration above 250 μmol L^{-1} (K_s = 30–70 μmol L^{-1}; Heidemann et al., 1998).

Fermentation process

After inoculation the perfusion fermentation process (Figure 2) was started with a dilution rate of D = 1.75 d^{-1}. When reaching a viable cell density of 1×10^6 cells mL^{-1} (24–48 hr after inoculation) the dilution rate was reduced (D = 0.26 d^{-1}) and the pulsed glutamine feed was started. Cell free supernatant was collected over 24 hr and harvested (HAR1-HAR8, 500 mL per day). Furthermore, 10 mL of cell containing sample were taken once a day, centrifugated and the supernatant was used for RT-PCR analysis (P1-P8). Throughout the complete fermentation process optimal nutrient supply was maintained by daily monitoring of substrates and concentration of metabolic products. Free amino acids were analysed by means of an automated reversed phase HPLC system DS 450 (Kontron GmbH) with precolumn derivatization using the OPA method (Büntemeyer et al., 1991) to determine all of the amino acids except for proline, cysteine and the dipeptide cystine. Inhibitory effects of ammonia were avoided by external consumption-dependent glutamine feed.

Purification

The harvests were concentrated by ultrafiltration (Ultrasette, cut off: 1 kDa, Pall). The molecular weight range of the growth inhibitory substances were determined by separation of the concentrated harvest (8 mL) on a Superdex 30 (separation range: <10 kDa) gelfiltration column C 16/100 with 120 mL gel volume and on a Sephacryl S-1000 (separation range: >600 kDa) C 16/100 with also 120 mL gel volume (both Amersham Pharmacia Biotech). The determination of the void volume (Vo) of Superdex 30 was follows by Blue Dextran (molecular weight: 2000 Da, Sigma). Determination of the void volume of Sephacryl S-1000 was not possible due to the separation range of 10^8 Da. The total volume of these two columns were specified by using aceton (molecular weight: 54 Da). Each column was equilibrated with 20 mM phosphate and 150 mM NaCl, pH 7.2. Fractions of 5 mL were collected starting with the beginning of sample application. All separations were performed at 7 °C with a BioPilot-System (Amersham Pharmacia Biotech).

Native PAGE

The inhibitory fractions were separated by native PAGE, PhastGel Gradient 4–15 (Amersham Pharmacia Biotech). Separation and the following silver staining (Heukeshoven and Dernick, 1988) were performed with the PhastSystem (Amersham Pharmacia Biotech). A high molecular weight marker (Thyroglobulin: 669 kDa, Ferritin: 440 kDa, Catalase 232 kDa, Lactate Dehydrogenase: 140 kDa and BSA: 67 kDa from Amersham Pharmacia Biotech) was used

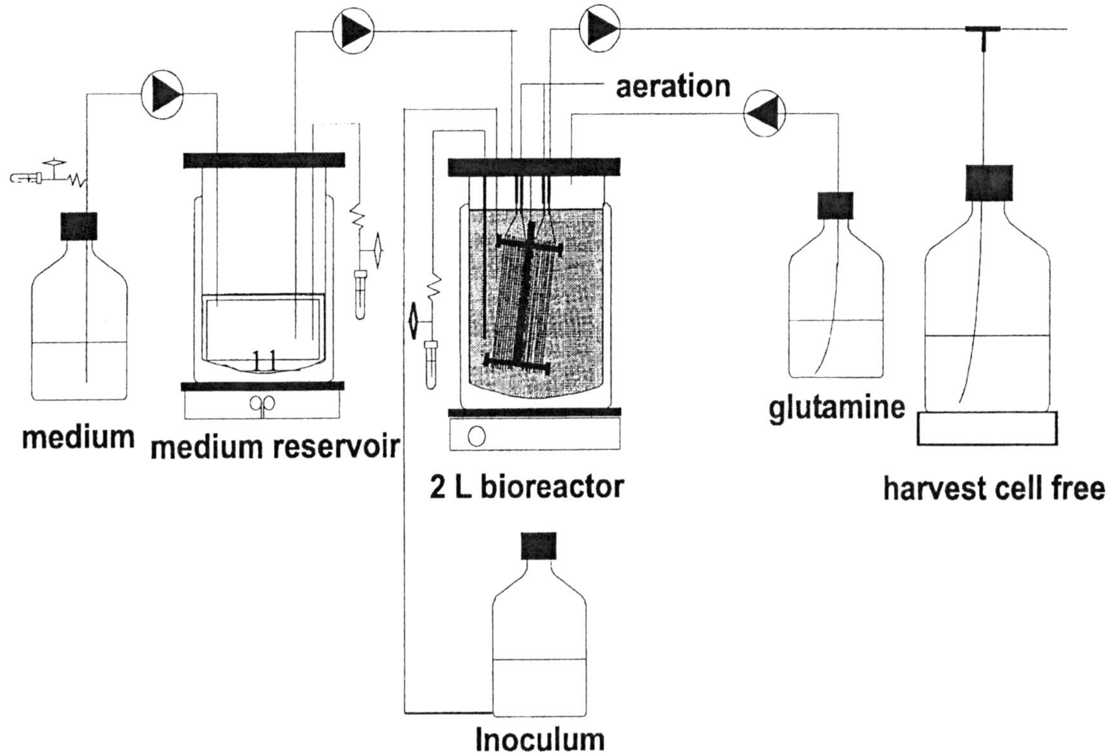

Figure 1. Setup of a 2 L perfusion bioreactor BIOSTAT BF 2 (B. Braun Biotech International). The reactor is equipped with a cooled medium reservoir and an additional consumption-dependent glutamine feed. The medium feed and the harvest of cell free supernatant are made possible by an internal perfusion membrane.

as a referance to determine the molecular weight of the inhibitory substances.

Western blot

The western blot was performed with Trans-Blot Electrophoretic Transfer Cell (Biorad). As blot membrane we used nitrocellulose (Schleicher and Schüll) with a pore size of 0.2 μm. The fransfer-buffer was: 25 mM Tris, 192 mM Glycine and 20% Methanol (pH 8.3). The blotting conditions were 120–140 V (0.8–0.9 A) for 3 hr.

Immunoblot

After the western blot procedure the membrane was first washed for 3 × 10 min with wash solution (2). In next step the membrane was blocked for 1 hr by using solution 3. Afterwards a washing step (2 × 10 min, solution 2) was followed. The incubation with the first antibody was performed for 2 hr. The washing step was repeated (3 × 10 min, solution 2) and the incubation (1 hr) with the second antibody followed. Another washing step (3 × 10 min with solution 2 and 10 min with solution 1) was performed. At least the membrane was treated with the developer.

Solutions:

1. 20 mM Tris-HCl, 500 mM NaCl, pH 7.5 (TBS)
2. Wash Solution: 20 mM Tris-HCl, 500 mM NaCl, 0.05% Tween 20 (TTBS)
3. Blocking Solution: TBS and 1% BSA
4. Antibody Buffer: TTBS and 1% BSA
5. Developer: 10 mg 4-chloro-1-naphtol, 3.3 mL methanol, 16.7 mL TBS and 10 μL 30% H_2O_2

Antibody-Dilutions:

1. Polyclonal-Antibody: Goat-Anti-Mouse-E-MuLV (1:500)
2. Monoclonal-Antibody: Anti-Goat-IgG-Peroxidase-Con jugate from mouse (Sigma), (1:1000)

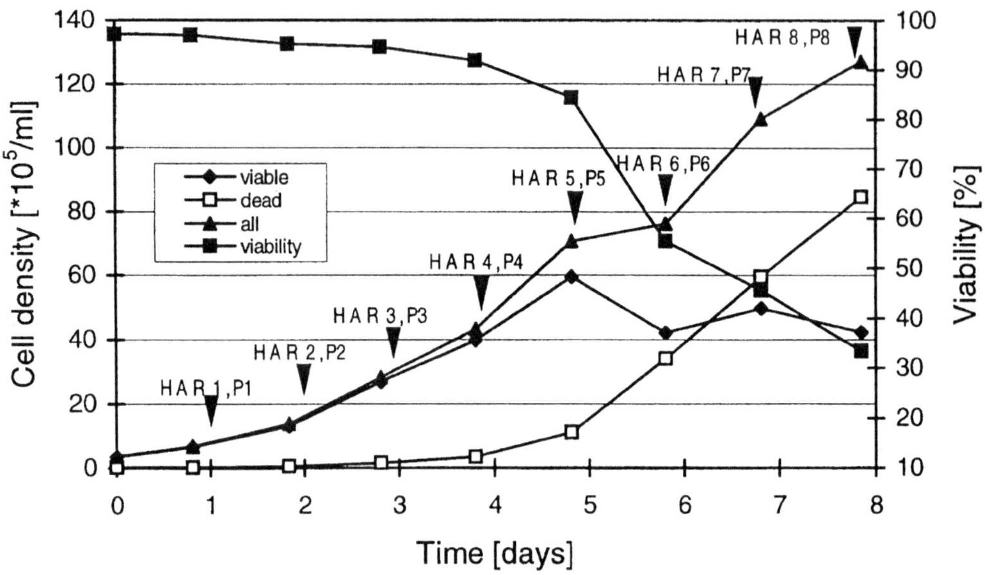

Figure 2. Time course of a continuous perfusion fermentation process with glutamine feed. After inoculation the perfusion process was started with a dilution rate of D = 1.75 d^{-1}. When reaching a viable cell density of 1×10^6 cells mL^{-1} (24–48 hr after inoculation) the dilution rate was reduced (D = 0.26 d^{-1}) and the pulsed glutamine feed was started. Cell free supernatant was collected over 24 hr and harvested (HAR1-HAR8, 500 mL per day). Furthermore, 10 mL cell containing sample were taken once a day, centrifugated and the supernatant used for RT-PCR analysis (P1-P8). The figure shows, that the viability decreased at day 5. Preliminary analysis demonstrated an inhibitory effect in HAR 5 to HAR 8.

Bioassay

To analyse inhibitory effects two test systems, the MTT-assay and the [^{3}H] Thymidinincorporation (Andersson et al., 1972) are established in our institute. For our examinations we chose the MTT-assay which determines the activity of dehydrogenases by colorimetric reaction (Mossmann, 1983; Hansen et al., 1989) and therefore, it is an indirect measure of cell growth activity. This measured activity is independent from the DNA-Synthesis (the S-Phase of cell cycle) in contrast to [^{3}H] Thymidinincorporation. The gelfiltration fractions (0.2 μm filtered) were diluted 1:1 with fresh medium and inoculated with cells (1×10^6 cells mL^{-1}) on microtiter plates (Nunc). After a proliferation period of two days, 10 μl MTT (3-(4,5-dimethylthiazol-2-yl)-2,5-diphenyl tetrazolium bromide, Sigma) solution (5 mg mL^{-1} in PBS) was added and the plate was incubated for 3 hr at 37 °C in a CO_2-incubator. Then 100 μL of 40 mL N,N-dimethylformamid, 40 mL dest. H_2O and 20 g SDS (Sigma) were added to each well to dissolve the blue crystals formed by the living cells. The intensity of the tinge was measured at 570 nm in an microplate-reader (Modul Biotek).

Isolation of RNA

First the cell supernatant (10 mL) samples P1 to P7 were ultracentrifugated (40000 × g, 20 min, 4 °C) and the pellets (1 mL) were used for RNA-Isolation with QIAamp Viral RNA kit (Qiagen GmbH). The inhibitory fractions 11 (Superdex 30) and 21 (Sephacryl S-1000) were used for RNA-Isolation without pre-preparation (no concentration with ultracentrifugation). Additional supernatant samples from Mink lung (ATCC: CCL64), Mus dunni (ATCC: CRL 2017) and a medium sample were used as negative controls.

RT-PCR

Specific primer (Life Technologies) for detection of murine leukemia viruses (E-MuLV and MCF) were used (Irving et al., 1993) in RT-PCR (PCR kit, Perkin Elmer). For visualization, 4 μL of the amplification products were separated by gel electrophoresis on 3% agarose (Sigma) in a TAE buffer containing ethidium bromide (Figures 9a and b).

Ultracentrifugation

The fractions 5 (control) and 12 (inhibitory) were ultracentrifugated (198000 × g, 5 hr, 4 °C) by using

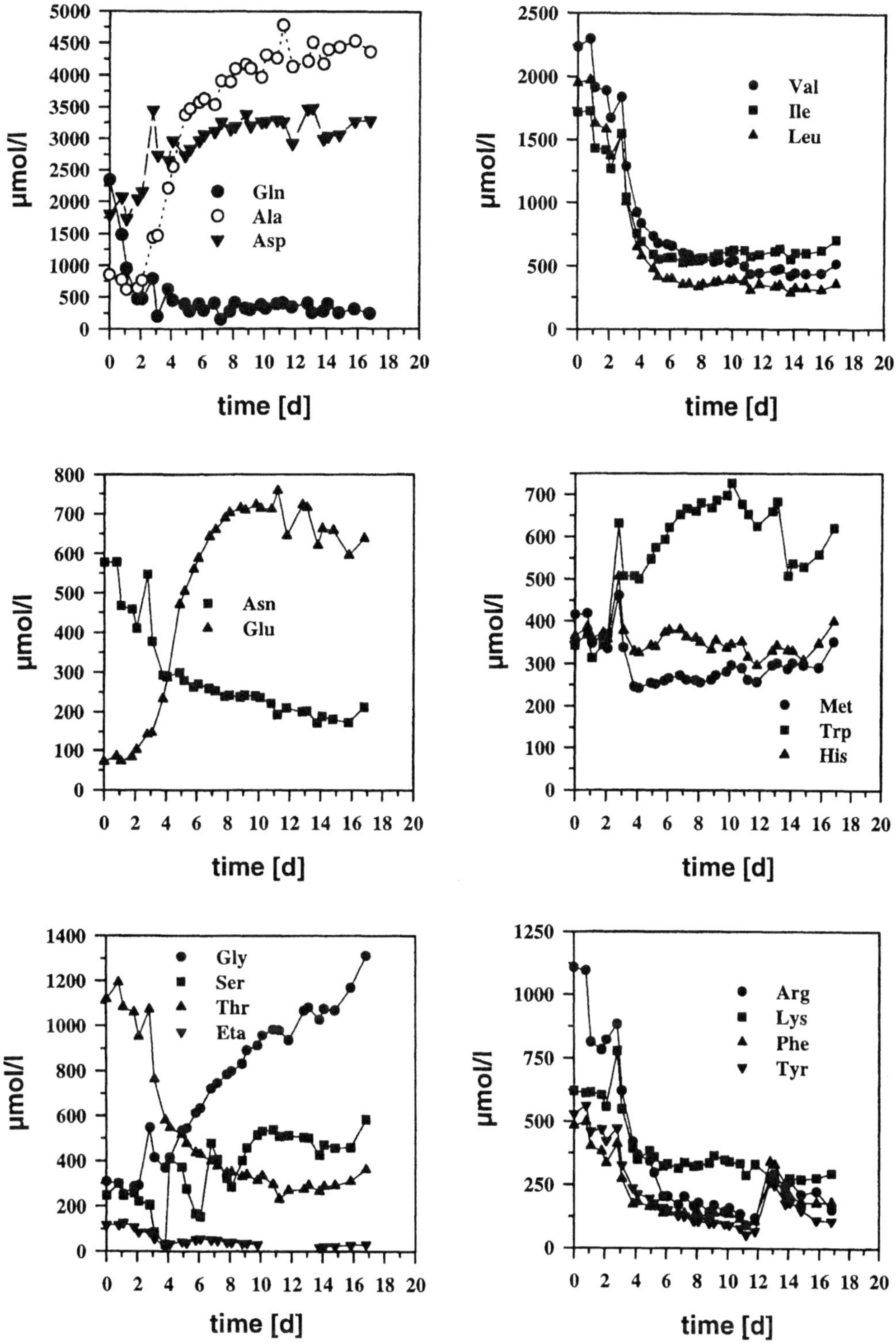

Figure 3. Amino acid concentrations in the 2-Liter bioreactor during perfusion fermentation with HB58 over a 17 days cultivation process. No limitation could be observed. Glutamine was always kept on a concentration above 250 μmol L^{-1}.

Fermentation Process

↓

Harvest (HAR5-HAR8)

↓

Ultrafiltration: 1:5 concentrated (cut off 1 kDa)

↓

Gelfiltration:

1. Superdex 30 (fractionation range: <10 kDa) ➔**MTT-assay** (Fig. 5)

2. Sephacryl S-1000 (fractionation range: >600 kDa) ➔**MTT-assay** (Fig. 6)

↓

Additional Concentration: Ultracentrifugation with Superdex 30 fractions:

Fraction 5 (non inhibitory, control)

Fraction 12 (inhibitory)

↓

MTT-assay (Fig. 8)

Figure 4. Purification scheme to determine the molecular weight range of the inhibitory substances.

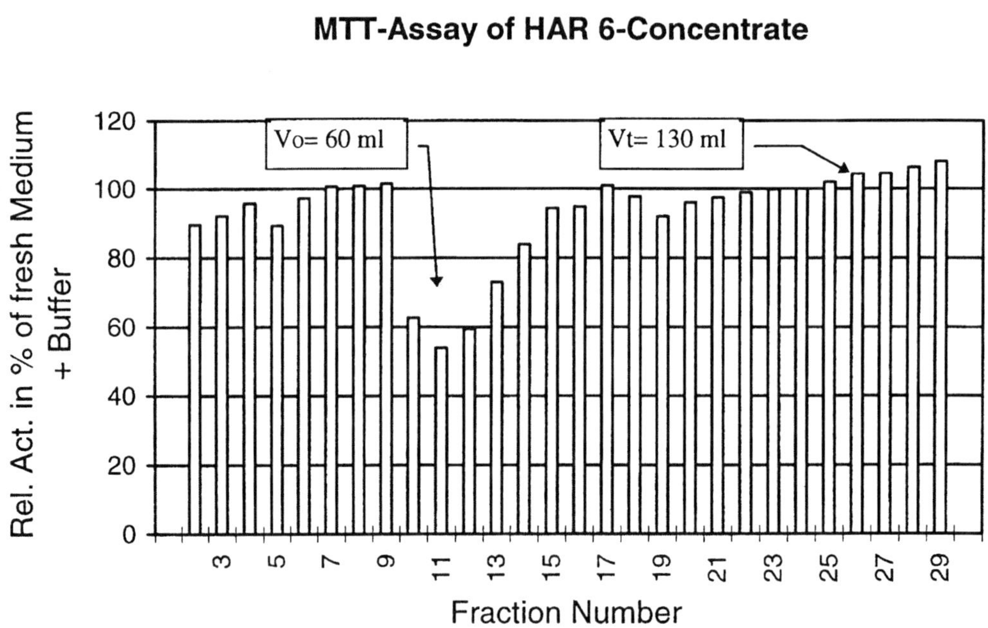

Figure 5. Results of the MTT-assay with Superdex 30 fractions of HAR 6-concentrate. This test detemines the metabolic state of the cells. Therefore, it is a indirect measure of cell growth activity in % relating to fresh medium and buffer. A drastic inhibitoric effect was observed by the fractions 10–14. These fractions were collected from the void volume (Vo = 60 mL) of the Superdex 30 column (Vt = total volume, 130 mL) which indicates a molecular weight of the inhibitory substances above 10 kDa.

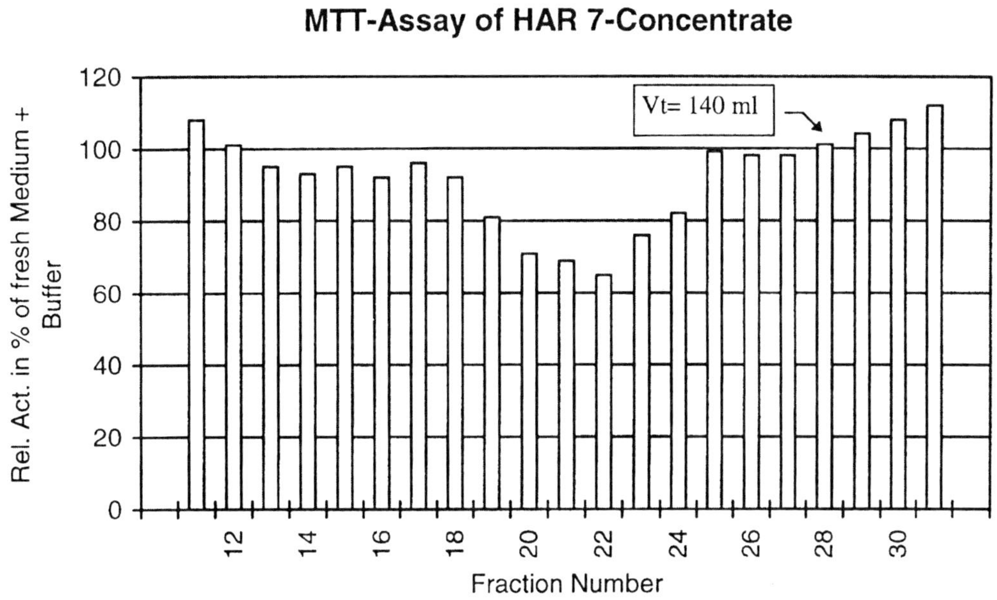

Figure 6. Results of the MTT-assay with Sephacryl S-1000 fractions of HAR 7-concentrate. A Sephacryl S-1000 column was used to determine the molecular weight range of the inhibitoring substances. The fractions 20–23 showed an cell growth inhibiting effect relating to fresh medium and buffer. The range between 100 and 125 mL is the same for the elution of inhibitory substances and Blue Dextran The possible molecular weight could therefore be reduced to the range between 600 and 2000 kDa.

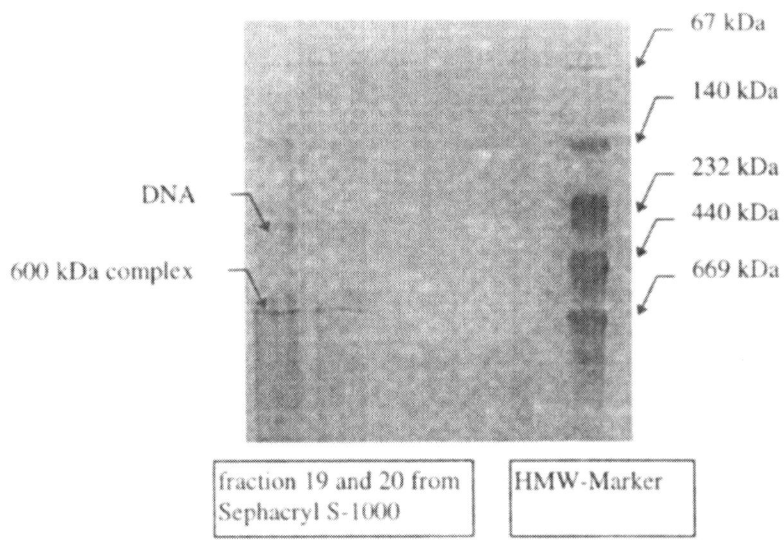

Figure 7. Native PAGE, PhastGel Gradient (4–15) of the inhibitory fractions 19 and 20 from the Sephacryl S-1000 column. The fractions show a approx. 600 kDa complex and DNA.

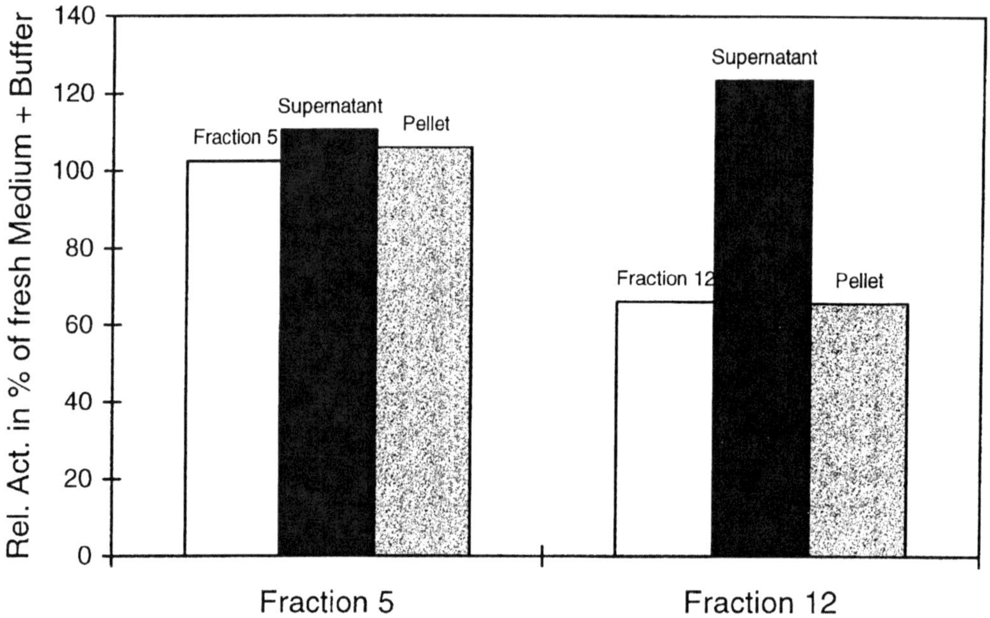

Figure 8. Activities of Superdex 30 fractions 5 and 12 before and after ultracentrifugation. Fraction 5 showed a constant activity of approx. 100%. At first the activity of fraction 12 was 65% of the control. After ultracentrifugation the supernatant showed an activity of 120%, whereas the pellet caused a significant inhibitory effect (65% activity).

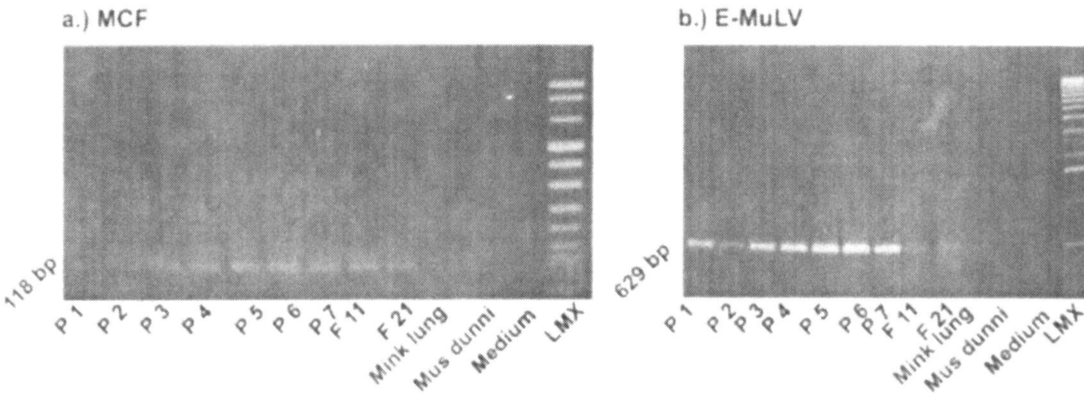

Figure 9. RT-PCR of fermentation samples P1 to P7 and the inhibitory fractions 11 (Superdex 30) and 21 (Sephacryl S-1000) for detection of E-MuLV and MCF. The figures shows that all supernatant samples, P1 to P7 and the inhibitory fractions (F 11 and F 21) were E-MuLV-(9b) and MCF-(9a) positive. From P1 to P7 an increasing signal could be observed, particularly for the E-MuLV. The samples from Mink lung, Mus dunni and the medium sample are retrovirus negative.

Sorvall OTD 50 (Rotor: TH 641). The resulting supernatants and pellets were analysed in the MTT-assay.

Results

Cell growth during perfusion fermentation

To examine growth inhibition effects and the expression of retrovirus-like particles we use HB 58 in continuous perfusion fermentation process. A typical growth curve (Figure 2) for this hybridoma cell line shows a viability rate of more than 90% during the first 4 days. After day 4 the viability decreased rapidly to 33% at day 8. The maximal viable cell density is reached at 6×10^6 cells mL^{-1} between day 4 and 5. From day 5 to day 8 the viable cell density is relatively constant between 4–5 $\times 10^6$ cells mL^{-1}, whereas the viability decreases. All substrates were monitored throughout the complete fermentation

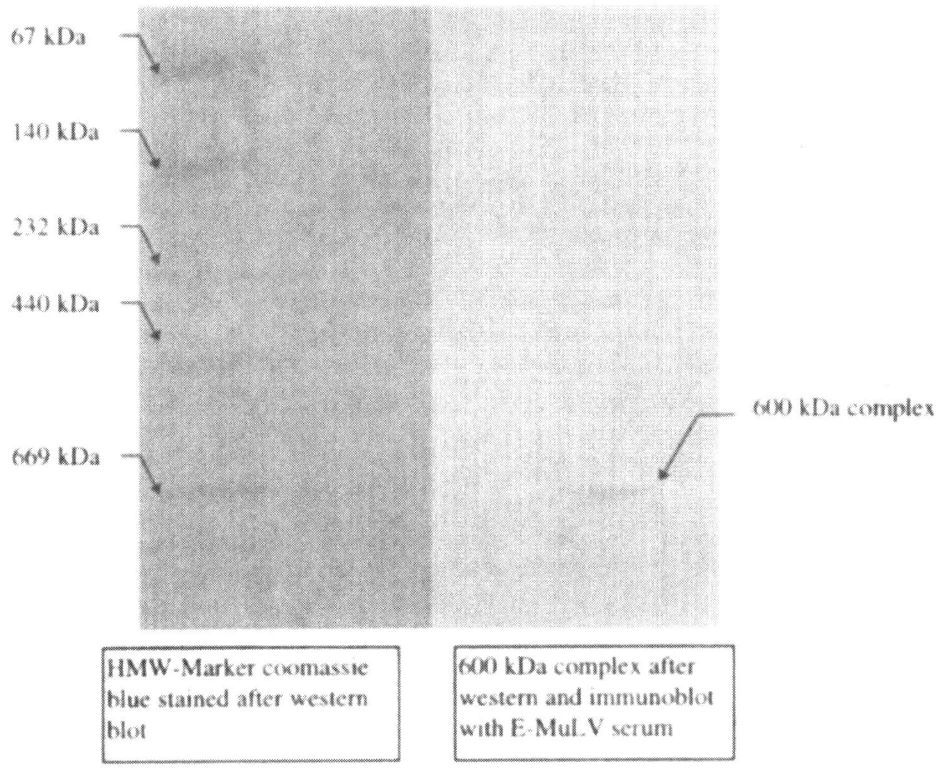

Figure 10. Results of the western blot and the following immunoblot with polyclonal serum against E-MuLV. The left lane of the figure shows the marker after western blot and coomassie blue staining. The marked approx. 600 kDa complex is shown on the right line.

process and no growth inhibition level of ammonia (<6 mmol L^{-1} at the end of the fermentation) and lactate (<30 mmol L^{-1} at the end of the fermentation) was measured. A growth inhibition effect caused by glucose (data not shown) and amino acid (Figure 3) limitations was not observed.

Determination of the molecular weight of the inhibitory substances

The growth inhibitors already described in literature belong to the class of peptides (Paukovits and Laerum, 1982; Elgjo and Reichelt, 1984) and proteins (Hsu and Wang, 1986; Wells and Mallucci, 1983) like TGF-β (Transforming growth factor-β) (Tucker et al., 1984; Sporn and Roberts, 1992) with a native molecular weight of 25 kDa. In accordance with these references we developed a purification strategy (Figure 4) to determine the molecular weight range of the inhibitory substances. Using this strategy the daily cell free supernatants (HAR5-HAR8) were first concentrated (1:5) by ultrafiltration and than separated on Superdex 30.

The resulting fractions of the Superdex 30 column (separation range: <10 kDa) were used in the MTT-assay to measure the cellular activity in each fraction. Figure 5 shows the results of this MTT-assay. A drastic inhibitory effect was observed in the fractions 10–14, collected from the void volume (Vo) of the Superdex 30 column.

This indicates that the inhibitory substance has a molecular weight above 10 kDa. For more exact determination we used a Sephacryl S-1000 column with a separation range of 5×10^5-10^8 Da. Figure 6 shows the activity of the Sephacryl S-1000 fractions of HAR 7-concentrate in the MTT-assay. The fractions 20–23 showed an inhibitory effect. The elution of the inhibitory substances and Blue Dextran are simultaneously between 100 and 125 mL. Therefore we can reduced to the molecular weight range between 600 and 2000 kDa. Apart from other biomolecules, the gel electrophoresis (native PAGE, PhastGel Gradient 4–15, Amersham Pharmacia Biotech) of the inhibitory fractions demonstrated (Figure 7) the existance of a complex with a molecular weight of approx. 600 kDa.

Further analysis (DNase treatment) show that these fractions also contain DNA (data not shown).

Further examination demonstrated the possibility of concentrating the inhibitory substances by using ultracentrifugation. The fractions 5 (control) and 12 (inhibitory) were ultracentrifugated. Afterwards the resulting supernatants and pellets were analysed in the MTT-assay. The results are shown in Figure 8. The fraction 5 had a constant activity of approx. 100%. The activity of fraction 12 was 65% of the control. After ultracentrifugation the supernatant showed an activity of 120%, whereas the pellet caused a significant inhibitory effect (65% activity). These results document the high molecular weight of the inhibitory substances and the possibility of concentrating by forced sedimentation.

Participation of retroviruses on growth inhibition effects

The cell line HB58 with the myeloma part P3X63Ag8.653 is known to contain E-MuLV, MCF and IAP's. To determine whether there are these type of particles present in the supernatant, fermentation samples P1 to P7 (see Figure 2) and the inhibitory fractions 11 (Superdex 30) and 21 (Sephacryl S-1000) were used for the detection of E-MuLV and MCF virus. Additional supernatant samples from Mink lung, Mus dunni and a medium sample were employed as retrovirus-negative references. Figures 9a and b show that all supernatant samples, P1 to P7 and the inhibitory fractions (fraction 11 and 21) were E-MuLV and MCF virus positive. From P1 to P7 an increase of the signal could be observed, particulaly for E-MuLV.

To indicate whether the approx. 600 kDa complex (native PAGE, PhastGel Gradient 4–15) is virus origin we use western blot and immunoblot with polyclonal goat-anti-mouse serum against E-MuLV membrane glycoproteins. Figure 10 shows the E-MuLV origin of this complex.

Summary and conclusion

During this type of perfusion fermentation process the viability decreased in spite of optimal nutrient supply. Inhibititory fractions could only be derived from harvests taken after this decrease in viability. Examination of the harvest demonstrated that an isolation of the inhibitory substances is possible. By using gel filtration and gel electrophoresis a complex with a molecular weight of approximately 600 kDa could be determined. It seems that this complex is E-MuLV origin. Furthermore two types of viruses, E-MuLV and MCF were detected in the supernatant. An increasing E-MuLV and MCF signal (RNA-content) could be observed in the supernatant throughout the whole fermentation. The inhibitory fractions were also MCF- and E-MuLV-positive. Therefore, we assume a correlation between the retrovirus-expression in continuous fermentation processes and the observed growth modulating effect. Further studies will aim to verify the involvement of retroviruses in growth inhibition processes.

Acknowledgements

The work was supported by Research Grant No. 0310754 of the German Ministry for Education, Science, Research and Technology (BMBF) and Boehringer Mannheim GmbH, Penzberg, Germany.

References

Andersson J, Möller G and Sjoberg O (1972) Selective induction of DNA synthesis in T and B lymphocytes. Cell Immunol 4: 381–393.

Büntemeyer H, Lütkemeyer D and Lehmann J (1991) Optimization of serum free fermentation processes for antibody production. Cytotechnology 5: 57–67.

Elgjo K and Reichelt KL (1984) Purification and Characterization of a Mitosis Inhibiting Epidermal Pentapeptide. Cell Biol Int Rep 8: 379–382.

Froud SJ, Birch J, McLean C, Shepherd AJ and Smith KT (1997) Viral contaminants found in mouse cell lines used in the production of biological products. In: Carrondo MJT, Griffiths JB and Moreira JLP (eds.) Animal Cell Technology: From Vaccines to Genetic Medicine. Kluwer Academic Publishers, Dordrecht pp. 681–686.

Hansen MB, Nielsen SE and Berg K (1989) Re-examination and further development of a precise and rapid dye method for measuring cell growth/cell kill. J Immunol Methods 119: 203–210.

Heidemann R, Lütkemeyer D, Büntemeyer H and Lehmann J (1998) Effects of dissolved oxygen levels and the role of extra- and intracellular amino acid concentrations upon the metabolism of mammalian cell lines during batch and continuous cultures. Cytotechnology 26: 185–197.

Heukeshoven J and Dernick R (1988) Improved silver staining procedure for fast staining in PhastSystem Development Unit: I. Staining of sodium dodecyl sulfate gels. Electrophoresis 9: 28–32.

Hsu YM and Wang JI (1986) Growth Control in Cultered 3T3 Fibroblasts Purification of an Mr 13000 Polypeptide Responsible for Growth Inhibitory Activity. J Cell Biol 102: 362–369.

Irving JM, Chang LWS and Castillo FJ (1993) A Reverse Transcriptase-Polymerase Chain Reaction Assay for the Detection and Quantification of Murine Retroviruses. In: BIO/TECHNOLOGY 11: 1042–1046.

Jäger V, Lehmann J and Friedel P (1988) Serum-free growth for the cultivation of a wide spectrum of mammalien cells in stirred bioreactors. Cytotechnology 1: 319–329.

Johnson L, Bryant J, Barford JP, Harbour C and Fletcher A (1997) Analysis of endogenous retrovirus production in a murine hybridoma by flow cytometry. In: Carrondo MJT, Griffiths JB and Moreira JLP (Eds.) Animal Cell Technology: From Vaccines to Genetic Medicine. Kluwer Academic Publishers, Dordrecht pp. 687–690.

Lehmann J, Vorlop J and Büntemeyer H (1988) Bubble-free reactors and their development for continuous culture with cell recycle. Animal Cell Biotechnology 3: 222–237.

Mossmann T (1983) Rapid colorimetric assay for cellular growth and survival: Application to proliferation and cytotoxicity assays. In: J Immunol Methods 65: 55–63.

Paukovits WR and Laerum OD (1982) Isolation and Synthesis of a Hemoregulatory Peptide. Z Naturforschung [C] 37: 1297–1300.

Rønning ØW, Schartum M, Winsnes A and Lindberg G (1991) Growth limitation in hybridoma cell cultures: The role of inhibitory or toxic metabolites. Cytotechnology 7: 15–24.

Sporn MB and Roberts AB (1992) Autocrine Secretion – 10 Years Later. Ann Int Med 117: 408–414.

Tucker RF, Shipley GD, Moses HL and Holley RW (1984) Growth Inhibitor from BSC-1 Cells Closly Related to Platelet Type Beta Transforming Growth Factor. Science 226: 705–707.

Wells V and Mallucci L (1983) Properties of a Cell Growth Inhibitor Produced by Mouse Embryo Fibroblasts. J Cell Physiol 117: 148–154.

Cytotechnology **28:** 31–42, 1998.
© 1998 Kluwer Academic Publishers. Printed in the Netherlands.

Isolation, characterization and recombinant protein expression in Veggie-CHO: A serum-free CHO host cell line

Brian Rasmussen, Ray Davis, James Thomas & Pranhitha Reddy*
Cell Sciences Department, Immunex Corporation, 51 University Street, Seattle, WA 98101, U.S.A.
E-mail: reddyp@immunex.com

Received 25 August 1998; accepted 25 August 1998

Key words: CHO, DHFR, gene expression, growth factor, recombinant protein, serum-free

Abstract

The dihydrofolate reductase-deficient Chinese hamster ovary cell line, DXB11-CHO, commonly used as a host cell for the production of recombinant proteins requires 7.5% serum-supplementation for optimal growth. Regulatory issues surrounding the use of serum in clinical production processes and the direct and indirect costs of using serum in large-scale production and recovery processes have triggered efforts to derive serum-independent host cell lines. We have successfully isolated a serum-free host that we named Veggie-CHO. Veggie-CHO was generated by adapting DXB11-CHO cells to growth in serum-free media in the absence of exogenous growth factors such as Transferrin and Insulin-like growth factor, which we have previously shown to be essential for growth and viability of DXB11-CHO cells. Veggie-CHO cells have been shown to maintain an average doubling time of 22 hr in continuous growth cultures over a period of three months and have retained the dihydrofolate reductase -deficient phenotype of their parental DXB11-CHO cells. These properties and the stability of its serum-free phenotype have allowed the use of Veggie-CHO as host cells for transfection and amplified expression of recombinant proteins. We describe the derivation a serum-free recombinant cell line with an average doubling time of 20 hr and specific productivity of 2.5 Units recombinant Flt-3L protein per 10e6 cells per day.

Introduction

Chinese hamster ovary (CHO) cells are widely used for the production of several complex recombinant proteins, e.g. cytokines, clotting factors, antibodies (Brasel et al., 1996; Kaufman et al., 1988; McKinnon et al., 1991; Wood et al., 1990). The dihydrofolate reductase (DHFR)-deficient mutant cell lines (Urlaub and Chasin, 1980), DXB-11 and DG-44, are the CHO host cell lines of choice because the efficient DHFR selectable and amplifiable gene expression system allows high level recombinant protein expression in these cells (Kaufman, 1990). In addition, these cells are easy to manipulate as adherent or suspension cultures and exhibit relatively good genetic stability. CHO cells and recombinant proteins expressed in them have been extensively characterized and have been approved for use in clinical manufacturing by regulatory agencies.

Serum is often used for the propagation of CHO cell lines, and there is increasing pressure to remove serum from the manufacturing process of therapeutics for several reasons. The expense of serum, variance between serum lots and serum quality, regulatory concerns regarding viral contamination in serum and the burden of removing serum proteins in downstream processing are some of the driving reasons to implement serum-free, cell-culture technology. Therefore, this technology has become an essential component of the production process and the ability to adapt adherent, serum-requiring recombinant CHO cell lines to grow as suspension cultures in serum-free media has been demonstrated (Adamson, 1994; Thomas and Fung, 1994). However, adaptation of recombinant cell lines to suspension and serum-free growth can be a time consuming step in process development

* Author for all correspondence.

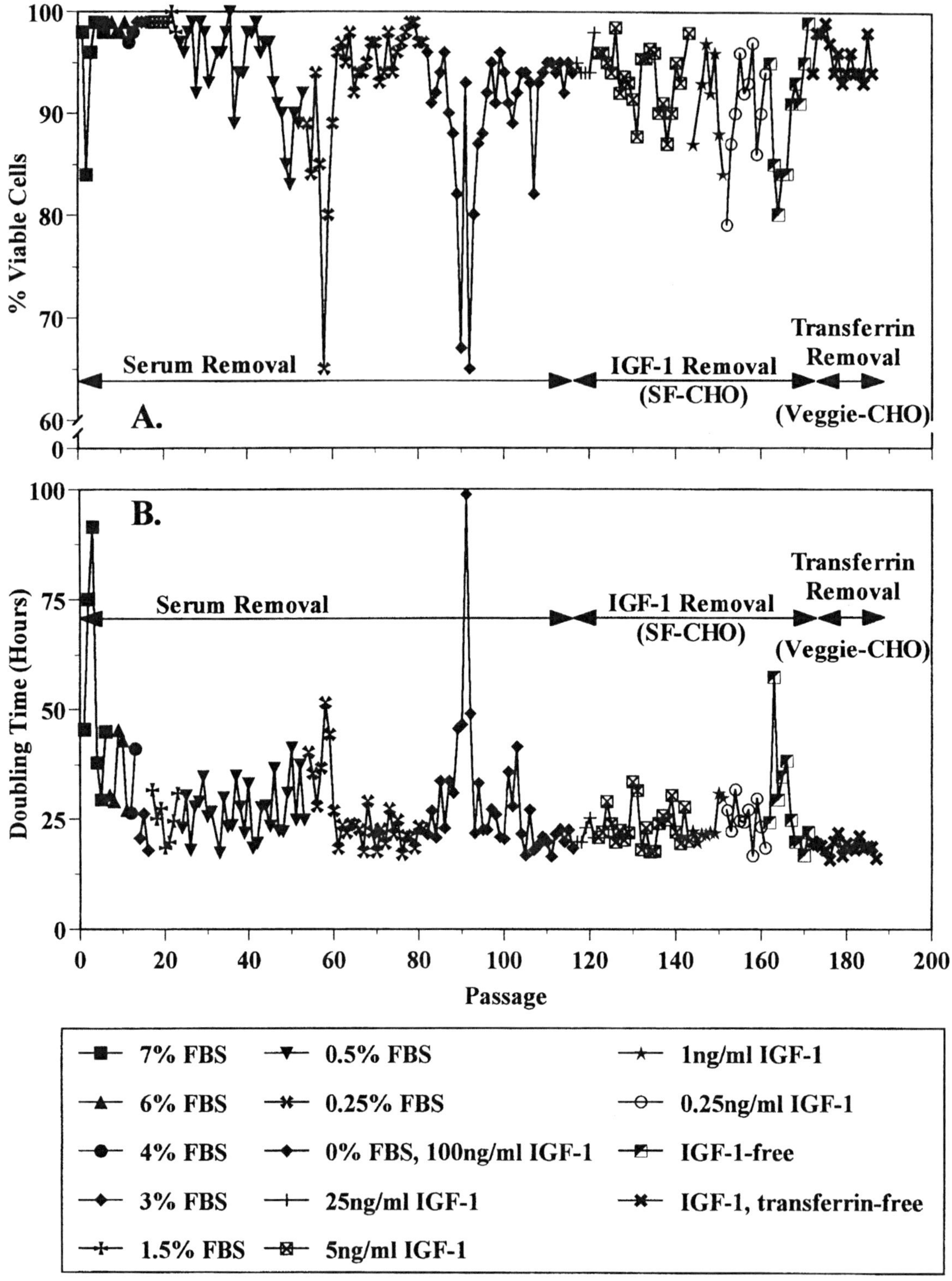

Figure 1. Isolation of Veggie-CHO cells by adaptation of DXB11-CHO cells to growth in serum-free media. Figures 1A and B show the average viability and cell doubling time respectively of DXB11-CHO cells during their adaptation to the gradual removal of serum, followed by the removal of growth factors: IGF-1 and transferrin.

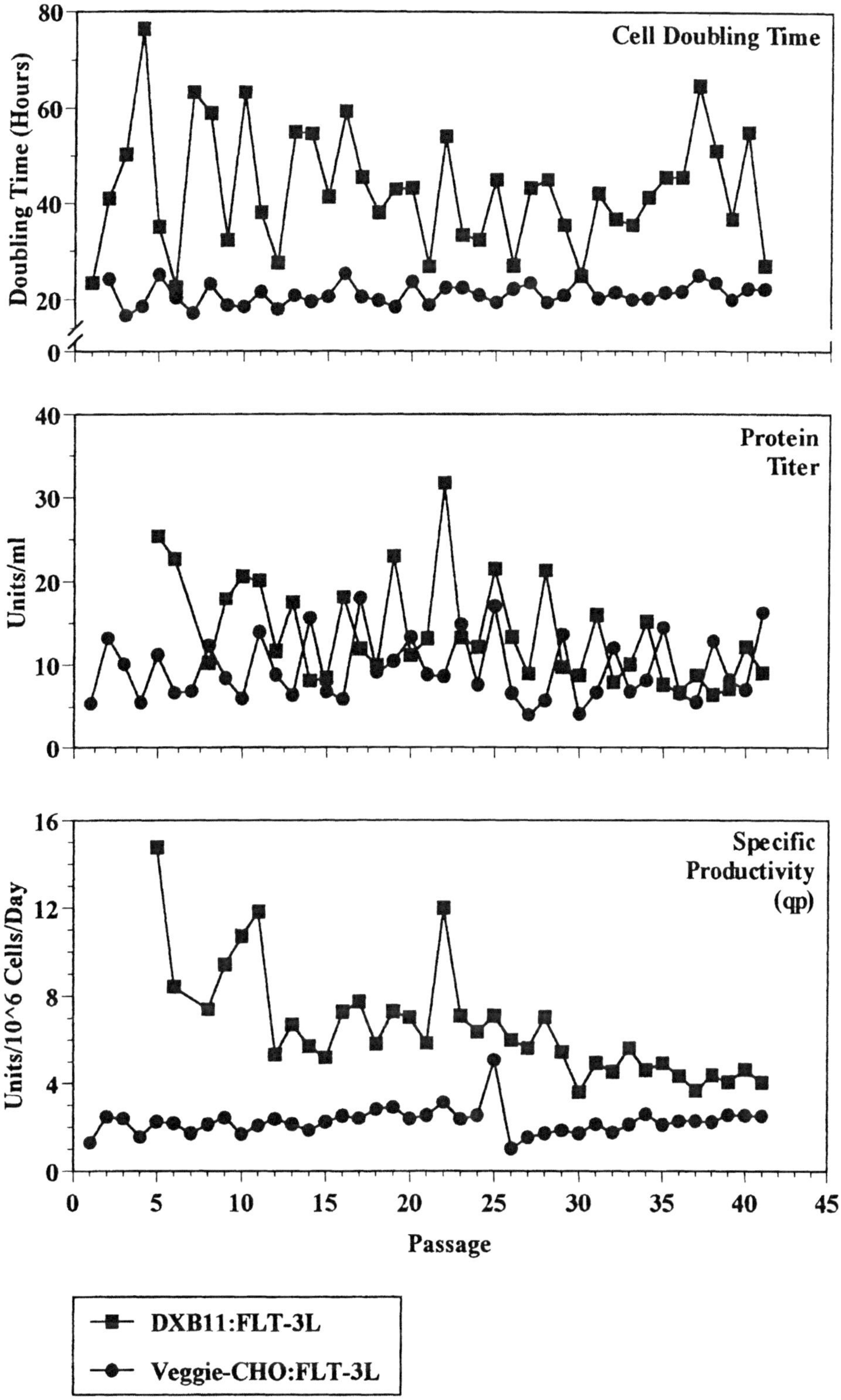

Figure 2. Growth and expression of a recombinant cytokine protein in Veggie-CHO and parental DXB11-CHO host cells. The doubling time, protein titers and specific productivity (qp) of the DXB11:FLT-3L and Veggie-CHO:FLT-3L cultures over 40 passages are shown.

with variable effects on recombinant protein expression and protein quality. Hence, a suspension and serum-free preadapted DHFR-deficient host cell line that could be used to generate recombinant cell lines expressing high levels of recombinant protein would greatly facilitate the process and enhance efficiency in the serum-free process development of recombinant proteins.

We used an adaptation strategy to derive the Veggie-CHO cells which grow in serum-free media in the absence of exogenous growth factors such as Transferrin and Insulin-like growth factor (IGF-1). This involved the gradual reduction of serum supplementation in the media and the replacement of serum with low levels of growth factors, IGF-1 and transferrin, in an enriched cell growth media. These serum-free cells were then weaned off these growth factors. The adaptation of cell lines to media with reduced or no serum is based on careful adjustment of medium components to provide cells with optimal nutrient levels and provide them with a mixture of growth factors essential for survival and growth in the absence of serum (Barnes and Sato, 1980; Evans et al., 1956; Hamilton and Ham, 1977). This approach has been demonstrated for several cell lines, including CHO cells (Gandor et al., 1995; Sinacore et al., 1996). The basal medium of a majority of serum-free formulations usually consist of Hams F-12, DMEM or a mixture of both. Over the past few years several serum-free, protein-free media and chemically defined media have become commercially available for use in development of serum-free cell lines. Some of these have been formulated for use with CHO cell lines and require an adaptation process to acquire optimal growth (Wyatt and Doak, 1995).

We demonstrate the convenience and advantage of using Veggie-CHO cells to generate recombinant CHO lines by using Veggie-CHO as host cells for the methotrexate (MTX)-amplified expression of the FLT-3L cytokine molecule. The FLT-3L molecule expressed in Veggie-CHO is active and is expressed at levels similar to those obtained using the conventional DXB11-CHO host cells.

Materials and methods

Materials

DMEM-F12, Nucleoside-free DMEM-F12 (–) and custom enriched formulations of these media, IMX 2.3 and IMX 1.2 respectively, were purchased from JRH Biosciences. Dialyzed and non-dialyzed fetal bovine serum (FBS) was obtained from Hyclone Laboratories. Tissue culture flasks were obtained from Corning and spinner flasks were obtained from Bellco. Recombinant Insulin-like growth factor (Long R3-IGF-1) (McKinnon et al., 1991) was purchased from GroPep and bovine were purchased from JRH and plant soy hydrolysate (NZ soy) was purchased from Quest. Pluronic F68 was purchased from BASF, and Trace elements A,B,C, were purchased from Cellgro. MTX was from Immunex.

Cell culture methods

DXB11 and Veggie CHO host cells

DXB11-CHO monolayers were maintained in DMEM-F12 media supplemented with 7.5% dialyzed FBS. Monolayers were subcultured three times weekly by trypsinization. Suspension CHO cultures were maintained in the DMEM-F12 media supplemented with decreasing amounts of serum as indicated in Figure 1. Serum-free IMX media (SF-IMX) consisted of the custom IMX2.3 base media containing pluronic and supplemented with trace elements, soy hydrolysates, intralipids, trace element mixes A, B, C, (to provide essential selenium and ferrous sulphate to supplement for iron carriers). Growth factors, IGF-1 and transferrin, were added at indicated concentrations. Glycine (2.52 mM), hypoxanthine (174 μM), thymidine (174 μM) were added to the DMEM F-12 and IMX2.3 media for the derivation and maintainence of Veggie-CHO cells in order to enrich for these components and minimize selection and outgrowth of any DHFR revertants. In addition, the peptone hydrolysates also provide additional amino acids.

Suspension adapted DXB11-CHO cell cultures were maintained as 100 mL cultures in 250 mL spinner flasks for the derivation of the Veggie-CHO cells. Cells were passaged three times weekly by centrifugation and 100% media replacement. Cells were seeded at a density of 2e5 cells mL^{-1} into enriched media (SF-IMX) with decreasing levels of serum or growth factor

Cell counts were determined by haemocytometer, and cell viability was determined by propidium iodide or trypan blue exclusion assay. DNA dyes, propidium iodide and Hoescht 33342 were used in flow cytofluorometry analysis to monitor viability and apoptosis of cell cultures (Dive et al., 1992; Sun et al., 1992). Cell doubling time (Td) was calculated using the following

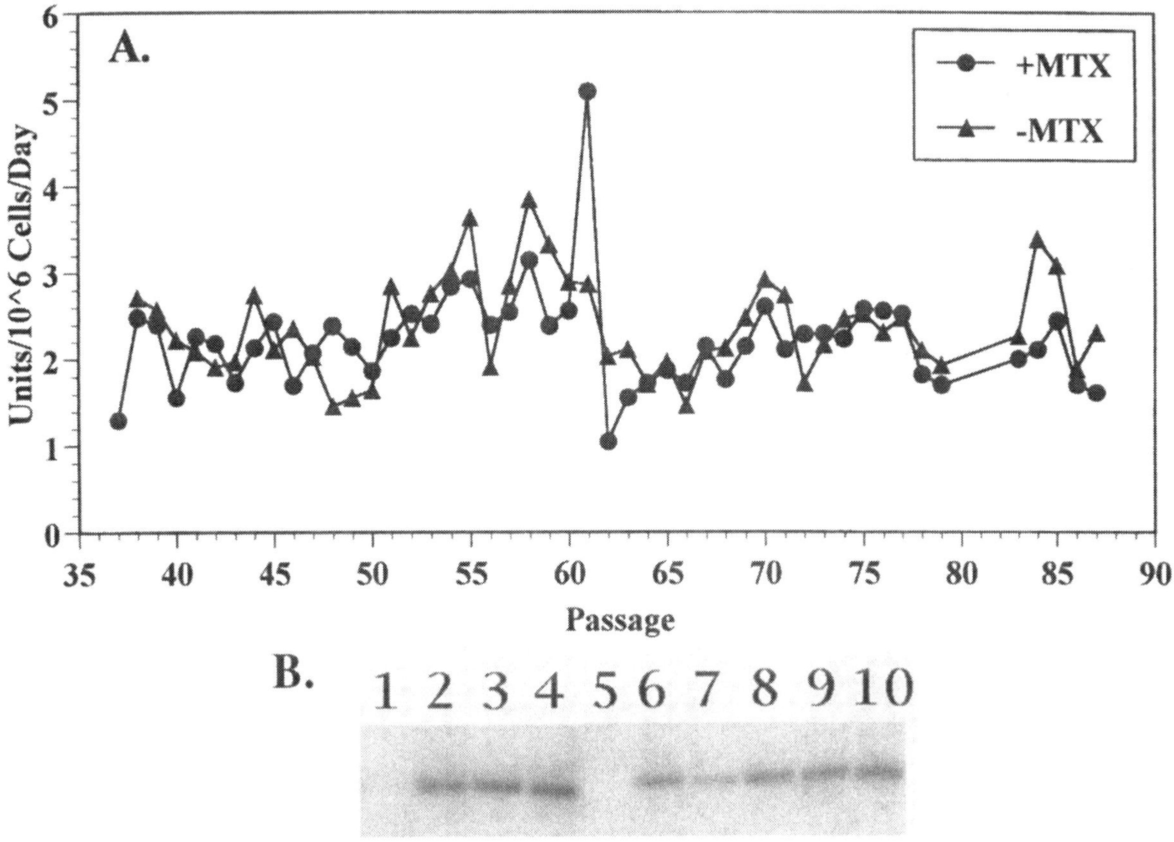

Figure 3a–b. Stability of recombinant FLT-3L expression in Veggie-CHO cells. A. Veggie -CHO:FLT-3L Cells were passaged in select media containing 150 nM MTX or in DHFR non-select media. FLT-3L protein levels in the culture media were determined at each passage using ELISA assays and used to calculate the specific productivity (qp). B. Chromosomal DNA was isolated from the cell cultures at various passages and FLT-3L gene copy number was determined using quantitative southern blot analysis (Sambrook et al., 1989). DNA was treated with restriction enzymes to release a single fragment of the FLT-3L coding sequence and probed with radiolabelled FLT-3L cDNA sequence. The signal was normalized for genomic DNA loading in each lane. DNA was loaded as follows: Lanes 1 and 5 control untransfected DXB-11, Lanes 2-4 Veggie-CHO:FLT-3L; Lane 2, 4 are DNAs from the +MTX culture at passage 37 and 63 respectively. Lane 3 is DNA from passage 63 no MTX. Lanes 6-10 are DNAs from DXB-11:FLT-3L. Lanes 6, 8, 9 and 10 are from +MTX DXB11 cultures at passage 66, 39, 62 and 82 respectively and lanes 7 is DNA from the nonselect (-MTX) DXB11 cultures at passage 66.

formula where t = days in culture, Nt is the final cell density and N0 is the initial seeding density.

$$Td = \ln 2/\mu .$$

$$\mu = \ln(\mathrm{Nt}/\mathrm{N0}) \times 1/\mathrm{t} .$$

Derivation and analysis of recombinant cell lines

The cDNA for the FLT-3L protein was subcloned into a bicistronic CHO expression vector with a CMV promoter and contained the selectable, amplifiable DHFR cDNA (Kaufman et al., 1991; Morris et al., 1997). Genomic DNA was isolated from recombinant cell lines and southern blot ananlysis was performed as described in the manual by Sambrook, Fritsch and Maniatis (1989). Recombinant protein expression in cell conditioned media was monitored using the Flt3L specific ELISA (Mckenna et al., 1995).

Effect of butyrate, IGF-1 and temperature on the recombinant Veggie-CHO and DXB11 cell lines was studied using 20 mL supension batch cultures in 125 mL agitated shake flasks. Cells were seeded at 1e6 to 2e6 cells mL^{-1} in the presence or absence of 1 mM butyrate and in the presence or absence of IGF1 (100 ng mL^{-1}) in serum free media (non selective SF-IMX). The lids of the shake flasks were slightly vented and cultures were incubated on a shaking platform at 31 or 37 °C for the period of the experiment. Cell counts, cell viability and protein expression were monitored as described above.

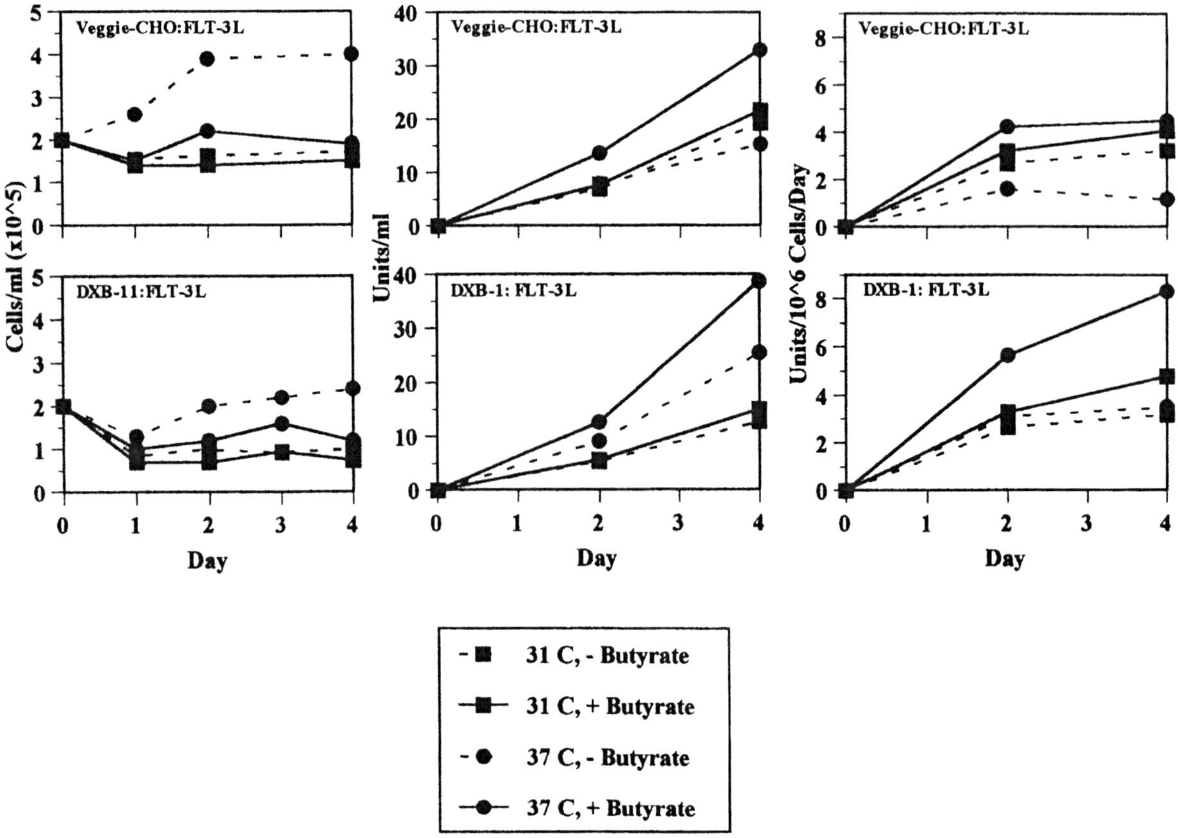

Figure 4. Effect of butyrate and temperature on growth and recombinant protein expression in Veggie-CHO cells. The cell density, protein titers and qp values of high density batch cultures of DXB-11:A and Veggie-CHO:FLT-3L in the presence of butyrate or absence of butyrate at 37 or 31 °C are shown.

Isolation of Veggie-CHO recombinants

Approximately 20×10^6 Veggie-CHO cells in mid-log phase of growth were plated in T75 tissue culture flasks on the day of transfection in serum-free media. Cells were left in a CO_2 incubator at 37 °C for one hour to allow the cells to adhere to the plates. Cells were then transfected with 50 μg of the FLT-3L cDNA expression vector in serum-free media using Lipofectamine reagent from Life Technologies following the general recommendations of the manufacturer. Transfection efficiency of Veggie-CHO cells with lipofectamine was several fold lower than that of DXB11 cells when monitored by transient expression of Lac Z protein. The transfected Veggie-CHO cells have a semi-adherent phenotype and were allowed to recover for 3 to 4 days and then gently trypsinized and selected for the selectable, amplifiable DHFR gene in the nucleoside-free, Selective SF-IMX1.2 media. The surviving cells or 'transfectant pools' were then selected in increasing concentrations of MTX (50 to 150 nM) in the Selective SF-IMX media to amplify recombinant protein expression. The amplified transfectant pool in 150 nM methotrexate expressed the highest level of recombinant FLT-3L molecule. This amplified transfectant pool was subcloned by limiting dilution in 96-well dishes using 3% dialyzed serum-supplemented, nucleoside-free DMEM-F12 (DHFR selective) media. The wells were visually screened for individual colonies and cells from these wells were transferred to 24-well dishes in serum-free media. Individual clonal colonies arising in the 96 wells are semi-adherent phenotype. Isolates were successively transferred to suspension growth in agitated shake flasks and then to spinner flasks in 10 to 12 passages and maintained as suspension cultures in Selective SF-IMX media. Clones were screened for recombinant FLT-3L expression using ELISA assay to select the final Veggie-CHO:FLT-3L recombinant clone. Although most of the clones assumed their serum-free and suspension growth phenotype, a small number

(20%) were terminated because of their long doubling times or low culture viabilites when passaged into serum-free and suspension growth

Isolation of DXB11 recombinants

Monolayers of adherent DXB11-CHO host cells were transfected using Lipofectamine reagent and plated as adherent cultures in nucleoside-free DHFR selection growth media (7.5% dialysed serum, DMEM-F12). Individual transfectant colonies were screened for FLT-3L expression using the ELISA assay. Colonies expressing the highest cytokine levels were further amplified in increasing amounts of methotrexate, up to a final concentration of 250 nM methotrexate. The 250 nM MTX pool was subcloned using the 96-well limiting dilution technique and the resulting colonies were screened for expression. The final DXB11:FLT-3L recombinant clone was identified as the highest expressing isolate. These cells were suspension adapted by passaging cells over several generations in serum-containing selective growth media in a spinner flask (100 mL culture in a 250 mL spinner flask)

Results

Isolation of Veggie-CHO Cells by adaptation of DXB11-CHO cells to growth in serum-free media and protein-free media

Since removal of serum leads to removal of cell attachment factors, we first adapted the adherent DXB11-CHO host cell line to grow in suspension in serum-supplemented DMEM-Hams F-12 media in stirred spinner flasks. These suspension adapted DXB11-CHO cell cultures were passaged every 2 to 3 days (by 100% media replacement) into decreasing levels of serum (Figure 1). We used a basal media (SF-IMX) consisting of a mix of Hams F-12 and DMEM enriched with amino acids and vitamins and supplemented with growth factors IGF-1 (100 μg L^{-1}) and transferrin (10 mg L^{-1}) to replace serum components. We have found that unlike insulin, IGF-1 acts as a growth and viability factor at physiological levels in DXB-11 cells (Thomas and Fung, 1994), and we used it initially, along with bovine transferrin, to replace serum.

Adaptation of the cells to the stepwise removal of serum from their growth media required approximately 100 passages (over approximately 160 cumulative cell doublings). These cells, designated SF-CHO, maintained stable growth (25 hour doubling time) and high cell viability (over 90%) in serum free media over 6 passages. This is indicative that the cells are completely adapted to grow in SF-IMX containing growth factors (IGF-1 and transferrin). Removal of transferrin from the serum-free media of SF-CHO could be accomplished in a few passages, without a significant drop in viability (below 90%) or growth (data not shown). Therefore, serum-free adaptation of the SF-CHO cells has also made these cells independent of transferrin as a growth factor. However, the SF-CHO cells continue to require IGF-1 for growth and viability. Sudden dramatic decrease in levels or removal of IGF-1 from the SF-CHO cell line lead to loss in culture viability and lack of growth. We chose to keep the transferrin in the SF-CHO media, to stabilize IGF-1, while we adapted these cells to lower concentrations of IGF-1. After the cells were adapted to be completely independent of IGF-1 for growth by passage 170 (approximately 383 cumulative cell doublings) the bovine transferrin was eliminated from the media. Cells rapidly adjusted to the removal of transferrin (Figure 1) to grow in protein-free media and were then designated as Veggie-CHO. Addition of iron-carriers, i.e. tropolone, to substitute for transferrin, did not enhance cell growth or viability.

Stepwise reduction of the serum or IGF-1 supplementation in the growth media often led to a transient decrease in cell viability at some of reduction steps, which was often accompanied by a transient increase in doubling time (see Figure 1). More dramatic reductions in the levels of serum and IGF-1 than those indicated in Figure 1 were attempted but led to loss of culture viability to below 50% (data not shown) and these cultures were not maintained further. The adherent DXB11-CHO cultures have been reported to grow with an average doubling time of 18 hr (Urlaub and Chasin, 1980). We found that our suspension and serum-free adapted cultures grew with an average doubling time of 28.6 hr. Removal of IGF-1 from this media led to cultures with an average doubling time of 25.7 hr and the Veggie-CHO cells have an average doubling time of 22.5 hr.

The serum-free adapted cultures and the parental DXB11 cells were analyzed for reversion of the DHFR phenotype using fluorescent methotrexate. FACS analysis of cells were incubated for 20 hr with fluorescent methotrexate showed similar profiles for both cell types with a single nonstaining population. In additon, cells were taken from different stages of the serum and growth factor weaning and tested for growth in serum-free and serum-containing DHFR-selective growth media. Lack of growth and the rapid loss of cell viability in the DHFR-selective media along with the fluorescent methotrexate analysis demonstrate the DHFR-deficient phenotype of the serum-free cell host.

Veggie-CHO and SF-CHO cells and the recombinant transfectant pools and clones generated from these cells (see below) can be cryopreserved in serum-free, protein-free media containing 10% DMSO. Cells were shown to recover and maintain growth and viability after 6 months of storage in liquid nitrogen.

Recombinant cytokine protein expression in Veggie-CHO

The Veggie-CHO serum-free host and the parental serum-requiring host DXB11-CHO were genetically engineered using the DHFR/MTX selection and gene amplification method to secrete high levels of recombinant FLT-3L (McKenna et al., 1995). The Veggie-CHO transfectant pools were selected for growth in 150 nM MTX to amplify recombinant protein expression. Selection of the Veggie-CHO transfectant pool in MTX levels higher than 150 nM concentration did not result in significant increase in the expression of this cytokine molecule. The DXB11 transfectant pools (i.e. colonies) were step amplified to a final 250 nM MTX concentration as described in the methods section.

Clonal FLT-3L expressing recombinant cell lines (Veggie-CHO:FLT-3L and DXB11:FLT-3L) were isolated from these final amplified pools using the 96-well limiting dilution technique. Growth and recombinant protein expression of the cultures were monitored and are shown in Figure 2. The Veggie-CHO:FLT-3L cells exhibit better growth; they have a shorter doubling time and reach higher cell densities than the DXB11:FLT-3L cells as illustrated in figure Figure 2 and Table I (also see Figure 4 below). These two cell lines produce similar final titers of recombinant protein.

Genetic stability of recombinant Veggie-CHO:FLT-3L cells

Genetic stability of the DXB11 and Veggie-CHO recombinant cell lines were evaluated by monitoring recombinant FLT-3L protein expression and integrated gene copy number of FLT-3L over 87 passages in cells grown in DHFR-selective or non-selective growth media. This corresponded to 133 cumulative cell doublings (CCD) in selective media and 124 CCD in non selective growth media. Protein expression and integrated gene copy number of FLT-3L were monitored

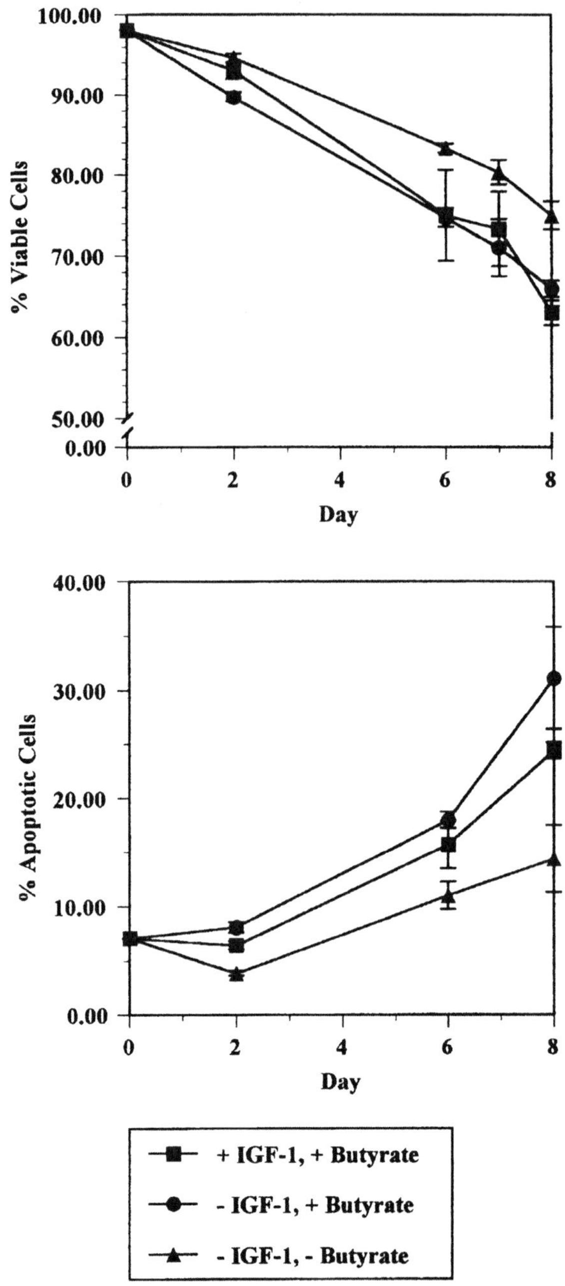

Figure 5. Effect of butyrate and IGF-1 on cell viability and apoptosis in batch cultures of Veggie-CHO. The viability and apoptosis in high density batch cultures of DXB-11:A and Veggie-CHO:FLT-3L in the presence of 1mM sodium butyrate or absence of butyrate with 100 μg mL^{-1} IGF-1 or absence of IGF-1 are shown.

Table 1. Specific protein productivity (qp) and growth rates (doubling time) of the two FLT-3L recombinant cell lines, Veggie-CHO:FLT-3L and DXB11:FLT-3L

Cell type	qp (Units 10e6·cells^{-1}·day^{-1}	Doubling time (hr)
DXB11:FLT-3L	5.58±2.39	42.1±11.8
Veggie-CHO:FLT-3L	2.27±2.94	21.2±2.9

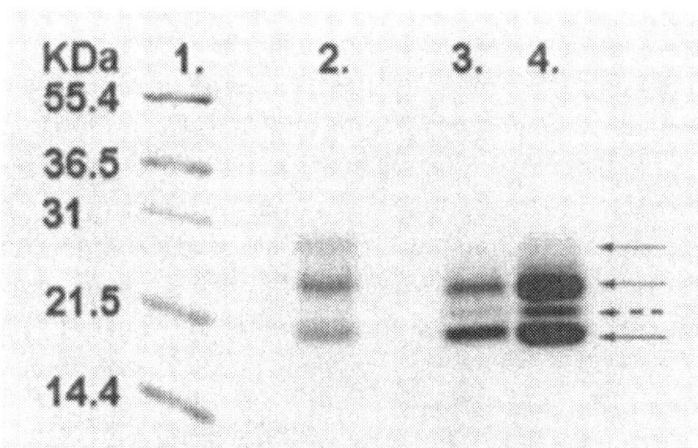

Figure 6. SDS-PAGE analysis of recombinant protein expressed in Veggie CHO and DXB11 cells. Cytokine protein purified from conditioned media derived from 10 day production cultures was analyzed on reducing 16% SDS PAGE gel electrophoresis. Positions of the three different glycoforms of the FLT-3L are indicated by solid arrows, position of the clipped FLT-3L is indicated by dashed arrow. Sample order is as follows: lane 1, molecular weight standards; lane 2, Veggie-CHO:FLT-3L (5 μg); lanes 3 and 4, DXB11:FLT-3L (5 and 10 μg, respectively).

using ELISA assays and southern blot analysis respectively (Figure 3). Results of these studies show that integrated copy number of FLT-3L in Veggie-CHO:FLT-3L cells was maintained at 8 copies/cell over the 87 passages in both, the presence or absence of selective pressure for the DHFR gene function (Figure 3, lanes 2–4).

The data in Figure 2 seem to suggest that the DXB11:FLT-3L recombinant cell line seems to be losing recombinant FLT-3L protein expression over 40 passages in selective growth media. We found however, that inoculum cultures derived from later passages (passage 38 and passage 80) of this culture do not show decreased protein expression when treated with sodium butyrate. We have also found that the integrated gene copy number of the FLT-3L cDNA stays stable at 12 copies and does not decrease in the later passages (Figure 3, lanes 8–10). It is possible that the apparent decrease in recombinant protein expression in the DXB11:FLT-3L may be a result of some reversible mechanism of gene down-regulation (i.e. histone acetylation, Chen et al., 1997), the effect of which seems to be reversed by sodium butyrate. We have not observed this butyrate-reversible decrease in protein expression in other DXB11 or Veggie-CHO cells recombinant cell lines. We also observed that both protein expression and integrated gene copy number of FLT-3L decreased over 40 passages in DXB11:FLT-3L cultures grown in DHFR-nonselect media (Figure 3, lane 7). This instability could be a function of the integration site in this cell line.

Effect of butyrate and temperature on growth, cell viability and recombinant protein expression in Veggie-CHO cells

Temperature reduction is a common manipulation of cell cultures which has variable effects on cell viability and protein expression (Bloemkolk and Gray, 1992; Weidemann et al., 1994). We have observed increased viability and decreased growth of the Veggie-CHO and DXB11-CHO batch cultures at lower temperatures (Figure 4). Reduction in temperature to 31 °C also leads to a small increase in specific protein expression (qp) in Veggie-CHO cultures but not in DXB11-CHO cultures as shown in Figure 4. We have found that the response in protein expression to changes in temperature varies for different proteins expressed in Veggie-CHO.

Sodium butyrate has been reported to have several different effects on cells; it has been shown to effect gene expression, cell cycle distribution of cells in culture and induce cell differentiation (Chen et al., 1997; Kruh, 1982). Both cell lines, DXB11-CHO and Veggie-CHO, respond similarly to sodium butyrate addition (Figure 4). Addition of sodium butyrate results in decreased growth and a significant increase in specific recombinant protein expression at 37 °C but not at 31 °C. The recombinant DXB11 cells however grow less than the Veggie-CHO:FLT-3L cells at 37 °C even in the absence of butyrate in these experiments. This is because the DXB11 cells have a long doubling time (40 hr) in serum and these cells are not adapted to grow in serum-free media used for these batch culture experiments (see Section Methods).

Butyrate was also found to have an adverse effect on viability of the cells (Figure 5). FACS analysis of the cells from these serum-free batch cultures to examine apoptosis revealed that 1mM sodium butyrate induces apoptosis in Veggie-CHO cultures (Figure 5) and DXB11 cultures (data not shown). Since we have previously found that IGF-1 suppresses butyrate induced apoptosis in DXB11-CHO cells we examined the effect of IGF-1 addition on cell viability and apoptosis in butyrate-treated Veggie-CHO cultures. Results of these experiments, shown in Figure 5, illustrate that addition of IGF-1 to the Veggie CHO batch cultures did not significantly enhance cell viability nor suppress butyrate-induced apoptosis in these cultures.

FLT-3L protein analysis

FLT-3L protein was purified from serum-free butyrate induced cultures of DXB11:FLT-3L and Veggie-CHO:FLT-3L with similar process parameters. Comparison of SDS PAGE analysis of the the purified Flt3L protein show that the protein from both sources has multiple banding pattern, with three FLT-3L protein species (Figure 6). Western blot and glycoform analysis demonstrates that the differently migrating species represent various glycoforms of the FLT-3L molecule. The FLT-3L molecule has two N-glycoslylated sites. The slowest migrating form represents most of the double glycosylated species and the middle band represents most of the single glycosylated species and the fastest migrating band represents the non-N-glycosylated species (C. Jochheim and L. Novick, personal communications). Although all three species are expressed in both cell types, the higher glycosylated species are more prevalent in the Veggie-CHO (Figure 6, lane 2) samples and least prevalent in the DXB11 samples (Figure 6, lanes 3 and 4). In addition the DXB11 sample has a fourth band (indicated with a dotted arrow in Figure 6) that represents a clipped version of FLT-3L. This fourth band is sometimes also seen in Veggie-CHO FLT-3L samples and can not be attributed to any one single culture parameter. Assay of the purified protein in '*in vitro*' bioassays show that the protein from both cell lines has similar biological activities. More detailed biochemical analysis aimed at better understanding of the post-translational modifications on this molecule and the effect of culture and purification conditons on these modifications are underway.

Discussion and conclusions

Veggie-CHO cells derived from DXB11-CHO cells maintain vigorous growth and high viability as well as their DHFR-deficient phenotype in serum-free, essentially protein-free media. We have demonstrated their utility as a host cell line for the stable, high level expression of active recombinant cytokine protein. Generation of serum-free recombinant cell lines directly from the preadapted Veggie-CHO host is both time and labor efficient since we have eliminated multiple rounds of adaptation, cloning and characterization that would be involved in isolating DXB11-CHO recombinants and then adapting them to growth in serum-free media. In addition the elimination of expensive growth factors and serum (except in 96-well cloning) during the generation of the recombinant Veggie-CHO lines and the recombinant protein production provided significant cost savings.

The Veggie-CHO cells that we derived by gradual reduction of serum and growth factors using an enriched mix of DMEM and HamsF-12 base media have retained their DHFR-deficient phenotype after extensive culture in serum-free conditions.

DHFR enzyme is involved in the synthesis of glycine, hypoxanthine and thymidine which are in turn required for DNA synthesis and cell division. Glycine (2.52 mM), hypoxanthine (174 μM), thymidine (174 μM) were added to the non-selective DMEM F-12 and SF-IMX medias used for the derivation and maintainence of Veggie-CHO cells in order to enrich for these components and minimize selection and outgrowth of any DHFR revertants. Although we tried more drastic reduction in serum and IGF-1 we choose to follow a very gradual removal of these in order to maintain the cell viability at over 60% in deriving the VeggieCHO cells. Two other groups (Gandor et al., 1995; Sinacore et al., 1996) reported the adaptation of DXB11-CHO cells to growth in serum-free media. The serum-free cells derived by these groups revert to a DHFR-positive phenotype during prolonged continuous culture in the serum-free media. Both these groups used more drastic removal of serum which resulted in drops in cell viability to 20% or lower. In addition the other investigators did not enrich their serum-free media with as high levels of nucleosides and glycine as we did in deriving the Veggie-CHO cells. Unlike the other groups, we also supplement our media with plant hydrolysates that also provide additional glycine and other nutrients. Any one or combination of these differences could have helped decrease the selective pressure and help maintain the DHFR-deficient phenotype in the Veggie-CHO cells.

Other investigators have found that previously adapted serum-free cultures reverted to serum-dependent phenotype when cultured in serum-containing media (Yao et al., 1991). The serum-free and suspension growth phenotype of the Veggie-CHO cells is stable after these cells have been maintained for several generations as adherent cultures in serum-containing media. This is demonstrated here in the cloning of the recombinant cell lines in 3% serum-containing media in 96 well-dishes and the subsequent culturing of the clones in serum-free media. Majority of the clones did not require a prolonged adaptation period to growth in serum-free media.

The recombinant Veggie-CHO line expressing FLT-3L can grow and maintain high viabilites in two other plant derived peptones as well as in the complete absence of peptone supplementation in the SF-IMX growth media. Experiments to study protein expression in these other media are under way. The observation that growth and viability of these Veggie-CHO cells are maintained even in the absence of peptone supplement suggests that these cells are not dependent on peptides present in the plant hydrolysates, which could function as growth factors.

To further evaluate their advantage and utility, the Veggie-CHO and the SF-CHO cells have been used as host cells for the stable expression of several other soluble proteins (receptors and ligands). Expression levels (titers) and functional activity of some of these proteins expressed in these serum-free hosts were comparable to those from suspension adapted recombinant lineages generated using the conventional, serum-requiring DXB11-CHO cells. In general, most of the recombinant serum-free CHO transfectant exhibit faster growth and attain higher cell densities when compared to the conventional DXB11-CHO derived, suspension adapted, recombinant cells expressing the same protein. The DXB11:FLT-3L cell line described here however exhibits an atypically long doubling time compared to other DXB11 cell lines. In addition, three other recombinant cell lines derived from the serum-free hosts and expressing different recombinant proteins, have been shown to retain stable recombinant protein expression over 30 passages.

It seems likely that the selection and amplification procedures used to derive clonal Veggie-CHO recombinant lines, i.e. selection and amplification in suspension pool cultures of the transfectants followed by clone isolation, allow for the selection and out-growth of the fastest growing transfectants. These fastest growing populations may not have the highest levels of specific recombinant protein expression. Hence the best expressing transfectants may be eliminated from the amplified pool before isolation or screening of final clones. The conventional DXB11-CHO cell lines were generated by isolating and screening individual colonies at most selection and amplification steps, thereby allowing transfectants with the best specific production, and often slower growth rates, to be screened and chosen. Cloning transfected Veggie-CHO pools at early stages of the selection and amplification steps may eliminate this bias toward the faster growing transfectants. Examining clones at earlier stages may lead to isolation of Veggie-CHO recombinants with a wider range in their growth and specific protein production phenotypes than we currently observe among the Veggie-CHO isolates derived from amplified pools. This may lead to the isolation of serum-free recombinant cell lines with higher specific production.

References

Adamson R (1994) Design and operation of a recombinant mammalian cell manufacturing process for rFVIII. Ann Hematol 68 Suppl 3: S9–S14.

Barnes D and Sato G (1980) Methods for growth of cultured cells in serum-free medium. Anal Biochem 102: 255–270.

Bloemkolk J-W and Gray MR (1992) Effect of temperature on hybridoma cell cycle and Mab production. Biotech Bioeng 40: 427–431.

Brasel K, McKenna HJ, Morrissey PJ, Charrier K, Morris AE, Lee CC, Williams DE and Lyman SD (1996) Hematological effects of Flt3 ligand *in vivo* in mice. Blood 88: 2004–2012.

Chen WY, Bailey EC, McCune SL, Dong JY and Townes TM (1997) Reactivation of silenced, virally transduced genes by inhibitors of histone deacetylase. Proc Natl Acad Sci USA 94: 5798–5803.

Dive C, Gregory CD, Phipps DJ, Evans DL, Milner AE and Wyllie AH (1992) Analysis and discrimination of necrosis and apoptosis (programmed cell death) by multiparameter flow cytometry. Biochim Biophys Acta 1133: 275–285.

Evans VJ, Bryant JC, Fioramonti MC, McQuilkin WT, Standford KK and Earle WR (1956) Studies of nutrient media for tissue and cells *in vitro*. A protein-free chemically defined medium for cultivation of strain L cells. Cancer Res 16: 77–86.

Gandor C, Leist C, Fiechter A and Asselbergs FA (1995) Amplification and expression of recombinant genes in serum-independent Chinese hamster ovary cells. FEBS Lett 377: 290–294.

Hamilton WG and Ham RG (1977) Clonal growth of chinese hamster cell lines in protein-free media. In Vitro 13: 537–547.

Kaufman RJ (1990) Selection and coamplification of heterologous genes in mammalian cells. Meth Enzymol 185: 537–566.

Kaufman RJ, Davies MV, Wasley LC and Michnick D (1991) Improved vectors for stable expression of foreign genes in mammalian cells by use of the untranslated leader sequence from EMC virus. Nucleic Acids Res 19: 4485–4490.

Kaufman RJ, Wasley LC and Dorner AJ (1988) Synthesis, processing, and secretion of recombinant human factor VIII expressed in mammalian cells. J Biol Chem 263: 6352–6362.

Kruh J (1982) Effects of sodium butyrate, a new pharmacological agent, on cells in culture. Mol Cell Biochem 42: 65–82.

McKenna HJ, de Vries P, Brasel K, Lyman SD and Williams DE (1995) Effect of flt3 ligand on the *ex vivo* expansion of human $CD34^+$ hematopoietic progenitor cells. Blood 86: 3413-3420.

McKinnon P, Ross M, Wells JR, Ballard FJ and Francis GL (1991) Expression, purification and characterization of secreted recombinant human insulin-like growth factor-I (IGF-I) and the potent variant des(1-3) IGF-I in Chinese hamster ovary cells. J Mol Endocrinol 6: 231–239.

Morris AE, Lee CC, Hodges K, Aldrich TL, Krantz C, Smidt PS and Thomas JN (1997) Expression augmenting sequence element (EASE) isolated from chinese hamster ovary cells. In: Carrondo MJT, Griffiths B and Moreira JLP (eds.) Animal Cell Technology, Kluwer Academic Publishers, The Netherlands, pp. 529–534.

Sambrook J, Fritsch E and Maniatis T (1989) Molecular Cloning: A Laboratory Manual, 2nd Edition. Cold Spring Harbor Press, Cold Spring Harbor.

Sinacore MS, Charlebois TS, Harrison S, Brennan S, Richards T, Hamilton M, Scott S, Brodeur S, Oakes P, Leonard M, Switzer M, Anagnostopoulos A, Foster B, Harris A, Jankowski M, Bond M, Martin S and Adamson SR (1996) CHO DUKX cell lineages preadapted to growth in serum-free suspension culture enable rapid development of cell culture processes for the manufacture of recombinant proteins. Biotech Bioeng 52: 518–528.

Sun XM, Snowden RT, Skilleter DN, Dinsdale D, Ormerod MG and Cohen GM (1992) A flow-cytometric method for the separation and quantitation of normal and apoptotic thymocytes. Anal Biochem 204: 351–356.

Thomas JN and Fung V (1994) Comparison of long R^3 IGF-1 with insulin in the support of cell growth and recombinant protein expression in CHO cells. In: Spier RE, Griffiths JB and Berthold W (eds.) Animal Cell Technology: Products of Today, Prospects for Tomorrow. (pp. ESACT Butterworth-Heinemann).

Urlaub G and Chasin LA (1980) Isolation of Chinese hamster cell mutants deficient in dihydrofolate reductase activity. Proc Natl Acad Sci USA 77: 4216–4220.

Weidemann R, Ludwig A and Kretzmer G (1994) Low temperature cultivation – A step towards process optimisation. Cytotechnology 15: 111–116.

Wood CR, Dorner AJ, Morris GE, Alderman EM, Wilson D, O'Hara Jr., RM and Kaufman RJ (1990) High level synthesis of immunoglobulins in Chinese hamster ovary cells. J Immunol 145: 3011–3016.

Wyatt DE and Doak J (1995) Performance characteristics of a protein-free medium for mammalian suspension cells. In: Beuvery EC, Griffiths JB and Zeijlemaker WP (eds.) Animal Cell Technology: Developments Towards the 21st Century, Kluwer Academic Publishers, The Netherlands, pp. 349–356.

Yao A, Huang EW and Rubin H (1991) Growth in high serum concentration leads to rapid deadaptation of cells previously adapted to growth in an extremely low concentration of serum. Proc Natl Acad Sci USA 88: 9422–9425.

Cytotechnology **28:** 43–48, 1998.
© 1998 Kluwer Academic Publishers. Printed in the Netherlands.

Collective experiences of adventitious viruses of animal-derived raw materials and what can be done about them

Stephen J. Wessman & Randall L. Levings
USDA, APHIS, VS, Center for Veterinary Biologics-Laboratory, P.O. Box 844, Ames, IA, 50010, U.S.A.

Received 26 August 1998; accepted 26 August 1998

Key words: animal-derived raw materials, bovine viral diarrhea virus, contamination

Abstract

Contamination of animal-derived raw materials with viruses, mycoplasmas, bacteria and fungi is common. These contaminants can interfere with the diagnosis of viral infection, and vaccines produced using infected cell cultures could lead to seroconversion or disease in the vaccinated animal. The purity, safety and efficacy of viral vaccines requires testing of the ingredients, cell substrates and final product. Methods for detection of viruses, especially bovine viral diarrhea virus, in nutrient serum, cell cultures, seed viruses and viral vaccines, and the frequency of their detection at the Center for Veterinary Biologics are discussed.

Abbreviations: BVDV: Bovine Viral Diarrhea Virus; FBS: Fetal Bovine Serum; FA: fluorescent antibody.

Introduction

Contamination of cell cultures used in the production of viral vaccines and in various diagnostic and serologic testing can be introduced in several ways. Primary and continuous cell lines may be contaminated if the original material is infected (tissues and eggs), if the dissociation material (usually trypsin) is virally contaminated, or if the nutrient serum contains virus. BVDV is probably the most common contaminant, as it is common in the bovine population.

Bovine viral diarrhea virus (BVDV) is currently classified as a member of the pestivirus genus of the Flaviviridae (Francki, 1991). Its two naturally occurring biotypes can be differentiated by the induction of cytopathic effects (CPE) in cell cultures (Lee, 1957; Gillespie, 1960). Isolates of the virus are diverse genetically and antigenically, although cross-reaction between strains by neutralization and other serologic methods is extensive (Gillespie, 1961; Fernelius, 1971; Castrucci, 1975; Horzinek, 1981). There is also significant serologic cross-reaction between BVDV and the other two members of the pestivirus group, hog cholera virus and border disease virus of sheep (Darbyshire, 1960; Osburn, 1973; Plant, 1973). The *in utero* infection of bovine fetuses can result in the contamination of serum processed from the fetuses. Contamination with BVDV of the fetal bovine serum (FBS) used in a test system can lead to false-positive pestivirus isolation by immunocytochemical or other techniques, or false-negative isolation of other viruses due to interference or masking. Contamination of test system FBS with antibody can lead to false-negative isolation, or false-positive bovine BVDV serology results. Persistent infection with non-cytopathic strains can occur.

Use of BVDV-contaminated FBS in cell cultures can lead to the infection of the cell cultures (Nuttall, 1977). Cell lines which have been reported to be infected with BVDV (Horzineck, 1981; Potts, 1989) include embryonic bovine lung (EBL), embryonic bovine kidney (EBK), Madin-Darby bovine kidney (MDBK), bovine turbinates (BoTur), embryonic bovine trachea (EBTR), porcine kidney lines PK-15,PK-2a and MVPK-1, rabbit kidney-13, Crandell feline kidney, feline lung (Fc2Lu), the Vero strain of African green monkey kidney and several mosquito cells. Infection of the cells with BVDV can alter the growth characteristics and viral yield of the cell lines, as well as leading to false- positive BVDV diagnosis by viral isolation.

Materials and methods

The techniques used by the Center for Veterinary Biologics Laboratory (CVB-L) to test fetal serum are those stated in 9 Code of Federal Regulations (9CFR) as 'Requirements for ingredients of animal origin used for production of biologics', plus additional testing. The testing is done sequentially, with any positive samples eliminated at each step. Initially, a virus neutralization test (1:2 dilution) of one to two animal pools is made. Negative samples are examined for the presence of virus by use in growth medium (15% by volume) for three passages on susceptible cells (MDBK, BoTur, EBK) with direct or indirect fluorescent antibody (FA) staining performed on the cells at each passage. Negative cultures are frozen and thawed three times and inoculated onto susceptible virus-free cells (MDBK, BoTur, EBK) for two passages and examined by FA for virus. Pools of four negative samples are ultracentrifuged sufficiently to pellet BVDV. The pellet is resuspended and inoculated onto susceptible cells as above and two cell culture passages are made, with a final examination by FA for BVDV. Only negative samples are retained. Samples meeting the criteria of being negative for antibody to BVDV by neutralization tests, negative to BVDV by FA on all tests (direct growth and examination, freeze-thaw and ultracentrifugation) and having satisfactory growth potential (non-toxic, normal cell morphology observed, no cytopathic effect and a normal time to confluency) for the indicator cells (usually MDBK and EBK), are then checked for antibody and interference to twenty viral agents (bovine viral diarrhea, infectious bovine rhinotracheitis, Parainfluenza 3, bluetongue and other bovine viruses; porcine rotavirus, pseudorabies, porcine parvovirus, transmissable gastroenteritis, rabies, canine parainfluenza, canine distemper and other canine viruses and feline corona, feline calicivirus and other feline viruses) and pooled for specific uses. Pools are checked for purity from bacteria, fungi and mycoplasma. Additionally, these pools are either heat-treated for 30 min at 56 °C, or gamma-irradiated at 25–35 kiloGrays at –40 °C (House, 1990; Hansen, 1993).

Testing of master cell stocks follows the 9CFR 'Requirements for cell lines used for the production of biologics', which includes a minimum of two subcultures over a period of twenty-one days while observing for CPE. The final culture is tested for cytopathogenic and/or hemadsorbing agents, and by fluorescent antibody staining for specific extraneous agents, including BVDV. An aliquot of the final cell culture is frozen and thawed three times and the disrupted cells are inoculated onto appropriate cell lines, including bovine cell lines susceptible to BVDV. The cells are observed for 14 days with at least one subculture, followed by tests for cytopathogenic and/or hemadsorbing agents, and a FA test for specific extraneous agents.

Master seed viruses and selected viral vaccines are tested according to 9CFR under 'Detection of extraneous agents in master seed virus'. Briefly, the test is the same as that described above for cell lines after disruption.

In the past few years an additional series of tests has been used to screen out contaminated lots of FBS. Our current supplier has used a combination of 1) initial indirect (IFA) and immunoperoxidase staining, 2) amplification by flask and roller bottle culture, and detection with IFA and immunoperoxidase staining and 3) serum neutralization testing for both BVDV antibody and virus, plus other selected viruses (IBR, PI3 and Reo).

Results

The frequency of virus or antibody detection in 1 L lots containing sera from no more than two fetuses in fiscal years from 1982 through 1990 ranged from 20 to 87% positives (Levings, 1990). A review of testing records for the period 1990 to 1997 shows a rejection rate ranging from 31.98 to 68.47% (see Table 1) for serum samples going through a prescreening consisting of viral isolation, immunoperoxidase staining and neutralization testing. Since 1988 the Cytology section of the CVB-L and the Diagnostic Virology Section of the NVSL have rejected 23 of 394 pretested 4–5 L serum lots for BVD virus; one for bovine herpesvirus-4, two for reovirus by IFA and one for BLV by agar-gel immunodiffusion, as well as three due to toxicity or poor growth, for a final rejection rate of 7.6%. In addition, 18 lots showed some interference with serum neutralization tests (15 for rabies, one for canine parainfluenza and two for feline panleukopenia).

One positive master cell stock was detected in 1986 out of 16 lines tested. None of 69 master cells were found positive in fiscal years 1987 to 1990 (Levings,

Table 1. Fetal bovine serum samples rejected by year (1990–1997)

	Year							
	1990	1991	1992	1993	1994	1995	1996	1997
Antibody	79	102	24	170	305	263	217	160
BVD virus	182	399	248	418	131	134	53	108
Rejected/total samples	266/621	506/919	275/860	658/961	447/853	422/854	309/734	272/516
Per cent rejected	42.8	55	32	68.5	52.4	49.4	42.1	52.7

1990). From 1991 to 1997 none of 81 cell lines were found positive. Prior to fiscal year 1986 this laboratory found EBK, PK2a,PK-15 and CRFK master cell stocks positive for BVDV.

In the last ten years various cell lines have been developed at the CVB or received as possible test substrates. All are treated as potentially contaminated until tested according to the protocol for master cell stocks. Among those found to be virally contaminated were MVPK-1, RK-13 and CRFK with BVDV, and ESN and Neuro2a with reovirus. Among primary cell cultures lamb testes and bovine primary lymph node and kidney were BVDV contaminated. In addition, bovine kidney, Vero, CRFK, MA104A, swine tonsil, swine lung and swine alveolar macrophages were found to be contaminated with mycoplasmas (*Acholeplasma laidlawii* and *Mycoplasma orale*).

One master seed virus was tested and found to be BVDV contaminated between 1986 and 1990. Extraneous BVDV tests of final product for fiscal years 1981 to 1990 had positive rates of 0 to 3.6% of the number tested, with no positives detected after 1983. Testing rates during that period dropped from 60% of eligible serials to 11%. During the past seven years vaccine lots have been found contaminated with bluetongue virus (Ianconescu, 1996) and BVDV, introduced by contaminated serum, and master seed viruses have been found contaminated with reovirus, also introduced by serum. Primary tissue culture derived from chicken eggs has been contaminated with avian reovirus, poxvirus and avian adenovirus, and reticuloendotheliosis virus and tenosynovitis virus have been isolated from avian vaccines produced on avian cell culture.

Discussion

Viral contamination of raw animal materials (FBS, eggs, tissues and other) is a significant problem for researchers, diagnosticians, and vaccine manufacturers. The results both prior to 1990 and since 1990 are similar to that reported by others (Horzineck, 1981) who have observed that from 20 to 50% of FBS was virus positive. Depending on the year up to 60% of sampled lots were either virus and/or antibody positive for BVDV. At the CVB-L, all lots of serum which are negative for both BVDV virus and antibody by the testing methods indicated above are still irradiated before being used in either testing or cell production.

Many cell types appear to be susceptible to BVDV contamination (Lee, 1957), including those from nonhuman and human primates, (Erickson, 1991) although viral replication may not occur (Xue, 1996; Black, 1997). Currently all cell lines are tested for BVDV by the techniques outlined in the Code of Federal Regulations. Others have suggested that this approach is only a minimum requirement (Black, 1997), and that enhanced testing will reduce the possibility that cell lines may be contaminated at a very low level, which might not be picked up by the above testing. BVDV can be isolated if the virus can be captured by inoculation onto a suitable susceptible substrate, not masked by antibody in the nutrient serum, and amplified by replication so that the increase in viral numbers produces a detectable effect in cell culture. For most materials the detection phase indicators are CPE, cell death, hemadsorption, giant cell and syncytia formation and vacuolation, or an immunoassay such as FA or immunoperoxidase staining. BVDV causes problems because although it captures and amplifies well, the detection phase can be variable. There are two biotypes of BVDV. The 1st type (for example,

NADL and Singer strains) causes a strong CPE, which can completely destroy the cell culture, but does not persist or chronically infect cell cultures. Included in this biotype is strain Oregon C-24V which causes a foamy, incomplete CPE, with no persistent infection. The second type is the one that is of most concern to vaccine producers and diagnosticians. It causes little or no CPE, and may cause persistent or chronic infections in cell culture. Examples of this biotype of BVDV are NY-1, Tn-131, and 890 strains. Among the additional enhancement techniques which may be used to detect these, and other strains of BVDV, are the use of the 9CFR as a minimum standard, plus inclusion of an FA stain at each passage, rather than just at 21 days; the use of a non-cytopathic strain (e.g. NY-1) as a positive control, rather than a cytopathic strain, or both; the use of a one to two day old, subconfluent cell culture of susceptible cells; and the use of a 100% acetone fixation, with no water rinse to remove buffer crystals, since the BVDV FA stain is of lower avidity than many other FA stains (Black, 1997). A good FA scope, with clean lenses and a mercury or xenon light source, along with a good FA conjugate which has not been freeze-thawed or exposed to light for extended periods of time will also enhance the ability to detect this virus. Additional techniques which may enhance a laboratory's ability to detect BVDV are IFA staining using monoclonal antibodies against both CPE and non-CPE strains and polymerase chain reaction (PCR) primers to detect virus as well as distinguish between the two BVDV genotypes and subtypes Ia and Ib. PCR is very sensitive, but sampling problems may make this a test that may not be biologically relevant, especially if it picks up inactivated particles. The threshold (Lower, 1991) for detection (1000 to 10 000 per mL^3) may also be above that contained in many materials (especially when only 25–50 μL are sampled) as opposed to the 15 to 100 mL of serum often used for cell growth and viral amplification before testing. It is possible to detect a nonCPE strain in a modified live virus CPE vaccine strain, using rapid passes on bovine turbinates to preferentially increase the nonCPE strain and then IFA and PCR to detect contaminants of different genotypes. For those contaminants of the same genotype as the vaccine strain, plaque purification and subsequent biotype testing of numerous isolates would be required. Currently the CVB has the expertise to do this kind of testing in special cases, but not the staffing to do this routinely. PCR testing should detect both live BVDV and fragments of BVDV nucleic acid, as may FA staining. The gold standard for detection of BVDV still remains virus isolation.

The routine testing of trypsin by the 9CFR also is probably a minimum standard. It has been reported that, while only 1 of 8 lots of trypsin was found positive by the 9CFR, an enhanced test detected porcine parvovirus in 5 of the same 8 lots (Black, 1997). This increase in detection rate was due to increasing the subculturing from a minimum of one time over 14 days to 5 times over six weeks (although later tests showed that 2 subcultures over 21 days was optimal), using a pH 9.5 buffer to solubilize and dissolve the powdered trypsin and by using roller bottles instead of stationary flasks. Together, these changes were reported to increase the sensitivity of the test 10-fold. The FA test showed a typical nuclear fluorescence.

There have been many changes noted in the past seven years. Quality assurance programs and filtration have improved the quality of the nutrient products available, better irradiation techniques have improved the radiation process, and improvements in staining and amplification techniques have increased our ability to detect virus in raw materials, cell and viral stocks, and vaccines. The ability to use techniques like immunoperoxidase staining, PCR, and monoclonal antibodies to increase specificity, the ability to differentiate between genotypes, and the ability to differentiate contaminant BVDV from product BVDV will improve quality control.

Quality assurance programs at firms that process raw materials and produce nutrient serums and other materials have improved. Serum lots that were previously submitted for testing are now quality controlled at the point of origin and/or processing, so there are fewer lots of serum to be rejected. Also irradiation is much more rigorously controlled, so more lots are satisfactory for use in cell production, resulting in less reliance on heat inactivation or chemical treatments. Biologics firms still have the primary responsibility for monitoring the ingredients used in the production of vaccines. If the serum used is antibody and virus negative (and irradiated), the master cell and master seeds are negative, and aseptic techniques are used in production, then pure vaccines should be produced.

Filtration has greatly improved the quality of serum and seems effective in removing almost all mycoplasmas, and reducing or eliminating many viruses (Art To Science, 1989). Porcine parvovirus, however, shows a limited titer decrease after filtration. It is still necessary to test any lots of serum before they are used in production or testing.

What lies ahead? As harmonization efforts continue (Hodgson, 1993; Sanhueza, 1996), panels of experts will probably be formed, to determine the testing and inspection strategies necessary. They may recommend that a) the current tests be continued, with sensitivity, reproducibility, and comparability agreed upon; b) the tests continue to be improved; or c) the current tests be eliminated and a new set of standards agreed upon. All of this involves risk assessment, (Sutmoller, 1997) and will involve such issues as test validation (Vicari, 1991) and the use of irradiation even after negative testing results. In addition, new tests, such as use of transgenic mice for prion isolation, continue to be developed. There are good testing methods already in place to screen raw animal products for the presence of viruses and other contaminants, such as the use of ELISA (Bock, 1997), PCR (Kappeler, 1996; Laamanen, 1997) and monoclonal antibodies, in conjunction with the standards of virus isolation and serum neutralization testing. Our challenge is to continue to improve them.

References

Art To Science (1989) 40 nm filtration method reduces FBS microbial contaminants. Art to Science 8(1): 1–2.

Black J (1997) Pitfalls in the detection of BVD virus by 9CFR APHIS Veterinary Biologics Public Meeting: 312–335.

Black J (1997) False negative PPV virus recovery from trypsin using minimum 9CFR protocol. APHIS Veterinary Biologics Public Meeting: 335–346.

Bolin SR, Mathews PJ and Ridpath JF (1991) Methods of detection and frequency of contamination of fetal calf serum with bovine viral diarrhea virus and antibodies against bovine viral diarrhea virus. Journal of Veterinary Diagnostic Investigation 3: 199–203.

Bock RE, Rodwell BJ and McGowan M (1997) Detection of calves persistently infected with bovine pestiviruses in a sample of dairy calves in south-eastern Queensland. Australian Veterinary Journal 75: 656–659.

Castrucci G, Avellini G, Cilli V, Pedini B, McKercher DG and Valente C (1975) A study of immunologic relationships among serologically heterologous strains of bovine viral diarrhea virus by cross immunity tests. Cornell Veterinarian 65: 65–72.

Code of Federal Regulations (1996) Title 9, Part 1, Sections 113.52–113.55. Office of the Federal Register, National Archives and Records Administration, Washington, D.C.

Darbyshire JH (1960) A serological relationship between swine fever and mucosal disease of cattle. Veterinary record 72: 331.

Deregt D, Bolin SR, Heckert RA and Loewen KG (1994) Monoclonal antibodies to bovine viral diarrhea virus:cross reactivities to field isolates and hog cholera virus strains. Canadian Journal of Veterinary Research 58: 71–74.

Erickson GA, Bolin SR and Landgraf JG (1991) Viral contamination of fetal bovine serum used for tissue culture: risks and concerns. Developments in Biological Standardization 75: 173–175.

Fernelius AL, Lambert G and Booth GD (1971) Bovine viral diarrhea virus-host cell interactions: serotypes and their relationships to biotypes by cross neutralization. American Journal of Veterinary Research 32: 229–236.

Francki RIB, Fauquet CM, Knudsen DL and Brown F (1991) Classification and nomenclature of viruses. Archives of Virology, Supplement 2: 223–233.

Gillespie JH, Baker JA and McEntee K (1960) A cytopathogenic strain of virus diarrhea virus. Cornell Veterinarian 50: 73–79.

Gillespie JH, Coggins L, Thompson J and Baker JA (1961) Comparisons by neutralization tests of strains of virus isolated from virus diarrhea and mucosal disease. Cornell Veterinarian 51: 155–159.

Hansen G and Wilkinson R (1993) Gamma radiation and virus inactivation:new findings and old theories. Art to Science 12(2): 1–6.

Hodgson J (1993) Fetal bovine serum revisited. Bio/Technology 11: 49–53.

Horzineck MC (1981) Non-arthropod borne Togaviruses. Academic Press Inc., London.

House C, House JA and Yedloutschnig RJ (1990) Inactivation of viral agents in bovine serum by gamma irradiation. Canadian Journal of Microbiology 36: 737–740.

Ianconescu M, Akita GY and Osburn BI (1996) Comparative susceptibility of a canine cell line and bluetongue virus susceptible cell lines to a bluetongue virus isolate pathogenic for dogs. *In Vitro* Cellular and Developmental Biology-Animal 32: 249–254.

Kappeler A, Lutz-Wallace C, Sapp T and Sidhu M (1996) Detection of bovine polyoma contamination in fetal bovine sera and modified live viral vaccines using polymerase chain reaction. Biologicals 24: 131–135.

Laamanen UI, Neuvonen EP, Yliviuhkola EM and Veijalainen PM-L (1997) Comparison of RT-PCR assay and virus isolation in cell cultures for the detection of bovine viral diarrhea virus (BVDV) in field samples. Research in Veterinary Science 63: 199–203.

Lee KM and Gillespie JH (1957) Propagation of virus diarrhea virus of cattle in tissue culture. American Journal of Veterinary Research 18: 952–955.

Levings RL and Wessman SJ (1990) Bovine viral diarrhea virus contamination of nutrient serum, cell cultures and viral vaccines. Developments in Biological Standardization 75: 177–181.

Lower J (1991) Virological aspects of the quality control of biologicals:quantitative considerations. Developments in Biological Standardization 75: 221–226.

Nuttall PA, Luther PD and Stott EJ (1977) Viral contamination of bovine foetal serum and cell cultures. Nature 266: 835–837.

Osburn BI, Clarke GL, Stewart WC et al. (1973) Border disease-like syndrome in lambs. Antibody to hog cholera and bovine viral diarrhea viruses. Journal of the American Veterinary Medical Association 163: 1165–1167.

Plant JW, Littlejohns JR, Gardiner AC et al. (1973) Immunological relationships between border disease, mucosal disease and swine fever. Veterinary Records 92: 455.

Potts BJ, Sawyer M, Shekarchi JC, Wismer T and Huddleston D (1989) Peroxidase labeled primary antibody method for detection of pestivirus contamination in cell cultures. Journal of Virology Methods 26: 119–124.

Rossi CR, Bridgman CR and Kiesel GK (1980) Viral contamination of fetal lung cultures and bovine fetal serum. American Journal of Veterinary Research 41: 1680–1681.

Sanhueza S (1996) International scientific conference on viral safety and evaluation of viral clearance from biopharmaceutical products. Vaccine 14(3): 244–246.

Smithies LK and Modderman E (1975) BVD virus in commercial fetal calf serum and 'normal' and aborted fetuses. Proceedings of the Annual Meeting of the American Association of Veterinary Laboratory Diagnosticians 18: 113–119.

Sutmoller P and Vose DJ (1997) Contamination of animal products: the minimum pathogen dose required to initiate infection. Rev. sci. tech. Office Internationale Epizooties 16(1): 30–32.

Vicari G (1991) Validation of purification procedures for removing and/or inactivating viruses in biologicals:points to consider. Developments in biological standardization 75: 227–232.

Xue W and Minocha HC (1996) Identification of bovine viral diarrhea virus receptor in different cell types. Veterinary Microbiology 49: 67–79.

Cytotechnology **28:** 49–52, 1998.
© 1998 *Kluwer Academic Publishers. Printed in the Netherlands.*

An overview of viral and viral-like agents in cell culture systems

John C. Petricciani
International Association of Biological Standardization, P.O.Box 1925, Palm Springs, CA 92263-1925, U.S.A.
E-mail: jpmdpsca@aol.com

Received 26 August 1998; accepted 26 August 1998

Key words: adventitious agents, biological products, contamination, viral contaminants

Abstract

The first recognition of seriously contaminated biological products occurred almost 100 years ago. More recently, in the second half of this century, the potential for contamination by viral and viral-like agents became obvious when the SV-40 virus was identified as an endogenous agent in the cell cultures used to produce polio vaccines. The history of major contamination incidents is reviewed, and current issues are discussed.

Introduction

The historical background of viral and viral-like agents that have contaminated biological products for human use can help to put into context some of the data and issues that will be presented in the papers that follow. The issue of microbial agents – both real and hypothetical – as contaminants of biological products is a very old one. Transmissible spongioform encephalopathies are the latest, and perhaps the most challenging, in a series that began almost 100 years ago. Whether there are lessons to be learned from the past that can be applied to contemporary issues is, of course, the big question which will be addressed at the end of this paper.

From the very beginning, the regulation of biological products in the U.S. and elsewhere has evolved in response to crises. It also has relied heavily on science and scientists in the the development of policy and decisions in response to those crises. The first incident involving a biological product which attracted federal interest in the U.S. occurred at the beginning of this century during a diphtheria epidemic in St. Louis in which 14 children inoculated with a diphtheria antitoxin died – not of diphtheria, but of a tetanus contaminant in the antiserum (Report, 1901). While that incident involved a bacterial contaminant, it is relevant to the issue of viral and viral-like agents because it resulted in the 1902 Federal Virus Toxin Act, much of which still remains intact as part of the Public Health Service legislation that FDA uses to regulate most biological products.

No similar incidents occurred until 1955 when the prospects seemed bright for controlling poliomyelitis with the new Salk vaccine produced in primary Rhesus monkey kidney cells. In that same year, the potential for creating a biological disaster of major proportions in national and global immunization programs was dramatized by a case, commonly referred to as the Cutter incident, in which over 260 children contracted polio after receiving the Salk vaccine of which 10 died because of the incomplete inactivation of the wild-type poliovirus used to produce the vaccine. While this was not, strictly speaking, an endogenous or exogenous contaminant of the product, that tragedy re-emphasized the necessity for maintaining strict control of biological products by the manufacturer and the national regulatory agency.

Viral agents

Subsequently, the first *viral* contaminant of potential major public health significance in a biological product was identified. The virus was SV40 and the products were again inactivated polio vaccines. SV40 is an endogenous virus of Rhesus monkeys and was a frequent contaminant of primary Rhesus monkey kidney cell cultures used to produce polio vaccines (Girardi, 1962). The public health issues revolved around the fact that SV40 was shown to be tumorigenic in hamsters. There was therefore justifiable concern that the SV40 contaminant in polio vaccines might lead to cancer in the vaccinated children. That concern was heightened with the finding that SV40 could transform normal human diploid fibroblasts, WI-38, *in vitro*.

There were two major responses to the SV40 incident. First, the National Cancer Institute funded a long-term follow-up study of a cohort of those children who had received SV-40 along with the polio vaccine. After 20 yr, there was no evidence of an increased frequency of cancer (Mortimer, 1981). A more recent 30-year followup study confirmed the conclusions of previous reports that there is not a significant correlation between SV40 exposure during the neonatal period and certain cancers (Strickler, 1998). Second, the cell substrate was switched from Rhesus to African Green monkey kidney because SV-40 is not an endogenous virus for that species.

Nevertheless, the issue of endogenous viruses in nonhuman primates as potential contaminants continued to swirl as more and more latent viruses were identified. The potential for serious consequences was accentuated in the 1970s with the deaths of a number of persons who came into direct or indirect contact with African monkeys that were infected with what became known as the Marburg virus (Gear, 1975).

Two solutions to the problems associated with using primary cells as substrates for the production of biologicals were advanced. One was to develop closed and carefully monitored colonies of animals, and in some cases specific pathogen-free colonies, from which primary cell cultures could be derived. The other approach was put forward by Dr. Hayflick and his colleagues (Hayflick, 1961). They developed an alternative to primary monkey kidney cells, namely, the WI-38 human fibroblast cell line which has normal phenotypic characteristics, such as a finite life, a diploid chromosomal constitution, and a lack of tumorigenic potential. In addition, a variety of tests showed no evidence of viral contamination. However, WI-38 was slow to be accepted in the U.S., primarily because of Dr. Albert Sabin's position that they might contain a latent human leukemia virus for which there were no tests (Sabin, 1969). Obviously, if you don't know how to test for an agent that *might* be present, about all one can do is provide as much indirect safety evidence as possible. Eventually, that approach worked, and products derived from WI-38 were approved by FDA, but long after they had been approved in Europe. In retrospect, Dr. Sabin was correct to have been concerned about human leukemia agents since they were discovered years later. However, he had focussed on the wrong cell type.

The next viral contamination episode came in the 1970s when we discovered bacteriophages in bovine sera and in live viral vaccines that had been derived from cell culture systems in which bovine sera had been used to propagate the cells (Chu et al., 1973; Petricciani et al., 1973). The finding of phages in those biological products had more regulatory implications than public health significance, since phages were simply markers of prior bacterial contamination in the bovine sera, and did not pose a health risk. That discovery, however, focused attention once again on the issue of controlling the quality of materials used in the manufacture of biological products.

Nevertheless, FDA regulations clearly prohibited the marketing of products with a live viral contaminant. A benefit risk analysis led to the conclusion that it would be better to allow vaccines on the market with small amounts of phages in them than to remove those products from the market for some extended period of time while phage-free vaccines were manufactured. The fix, obviously, was to change the FDA regulations. In what is probably the record for a fast-track publication of a regulation, FDA allowed what was termed 'unavoidable bacteriophage' in live viral vaccines while steps were taken to produce bovine sera that were phage-free.

World Health Organization (WHO)

In the mid-1980s, a WHO expert group reviewed the entire area of cell substrates, and made it clear that viral and viral-like contamination was the major concern, and that the manufacturing process for biological products must be shown to eliminate any potential contaminating agents pathogenic for humans, taking into account the type of cell being used as well as the raw materials used in production (Report, 1987).

When the Study Group examined this area, they divided cells into three risk categories with respect to their potential for carrying infectious agents *pathogenic* for human beings, namely, low, medium, and high. In making these risk distinctions, they noted that human diploid cells had been used successfully for over 20 yr and that there was no evidence to support Dr. Sabin's concern about a latent human oncogenic agent. Similarly, even though avian tissues and cells are well-known to harbor viruses, there is no evidence for transmission of disease to human beings by products from these sources. For example, large quantities of yellow fever, measles, and live influenza vaccines have been produced for many years in eggs that contain avian leukosis viruses, but there is no evidence that these products have caused leukemia or lymphoma during their long history of use for human immunization. However, it is important to note that in 1942, approximately 50 000 U.S. military personnel developed hepatitis B from Yellow Fever vaccine due to contaminated human serum which was used as a stabilizer in the vaccine.

In considering mammalian cells, the WHO Study Group suggested that medium risk is associated with cells derived from non-hematogenous cells. Here, they recognized that rodent cells, like avian cells, are well-known to contain viruses or virus-like structures that are visible on electron microscopy. But products have been produced from such cells without any evidence of transmitting infectious diseases to humans. For example, primary hamster kidney cells have been used to produce rabies vaccine, and Chinese hamster ovary cells have been used extensively to manufacture recombinant DNA products.

And finally, the WHO group concluded that blood and bone marrow cells of human and nonhuman primate origin were high risk based on clear evidence that such human cells frequently transmit infectious diseases. Similarly, there was heightened concern for the potential of transmitting diseases by hybridomas where one partner was of human or nonhuman primate origin.

The primary reason the Study Group viewed viral or viral-like contaminants as the most significant issue from a public health and safety point of view is because they have been the single most important risk in the past, if one looks at the history of actual transmission of infectious diseases by biological products. It is strikingly evident that *primary tissues* – especially human – and *primary cells* have been the source of several problems that occurred in the past. Specifically, as already mentioned, polio vaccines carried SV-40 from primary Rhesus monkey kidney cells. Fortunately, there is no evidence to suggest that the children who received SV-40 in polio vaccines were at greater risk for cancer than those who did not get SV-40 (Strickler, 1998).

Human Plasma Protein Fraction (PPF) from one producer transmitted hepatitis B in the 1970s, probably as the result of a manufacturing flaw, and some deaths occurred. Human corneas derived from a cadaver transmitted rabies after they were transplanted because the medical history of the cadaver was ignored. Similarly, growth hormone derived from the pituitaries of human cadavers led to the development of Creutzfeld-Jacob disease (CJD) in many of the children who received that product. In addition, dura mater transplanted from cadavers has transmitted CJD. Factor VIII (Antihemophilia Factor) and Factor IX Complex derived from human plasma were licensed in the U.S. in the 1960s. Because neither tests for donor screening nor virus inactivation procedures for these products were available, they transmitted hepatitis viruses. When HIV entered the plasma supply in the late 1970s or early 1980s, it also was transmitted to hemophilia patients, with over 4000 of them in the U.S. having developed AIDS since that time. Currently, we are worried about: 1) the agent responsible for bovine spongioform encephalopathy as a contaminant of beef, especially in Europe; 2) prions as a potential contaminant of blood and blood products; and 3) retroviruses as potential pathogens in transplants of organs from various species to humans.

In contrast to these examples of actual transmission of disease, or concern about possible transmission, cell lines and raw materials can be tested extensively to rule out the presence of potential pathogens for humans. The ability to pre-test cell lines – whether diploid or continuous – is the single most significant factor that has resulted in the *absence* of any documented case of infectious disease transmission by *any* product of which I am aware that was derived from well-characterized and tested cell substrates.

The most recent actions related to potential viral contaminants in biological products occurred in 1996 when the WHO Expert Committee on Biological Standardization and the International Conference on Harmonization revised their respective guidance documents for both diploid cell and continuous cell lines. These updated recommendations recognize the evolution in thinking about cell substrates that has taken place over the past 30 yr, as well as the technology that is currently available, and they emphasize state-of-the-art tests for viruses endogenous to the cells of the species being used for production. And, of course, improved tests also are used to detect *exogenous* contamination of both the raw materials and the product. For example, if a new human cell line were proposed as the cell substrate for various products, and since we know quite a bit about human endogenous viruses, the polymerase chain reaction could be used in the characterization of the cell line to test for potential latent agents such as CMV, EBV, HTLV-1 and 2, HIV-1 and 2, and hepatitis A, B, and C.

Summary and conclusions

That brings us up to the present, and I think it's fair to ask what we have learned in the past century that could help us in facing the future when we consider the risk of products transmitting infectious viruses and viral-like agents. In a sense, we have been both frightened and reassured. Extraneous agents have, on a few occasions, been administered inadvertently to thousands of recipients of biological products. On balance, however, considering the hundreds of millions, if not billions, of doses and exposures that have occurred during this century, it is remarkable that there have been relatively few problems. However, we need only to think back to the early 1980s when HIV first appeared, or more recently to the difficult issues that prions present, to understand that, as Dr. Fred Robbins often pointed out, we should be prepared for surprises and not be deluded into thinking that we have all the answers or that everything is under perfect control.

If we look at the history of the *actual* transmission of infectious diseases by biological products, there are several points that emerge. First, problems are almost inevitable when strict control of the manufacturing process (including design, validation, and routine execution) or of the starting materials is not maintained, and when up-to-date technology is not applied. Second, a safe product can be made unsafe by the addition of a contaminated excipient (diluent, stabilzer, etc.). Third, clearly the biggest risk of the past has been associated with human agents that have contaminated primary human cells and tissues. The lesson here is that if it is possible to avoid using primary human material, it is safer and wiser to do so. Obviously, that is easier said than done, especially if one has no other choice, at least for the present, such as with whole blood and many blood derivatives. The final point that becomes obvious from past experience is that conventional wisdom does not necessarily lead us down the right path. For example, the initial anxiety and apprehension concerning the safety of human diploid cells as well as the bias towards using primary tissues and cells did not prove to be well-founded. As we move into the future, we should look carefully at these experiences of the past to guide us in making appropriate decisions about risk.

Acknowledgment

Dr. John Finlayson provided particularly helpful advice and examples of past viral transmissions.

References

Chu FC, Johnson JB, Orr HC, Probst PG and Petricciani JC (1975) Bacterial virus contamination of fetal bovine sera. In Vitro 9: 31–34.

Gear JSS, Cassel GA, Gear AJ et al. (1975) Outbreak of Marburg virus disease in Johannesburg. British Med J 4: 489–493.

Girardi AJ, Sweet BH, Slotnick VB et al. (1962) Development of tumors in hamsters inoculated in the neonatal period with vacuolating virus, SV40. Proc Soc Exp Biol Med 109: 649–660.

Hayflick L and Moorhead PS (1961) The serial cultivation human diploid cell strains. Exp Cell Res 25: 585–621.

Mortimer Jr, EA, Lepow MI, Gold E, Robbins FC, Burton GJ and Fraumeni Jr, JF (1981) Long-term followup of persons inadvertently inoculated with SV40 as neonates. N Eng J Med 305: 1517–1518.

Petricciani JC, Chu FC and Johnson JB (1973) Bacteriophages in live virus vaccines. Proc Soc Exp Biol Med 144: 789–792.

Report (1901) NY Med J. 74: 977.

Report of the meeting of the study group on biologicals (1987) WHO Technical Report Series, No. 747.

Sabin A (1969) Discussion, Session V of the International Conference on Rubella Immunization. Amer J Dis Child 118: 378–381.

Strickler HD, Rosenberg PS, Devesa SS, Hertel J and Fraumeni Jr, JF and Goedert JJ (1998) Contamination of poliovirus vaccines with simian virus 40 (1955–1963) and subsequent cancer rates. J Am Med Ass 279: 292–295.

Cytotechnology **28:** 53–64, 1998.
© 1998 *Kluwer Academic Publishers. Printed in the Netherlands.*

New adenovirus vectors for protein production and gene transfer

Bernard Massie[1*], Dick D. Mosser[1], Maria Koutroumanis[1], Isabelle Vitté-Mony[1†], Linda Lamoureux[1], France Couture[1‡], Luc Paquet[1§], Claire Guilbault[1,2], Julie Dionne[1], Dounia Chahla[1], Pierre Jolicoeur[1¶] & Yves Langelier[2]

[1] *Institut de Recherches en Biotechnologie, 6100 Royalmount Avenue, Montréal, Québec, H4P 2R2 Canada E-mail: bernard.massie@nrc.ca*

[2] *Institut du Cancer de Montréal, Centre de Recherche Louis Charles Simard, 1560 Sherbrooke Est, Montréal, Québec, H2L 4M1 Canada*

Received 25 August 1998; accepted 25 August 1998

Key words: adenoviral recombinant, functional genomics, gene therapy, green fluorescent protein, inducible gene expression, protein production

Abstract

Based on two new adenovirus expression cassettes, we have constructed a series of Ad transfer vectors for the over-expression of one or two genes either in a dicistronic configuration or with separate expression cassettes. Inclusion of the green or blue fluorescent protein in the vectors accelerates the generation of adenovirus recombinants and facilitates the functional characterization of genes both *in vitro* and *in vivo* by allowing easy quantification of gene transfer and expression. With our optimized tetracycline-regulated promoter (TR5) we have generated recombinant adenoviruses expressing proteins, that are either cytotoxic or which interfere with adenovirus replication, at levels of 10–15% of total cell protein. Proteins that are not cytotoxic can be produced at levels greater than 20% of total cell protein. As well, these levels of protein production can be achieved with or without adenovirus replication. This yield is similar to what can be obtained with our optimized human cytomegalovirus-immediate early promoter-enhancer (CMV5) for constitutive protein expression in non-complementing cell lines. Using the green fluorescent protein as a reporter, we have shown that a pAdCMV5-derived adenovirus vector expresses about 6-fold more protein in complementing 293 cells and about 12-fold more in non-complementing HeLa cells than an adenovirus vector containing the standard cytomegalovirus promoter. Moreover, a red-shifted variant of green fluorescent protein incorporated in one series of vectors was 12-fold more fluorescent than the S65T mutant, making the detection of the reporter protein possible at much lower levels of expression.

Abbreviations: Ad – adenovirus; AdV – adenovirus vector; AES - adenovirus expression system; CMV-IE – cytomegalovirus immediate early; GFP/BFP – green/blue fluorescent protein; MLP – major late promoter; MOI – multiplicity of infection; TCP – total cell protein; tTA - tetracycline-controlled transactivator; rtTA – reverse tTA.

Introduction

The adenovirus expression system (AES) is an efficient tool both for high level protein production and for gene transfer experiments. Recombinant adenovirus vectors (AdV) can be produced at high titers, infect a wide variety of cell types both *in vitro* and *in vivo*, including non-replicating cells, and can permit high levels of gene expression (reviewed in: Acsadi et

* Author for all correspondence. Other affiliations are: Centre de Recherche en Virologie, Institut Armand-Frappier, Université du Québec, Laval, Québec, Canada H7N 4Z3; Département de Microbiologie et Immunologie, Université de Montréal, Montréal, Québec, Canada H3C 3J7

† Present address: Diagnocure Inc., Québec, Québec Canada

‡ Present address: Centre de Recherche Hospitalier de l'Université Laval, Québec, Québec, Canada

§ Present address: Exogen Neuroscience Inc., Montréal, Québec, Canada

¶ Present address: Biophage Inc., 6100 Royalmount Avenue, Montréal, Québec, H4P 2R2 Canada

al., 1995; Berkner, 1988, 1992; Graham and Prevec, 1992; Stratford-Perricaudet and Perricaudet, 1994; Trapnell and Gorzilla, 1994). Recent improvements to the expression cassette in the transfer vector (pAd-CMV5) have resulted in attaining expression levels greater than 20% total cell protein (TCP) in 293 cells and in non-permissive cells (Massie et al., 1998). The inability to generate AdV expressing proteins that have cytotoxic effects or which interfere with AdV replication has been overcome by the development of a transfer vector (pAdTR5) that utilizes a tetracylcine-regulated expression cassette (Massie et al., 1998). A similar vector was also recently described by Hu et al. (1997). Infection of cells that express the tetracycline-controlled transactivator tTA (Gossen and Bujard, 1992) or rtTA (Gossen et al., 1995) or co-infection together with an AdV that expresses tTA can allow production of toxic proteins at levels approaching 10% TCP (Massie et al., 1998).

In the present work, to improve the utility of the AES, we have included coding elements in the optimized transfer vectors for the co-expression of GFP from *Aequorea victoria*. This marker of gene expression has the unique feature of being detectable in living cells and whole organisms. Its inclusion assists in the process of identifying recombinant AdV plaques and can be used to monitor gene transfer efficiency. Using GFP, we have recently developed a rapid and accurate titration method based on the measurement of gene transfer units by cytofluorometry (Couture et al., manuscript in preparation). The usefulness of this marker to monitor AdV-mediated gene transfer *in vivo* was illustrated by the co-detection of GFP and dystrophin in fixed muscles by histo-immunostaining as well as the development of a protocol to quantify GFP expression in muscle cells extract with a fluorometer (Couture et al., manuscript in preparation). Expression of GFP can occur independently of a gene of interest or can be linked through expression of a dicistronic message in which GFP is encoded as the second cistron (Mosser et al., 1997a). Using mutants of GFP with altered fluorescence properties (Palm et al., 1997; Stauber et al., 1998) increases the sensitivity of detection such that GFP expression can be easily detected in non-replicating cells that have been infected by a single AdV. Transfer vectors with expression cassettes for either GFP or the blue fluorescent protein (BFP) allow the independent identification of cells expressing more than one transgene.

Materials and methods

Cell culture and virus infection

HeLa cells expressing rtTA (HeLa-rtTA) and A549 cells expressing tTA (A549-tTA) have been described previously (Massie et al., 1998). A KB cell line, obtained from the American Type Culture Collection, expressing rtTA was generated by the same method described for the HeLa-rtTA cell line. Cell culture and virus production was performed as described (Garnier et al., 1994; Massie et al., 1995). Recombinant viruses were generated by *in vivo* homologous recombination between overlapping sequences of linearized transfer vectors and Ad5/ΔE1ΔE3 genomic DNA as detailed in Jani et al., (1997), with the following modifications. After the positive identification of Ad recombinants, they were plaque purified and amplified on BMAdE1 cells, an A549 cell line expressing AdE1 proteins from a vector designed to eliminate the generation of replication competent Ad (RCA) (Massie, 1998). BMAdE1 clone 78 was used for plaque purification for one or two rounds depending on the purity of the AdV clones. Subsequently, selected viral clones were amplified on BMAdE1 clone 220 up to 3×10^7 cells (passage # 3). Since the yield on BMAdE1 is about 5-fold lower than in 293 cells, AdV stocks were then further amplified in suspension culture of 293 cells (up to 3 to 6×10^9 cells) for only 2 passages, thereby minimizing the occurrence of RCA. The absorption conditions for infection and titration, either by plaque assay or by the gene transfer units method (Couture et al., manuscript in preparation), were done following the protocol of Mittereder et al., (1996) in minimal volumes (ex., 1.5 mL in 60 mm plates) for 6 to 16 hr in order to insure optimal entry of the virus. In our experience, this adsorption method can increase the uptake of Ad by a factor of 5 to 10 as compared to the previously described infection conditions of 2 hr incubation in a larger volume (ex., 3 mL in 60 mm plates). Thus, using the gene transfer units method, more than one third of Ad physical particles can be shown to be infectious (Couture et al., manuscript in preparation).

Transfer vectors and recombinant viruses

The transfer vector pAdTR5, which contains a modified tetracycline-regulated expression cassette, and pAdCMV5, which has an optimized human CMV-IE promoter expression cassette, have been described (Massie et al., 1998). Recombinant adenoviruses with tetracycline-regulated expression of the herpes simplex virus ribonucleotide reductase large subunit protein R1 (AdTR5-R1), an N-terminal deleted R1 protein (AdTR5-ΔR1) were described previously (Massie et al., 1998).

AdV's expressing either the adenovirus E1B-19K protein or the human methenyl-tetrahydrofolate synthetase (MTHFS) protein was generated using the pAdTR5 transfer vector into which was cloned, in the BglII site, a PCR fragment containing either the E1B-19K coding region from pXC38 (McKinnon et al., 1982) or the MTHFS cDNA (Dayan et al., 1995) with BamHI ends and an optimized Kozak consensus sequence flanking the ATG initiation codon. All recombinant DNA molecules were constructed by standard cloning procedures. Restriction enzymes, modification enzymes and Vent polymerase were purchased from New England Biolabs (Beverly, Mass.)

The transfer vector pAdTR5-DC/GFP was constructed from pAdTR5 through a number of intermediate modification and cloning steps. First, a unique FseI site was added at the position corresponding to map unit 0 of the Ad5 genomic sequence. This creates a convenient 8 base pairs cleavage site for plasmid linearization prior to transfection. This was accomplished by digestion of pAdTR5 with EcoRI and BglII followed by ligation with a PCR fragment derived using the pAdTR5 plasmid as a template and primers that retained EcoRI and BglII ends but included an FseI site downstream from the EcoRI site to generate the plasmid pAdTR5F. The BglII site in this plasmid, which is located close to the Ad5 ITR, was next destroyed by digestion with BglII followed by blunt ending with Klenow and re-ligation. Subsequently a unique BglII site was added 5′ of the TR5 promoter. For this the plasmid was cleaved with BamHI and XhoI and the fragment that was removed was replaced with a PCR fragment derived using pAdTR5 as a template and primers that retained the XhoI site at the 5′ end but generated a 3′ end with BamHI, PmeI and BglII sites. This plasmid, pAdTR5FMCS, was used to finally generate pAdTR5-DC/GFP and the K7 series of plasmids described below. A plasmid, pTR5-DC/GFP, was used as the source of a fragment containing the encephalomyocarditis virus internal ribosome entry site (IRES) and GFP. This plasmid was derived from pTR-DC/GFP (Mosser et al., 1997a) by the insertion of a PCR fragment containing the adenovirus tripartite leader sequence using pAdBM5 (Massie et al., 1995) as a template and primers that added SacII ends such that the 3′ end also contains BglII site upstream from the SacII site. The pTR5-DC/GFP plasmid was digested with BglII and partially digested with BamHI to remove the IRES-GFP encoding containing fragment. This fragment was cloned into BamHI-BglII digested pAdTR5FMCS to generate pAdTR5-DC/GFP (Figure 1A). A described earlier (Mosser et al., 1997a) the GFP in pTR-DC/GFP is the S65T mutant (Heim et al., 1995).

pAdTR5-DC/HSP70-GFP was constructed by combining sequences from the plasmids pAdTR5F and pTR5-DC/HSP70-GFP (Mosser et al., 1997b). First, the unique EcoRI site in pAdTR5F was destroyed by digestion with EcoRI and blunt ending with Klenow followed by ligation. An EcoRI fragment from pTR5-DC/HSP70-GFP, containing the IRES sequence and including a BamHI site was excised and retained. The remainder of the plasmid was re-ligated and then a BamHI fragment encoding hsp70 and GFP was removed and cloned into the unique BamHI site in pAdTR5F. The resulting plasmid was then digested with EcoRI and the EcoRI fragment encoding the IRES was returned in its proper orientation to produce pAdTR5-DC/HSP70-GFP.

All of the K7 series of plasmids (Figure 1B) were derived from pAdTR5FMCS. Initially, this plasmid was digested with BglII and BamHI and a fragment containing the CMV-IE promoter and either the red-shifted GFP gene (GFPrg25) or the BFP gene (BFPrg50) from the plasmids QBI25 and QBI50 (Quantum Biotechnologies Inc., Montréal, Canada) were inserted (we refer to GFPrg25 and BFPrg50 as GFP_Q and BFP_Q respectively). These fragments have 5′ BglII and 3′ BamHI ends. To add cloning sites for insertion of a gene of interest and a polyA sequence downstream from the TR5 promoter a PCR fragment was generated containing the β-globin polyA sequence with 5′ BglII and PmeI sites and a 3′ BamHI site using the plasmid pAdCMV5 as a template. After digestion with BglII and BamHI the PCR product was cloned into the BglII cut GFP_Q and BFP_Q Ad plasmids to generate pAdTR5-K7/GFP_Q and pAdTR5-K7/BFP_Q (Figure 1B). The K7 plasmids with the constitutive CMV5 and MLP5 promoters were created by replacing the TR5 promoter in the pAdTR5-K7/GFP_Q and pAdTR5-K7/BFP_Q, after removal by digestion with AflII and XhoI, with a 1134 bp AflII/XhoI fragment containing the CMV5 promoter from pAdCMV5 or a 811 bp AflII/XhoI fragment containing the MLP promoter from pAdBM5 (Massie et al., 1995). This resulted in the generation of the transfer vectors pAdCMV5-K7/GFP_Q, pAdCMV5-K7/BFP_Q,

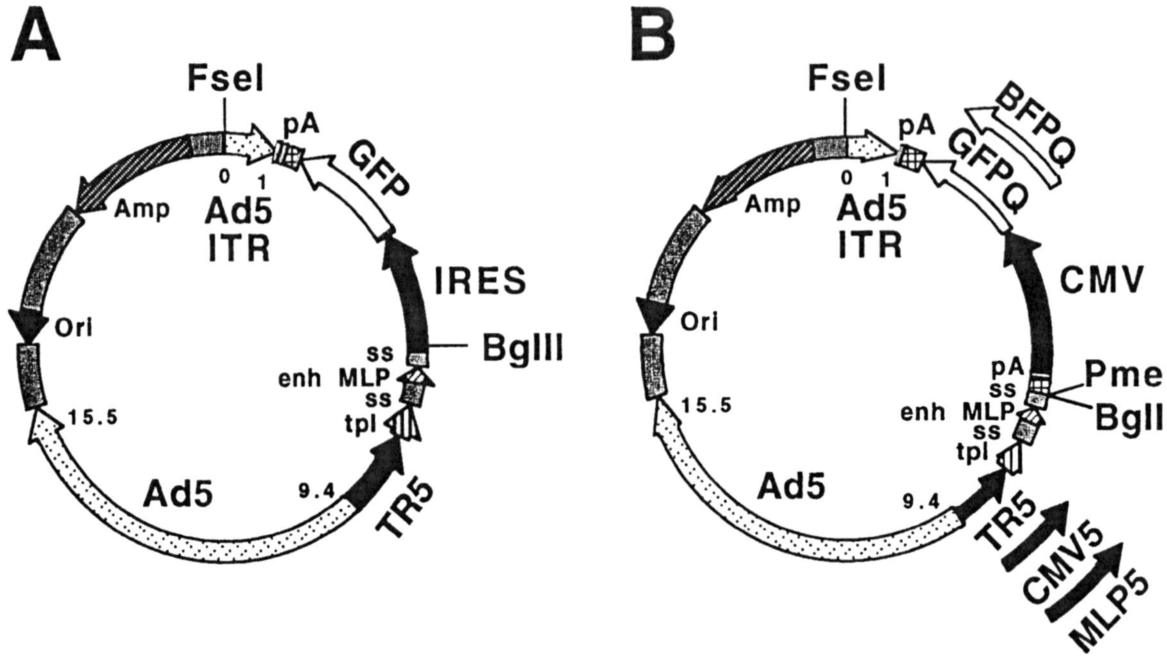

Figure 1. Schematic representations of the Ad transfer vectors encoding GFP. The dicistronic GFP expression cassette (A) and the independent GFP expression cassette (B) were derived from pAdTR5 and pAdCMV5 (Massie et al., 1998) and from pAdBM5 (Massie et al., 1995). The dicistronic cassette encodes the S65T mutant of GFP placed downstream of the encephalomyocarditis virus internal ribosome entry site (IRES). The independent cassettes encode either a red shifted mutant of GFP, GFP_Q, or a blue fluorescent protein, BFP_Q. Expression of an inserted gene of interest (into either the BglII in A or either the BglII or PmeI sites in B) is under the control of either a tetracycline-regulated promoter (TR) or the constitutive promoters CMV5 and MLP5. Each of the promoters is followed by the Ad5 tripartite leader sequence (tpl) and the MLP enhancer sequence (enh MLP) flanked by splice donor and acceptor sequences (ss). The Ad5 (Ad5 genome map units 9.4–15.5) and Ad5 ITR (m.u. 0–1) regions are Ad5 subgenomic sequences involved in homologous recombination used to generate Ad recombinants. A unique 8 base pairs FseI site can be used to linearize the plasmid prior to transfection.

pAdMLP5-K7/GFP_Q, and pAdMLP5-K7/BFP_Q (Figure 1B).

The AdCMV5-GFP virus, which expresses the S65T mutant of GFP under the control of the optimized CMV-IE promoter will be described elsewhere (Couture et al., in preparation). The plasmid pAdCMV5-GFP_Q was prepared by cloning a PCR product with BamHI ends encoding the red-shifted mutant of GFP (GFP_Q) from the plasmid pQBI25 (Quantum Biotechnologies Inc., Montréal, Canada) into BamHI cut pAdCMV5 (Massie et al., 1998) and was used to generate the virus AdCMV5-GFP_Q. The virus AdCMV-GFP_Q was generated using the transfer vector pAdMLP5-K7/GFP_Q. An adenovirus expressing β-galactosidase, AdCMV-LacZ (Ascadi et al., 1994), was used as a control in experiments with cells infected with adenoviruses expressing GFP.

Measurements of protein expression

For analysis of recombinant protein synthesis, 60 mm petri dishes of subconfluent cells were infected at a density of 0.5–1 × 10^6 cells/dish. At different times post-infection (p.i.), total protein extracts were prepared by lysing PBS washed cells with 2% SDS in 80 mM Tris-HCl pH 6.8 and 10% glycerol. Protein analysis by SDS-PAGE and Western blotting was performed as described previously (Lamarche et al., 1990). GFP levels were quantitated by western blotting using an anti-GFP monoclonal antibody (11EB, Quantum Biotechnologies Inc., Montréal, Canada) followed by incubation with a horseradish peroxidase conjugated goat anti-mouse antibody (Amersham Corp., Arlington Heights, Ill.) and ECL detection reagents (Amersham).

GFP fluorescence was measured by flow cytometry. Exponentially growing cells (1 × 10^6), seeded in duplicate 60 mm plates, were infected with AdCMV5-GFP, AdCMV5-GFP_Q or AdCMV-GFP_Q at various

MOIs. After 48 hr at 37 °C the cells were fixed with 2% paraformaldehyde for 30 min at 4 °C. GFP emission was analyzed using an EPICS XL-MCL flow cytofluorometer (Coulter, Miami, USA) equipped with a 15-mW argon-ion laser and the following filters: 488-nm laser-blocking, 488-nm long-pass dichroic, 550-nm long-pass dichroic and 525-nm band-pass. The fluorescence index was calculated as the product of the percentage of GFP positive cells by the mean fluorescence value.

Results and discussion

Features of the adenovirus transfer vectors with positive selection

Previously we described two new adenovirus expression cassettes that allow for either constitutive or inducible expression of a gene of interest in non-permissive cells (Massie et al., 1998). The transfer vector pAdTR5 utilizes a tetracycline-regulatable promoter (Gossen and Bujard, 1992) while the pAdCMV5 vector contains an optimized CMV-IE promoter/enhancer. To increase the utility of these transfer vectors we have included elements for the co-expression of GFP as described in Material and Methods (Figure 1). In the plasmid pAdTR5-DC/GFP (Figure 1A) the GFP gene, containing the S65T mutation (Heim et al., 1995), is encoded as the second cistron in a dicistronic expression cassette. A gene of interest can be cloned into the position of the first cistron by insertion into the unique BglII site. The two cistrons are separated by the encephalomyocarditis virus internal ribosomal entry site (IRES). We have previously shown that this dicistronic cassette can be used to facilitate the screening and selection of transfected cell lines stably expressing inducible gene products (Mosser et al., 1997a, b). Another series of transfer vectors that allow for the independent expression of GFP and a gene of interest were also assembled by including a GFP expression cassette (K7-GFP$_Q$) into either the pAdTR5, pAdCMV5 or pAdBM5 plasmids (Figure 1B). In these plasmids the expression of a red-shifted variant of GFP (GFP$_Q$), or a blue fluorescent protein (BFP$_Q$), is placed under the control of the standard CMV-IE promoter. Two unique cloning sites, BglII and PmeI, are available for insertion of a gene of interest. Three pairs of plasmids offer either tetracycline-inducible expression (pAdTR5-K7-GFP$_Q$/BFP$_Q$) or constitutive expression (pAdCMV5-K7-GFP$_Q$/BFP$_Q$, pAdMLP5-K7-GFP$_Q$/BFP$_Q$). All of the above transfer vectors can be used for (i) transient expression assays, (ii) to establish stable cell lines and (iii) to generate recombinant Ad's. In combination they offer the ability to express two different proteins with either constitutive or inducible expression and can be detected separately with co-expression of the blue or green fluorescent proteins.

Efficient protein production in absence of Ad replication

Although equally high levels of expression can be obtained with either the constitutive (AdCMV5) or the tetracycline-regulated (pAdTR5) AdV's, only the inducible system allows for the expression of potentially toxic proteins. This is because the generation of a recombinant Ad is accomplished *in vivo* in permissive cells transfected with the transfer vector and E1-deleted viral DNA. Constitutive expression of proteins, encoded by the transfer vector, that can interfere with viral replication or assembly will prevent the successful generation of a recombinant virus. This was assumed to be the reason for our previous unsuccessful attempts to generate an Ad expressing N-terminal deletions of the herpes virus ribonucleotide reductase protein R1 (ΔR1). However, a recombinant Ad with tetracycline-regulated expression of ΔR1 was readily obtained (Massie et al., 1998). Figure 2 shows examples of high level regulated expression of a number of proteins using AdTR5 vectors in KB cells that stably express rtTA. Levels of expression range from 30% TCP for AdTR5-R1 to 10–15% TCP for the others. Expression is undetectable in cells infected in the absence of the inducer, doxycycline (compare lanes 2 and 3). As we have shown previously (Massie et al., 1998), high level co-expression of two proteins can be easily achieved by co-infection of cells with each of the AdV. This is shown in lane 4 where cells were infected with an equal MOI (400 pfu) of the AdTR5-ΔR1 and AdTR5-DC/hsp70-GFP viruses. Expression levels obtained after infection with either of these viruses singly is shown in lanes 5 and 6. Two other examples of proteins that were expressed at high levels using the tetracycline-inducible AdV are the human enzyme MTHFS (lane 7) and the Ad E1B-19K protein (lane 8). Another AdV expressing the MTHFS under the control of the CMV5 promoter was also generated and it produced the enzyme at the same level as AdTR5-MTHFS (data not shown) like this was the case for R1 expressed in both vectors (Massie et al., 1998).

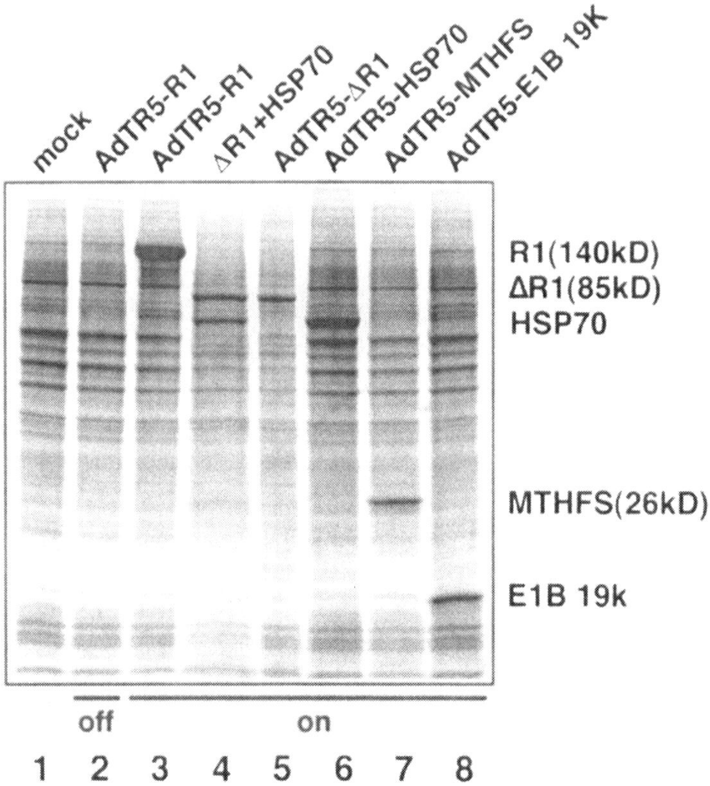

Figure 2. High-level protein production with AdTR5 vectors in KB-rtTA cells. Shown is a Coomassie blue stained SDS-PAGE gel of protein extracts (10 μg/lane) from cells infected at an MOI of 400 and collected 40 hr. later. Cells were infected with recombinant Ad's encoding the herpes simplex virus ribonucleotide reductase large subunit R1 (lanes 2 and 3), an R1 N-terminal deletion ΔR1 (lanes 3 and 4), the human 70 kDa heat-inducible protein hsp70 (lanes 4 and 6), the human methenyltetrahydrofolate synthetase enzyme MTHFS (lane 7) and the adenovirus E1B-19 K protein (lane 8). The rtTA inducer, doxycycline (1 μg mL^{-1}) was added to the culture medium at the time of infection. Lane 2 shows the absence of R1 protein synthesis in cells infected with the AdTR5-R1 virus without addition of doxycycline. An extract from non-infected cells (mock) is shown in lane 1.

The expression of R1 in KB-rtTA cells at 30% TCP is slightly higher than what was obtained in A549-tTA or HeLa-rtTA with AdTR5-R1 but is equivalent to the expression level obtained in the aforementioned cell lines following co-infection with an AdV encoding tTA (Massie et al., 1998). This suggest that this particular clone of KB cells is expressing rtTA at a level higher than tTA or rtTA are expressed in A549 or HeLa cells respectively, and also that its expression level rivals the one obtained following AdVCMVtTA co-infection. Taken together our results show that the AES is a very efficient protein production system allowing to express both toxic and non-toxic proteins at near saturation levels in absence of viral replication in a wide variety of mammalian cells. The absence of AdV replication is a feature that facilitates protein purification since no other abundant viral proteins are produced. Moreover, owing to the absence of viral replication and ensuing host protein shut-off, the production host can be maintained in good physiological state for prolonged periods of time (weeks) and continue to synthesize the recombinant protein, that could be harvested periodically, if secreted in the medium. Indeed, we have shown by radiolabeling with [^{35}S]methionine that protein synthesis reaches a maximun at 24 hr pi and can be maintained for at least one week (Massie et al., 1998; and unpublished data).

Cytotoxicity following overexpression of E1B-19K

The E1B-19K protein plays an essential role in adenovirus replication by inhibiting the apoptotic effects that are associated with the expression of E1A (White et al., 1991). Constitutive expression of E1B-19K in transfected cell lines confers upon them an increased

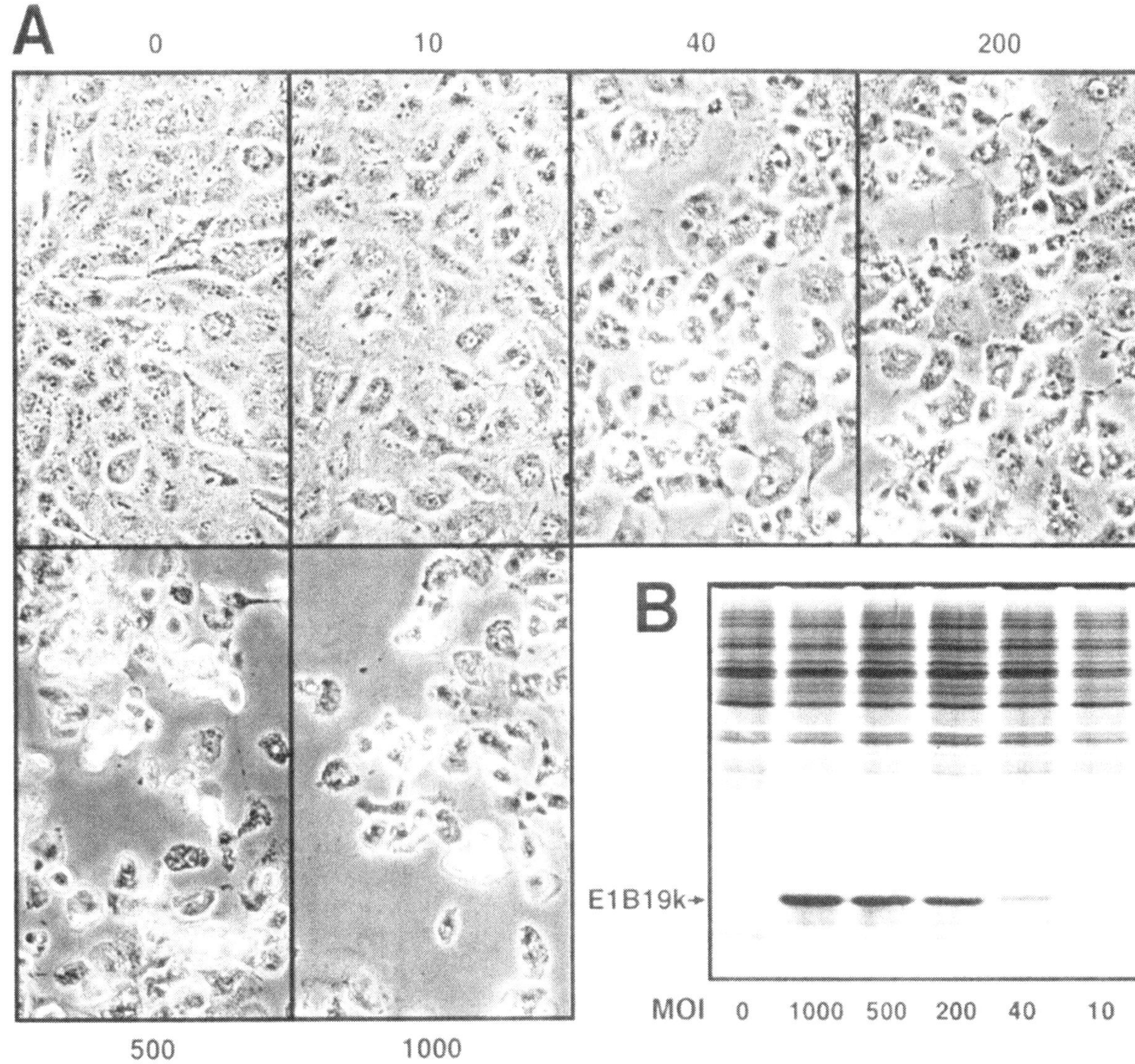

Figure 3. High level expression of the anti-apoptotic protein E1B-19K is lethal in A549-tTA cells. Morphological appearance and E1B-19K expression levels in cells infected with the AdTR5-E1B 19K AdV at MOI's of 10, 40, 200, 500 and 1000. Doxycycline (1 μg mL^{-1}) was added to each of the cultures, and to the non-infected cells (0), at the time of infection. Cells were photographed and then collected for SDS-PAGE 40 hr. after infection (15 μg of protein cell extract/lane). Cytotoxicity is first seen by 24 hr. in cells infected at an MOI of 40. Expression of E1B-19K represents approximately 2% of total cell protein in cells infected at this MOI.

resistance to apoptosis in response to a number of diverse stimuli (reviewed in Teodoro and Branton, 1997). Surprisingly, we were unable to generate a recombinant adenovirus expressing this protein under the control of the strong CMV5 promoter. Our ability to readily generate an E1B-19K expressing adenovirus with the tetracycline-inducible AdV led us to suspect that high level expression of E1B-19K was cytotoxic, unlike the anti-apoptotic function that it possesses when expressed at more moderate levels. This is clearly shown in Figure 3 where infection of A549-tTA cells at an MOI of 40 pfu or greater led to intense vacuolization and a loss of attached cells from the culture dish. Cytotoxicity was evident at expression levels that were approximately 2% of TCP whereas expression at lower levels resulted in protection against apoptosis induced by various treatments (Langelier et al.; manuscript in preparation). No cytotoxicity was observed when cells were infected in the presence of anhydrotetracycline which prevents tetracycline-regulated expression in tTA-expressing cells (not shown). The cytotoxicity occurring as a result of high level E1B-19K protein expression is not simply due to a general effect related to such high levels of protein production since this level of protein expression was well tolerated for a number of other proteins (Figures 2 and 5). E1B-19K is targeted to membranes of the nucleus and endoplasmic reticulum, and is associated with lamins (Rao et al., 1997). Toxicity may be the result of mislocalization of the E1B-19K protein or inappropriate or unregulated associations with essential cellular proteins when expressed at high levels. This is another example of a cytotoxic protein expressed using the tetracycline-inducible system in AdV (Hu et al., 1997; Massie et al., 1998). We are also expressing the pro-apoptotic proteins Bad and Bax and preliminary results indicates that only the inducible vector can generate viable Ad recombinants.

Autofluorescent protein expression for positive selection of Ad recombinants

Co-expression of GFP in the AdTR5-DC based cassettes simplifies the process of recombinant virus identification and greatly increases the utility of the recombinant Ad. Generally, recombinant Ad's are identified as plaques on monolayer cultures of 293 cells that have been overlaid with agarose. Individual plaques are verified as being genuinely recombinant by screening methods that involve plaque elution and amplification followed by dot blotting, PCR, or SDS-PAGE and western blotting to verify the presence of the recombinant protein (Jani et al., 1997). By contrast, recombinant Ad's that co-express GFP can be easily identified by direct observation with an inverted fluoresecent microscope. Since GFP expression is detectable in a large percentage of cells (typically 10 to 50%) one day after the transfection, it can be used to monitor the efficiency of transfection, a crucial parameter for the successful generation of recombinant Ad's. By the time the first plaques are visible, 7 to 10 days post-transfection, the percentage and intensity of fluorescent cells has diminished substantially due to the transient nature of GFP expression. Although, pockets of green cells are still present and can be confused with real plaques, a careful morphological examination allows to readily discriminate between them. Indeed, the cells within a plaque display a typical cytopathic effect and are more fluorescent. Plaques that are green can be rapidly identified and purified by subsequent rounds of infection and selection of individual green plaques, allowing to save as much as 3 to 4 weeks on the selection and plaque purification process.

Expression from the TR5 promoter in the 293 cell line is sufficient to identify green plaques in the uninduced state, although the level of GFP fluorescence is about 25-fold less than when induced. This is due to the leakiness of the TR5 promoter in 293 cells, likely resulting from the transactivation of the minimal CMV TATA box by the AdE1A proteins in combination with the high copy number of the vectors following transfection or Ad replication as discussed in our previous report (Massie et al., 1998). Obviously, this intrinsic leakiness might preclude the generation of AdV's expressing very toxic proteins, and so far we have found at least one gene that we were unable to rescue in AdV. Work is in progress to minimize the leakiness of the uninduced state while preserving the strong activity of the fully induced promoter. Despite this limitation, we have successfully produced recombinant Ad's with the AdTR5-DC/GFP transfer vector expressing the chaperone hsp70; the cytokines EPO, IL-2 and TGFβ; the G-coupled receptor protein hDOR; the adenovirus protein E1B-19K; and the pro-apoptotic proteins Bad and Bax using this strategy. By comparison to a vector in which the GFP gene is next to the TR5 promoter, the expression of GFP in the second cistron is reduced by 2- to 4-fold depending on the first gene in the dicistronic construct (not shown). The reason for this reduced expression is currently under investigation, but it is not a major problem since GFP is readily detectable at that level. However, since bright green fluorescence can only be seen when the dicistronic message is induced to high levels, and since this would prevent the generation of recombinant Ad's expressing potentially toxic proteins, we have also constructed transfer vectors in which the expression of GFP is independent of the gene of interest (pAdTR5-K7/GFP$_Q$). Recombinant Ad's with tetracycline-regulated expression of HSV ribonucleotide reductase subunit R1, the p53-regulated protein p21, the G-coupled receptor protein hDOR, the adenovirus proteins E1A, E1B-19K, E3-gp19K and E3-14.7K, as well as the murine cytokines IL-6 and IL-10 have been generated with the pAdTR5-K7/GFP$_Q$ transfer vector. Preliminary characterization of some of these recombinant Ad's indicated that not only the GFP reporter, but also the other gene in the construct was expressed at higher levels in the independent double cassettes as compared to the dicistronic vectors (not shown). Interestingly, so far every GFP positive plaque tested also produced the other transgenes showing that selection of AdV with this procedure is reliable. Given the structure of the tandem expression cassettes in the vector, it is formally possible that rearrangements of the AdV could occur through recombination between homologous sequences in the transfer vector generating unstable AdV stocks. However, this has not been observed with the limited number of Ad recombinants we have analyzed at this point.

An important consideration in the use of GFP for the identification of recombinant Ad's is the level of GFP expression that is minimally required for detection. Since high sensitivity would be a beneficial attribute, we compared the expression level and fluorescence intensity of GFP in 293 cells infected with AdV's expressing either the S65T mutant of GFP (Heim et al., 1995) or a red-shifted variant GFP_Q (Palm et al., 1997; Stauber et al., 1998). Figure 4 shows that while expression levels of both GFP S65T and GFP_Q are similar when expressed from the Ad-CMV5 viruses (Figure 4A and B, lanes 2 and 3) the fluorescence index of the GFP_Q is almost 12-fold higher than that of GFP S65T (93 vs. 8). When expressed under the control of the weaker promoter in the AdCMV virus, GFP_Q expression is approximately 6-fold less than that obtained with AdCMV5/GFP_Q (compare Figure 4B lane 5 which shows a 1:9 dilution of the extract from AdCMV5-GFP_Q infected cells with lane 6 which is the undiluted extract from AdCMV-GFP_Q infected cells). However, the fluorescence index is still about 2-fold higher than that of GFP S65T expressed under the control of the strong CMV5 promoter (17 vs. 8). Therefore, using a GFP variant that has been optimized for GFP fluorescence makes it possible to detect GFP positive cells without the need to express high amounts of the reporter protein.

Functional studies in non-permissive cells with autofluorescent proteins

We performed similar experiments in cells that do not support virus replication to determine whether GFP fluorescence could be detected in cells that received a single copy of the gene and also whether the level of GFP fluorescence could be directly correlated with the copy number delivered. This is an important consideration in applying AdV's for functional studies *in vitro* or *in vivo* gene therapy. Figure 5A shows GFP_Q protein levels in HeLa cells infected with AdCMV5GFP_Q at MOI's of 100, 1000 and 5000 and is compared to levels of GFP_Q in HeLa cells infected with AdCMV-GFP_Q at an MOI of 5000 and 293 cells infected with AdCMV5-GFP_Q at an MOI of 10. Equivalent amounts of GFP_Q can be produced in non-replicating cells at a high MOI (5000) as compared to 293 cells where the virus replicates to a high copy number (compare lanes 2 and 6). GFP_Q is undetectable by Coomassie staining in HeLa cells infected with AdCMV5-GFP_Q at an MOI of 100 (lane 4) or with AdCMV-GFP_Q at an MOI of 5000 (lane 7). Western blot analysis (Figure 5B) shows that the amount of GFP_Q produced by the standard CMV-IE promoter (lane 7) was similar to that produced using a 50-fold lower MOI of the stronger CMV5 promoter (lane 4). Increasing the MOI in both AdCMV5-GFP_Q and AdCMV-GFP_Q infected cells gives a proportional increase in the amount of GFP fluorescence (Figure 5C). The fluorescence index of GFP_Q in AdCMV5-GFP_Q infected cells is far superior to that of AdCMV-GFP infected cells (approximately 12-fold higher over an MOI range of 0.25 to 10). This is somewhat higher than the difference measured in 293 cells where the fluorescence index of AdCMV5-GFP_Q infected cells was about 6-fold higher than in AdCMV-GFP_Q infected cells (Figure 4B, 93 vs. 17). Although the fluorescence intensity is weaker in HeLa cells infected with the AdCMV-GFP_Q virus it is sufficient to detect GFP_Q in cells containing a single copy of the GFP_Q gene. Infection with an MOI of 1, as normalized in 293 cells (where an MOI of 1 gives 50–60% positive cells), resulted in the detection of 28% GFP positive cells with the AdCMV5-GFP_Q virus and 16% positive cells with the AdCMV-GFP_Q virus. The sensitivity is therefore about 2-fold less for AdCMV5-GFP_Q and 4-fold less for AdCMV-GFP_Q at an MOI of 1 in HeLa cells relative to 293 cells where the virus replicates and reaches saturation levels for GFP.

A proportional increase in the fluorescence index relative to MOI can be measured up to an MOI of 1000 in AdCMV5-GFP_Q infected HeLa cells (Figure 5D). For example, increasing the MOI from 10 to 100 or 1000 gives a 10- and 100-fold increase in fluorescence index. However, increasing the MOI from 1000 to 5000 or 10 000 does not give a proportional 5- and 10-fold increase but rather less than 2-fold going from 1000 to 5000. This may be because the uptake of the virus is near saturation at such high MOI's, or because GFP_Q protein synthesis levels are nearing saturation, or because of alteration of cell physiology at such high MOIs. The fact that in complementing

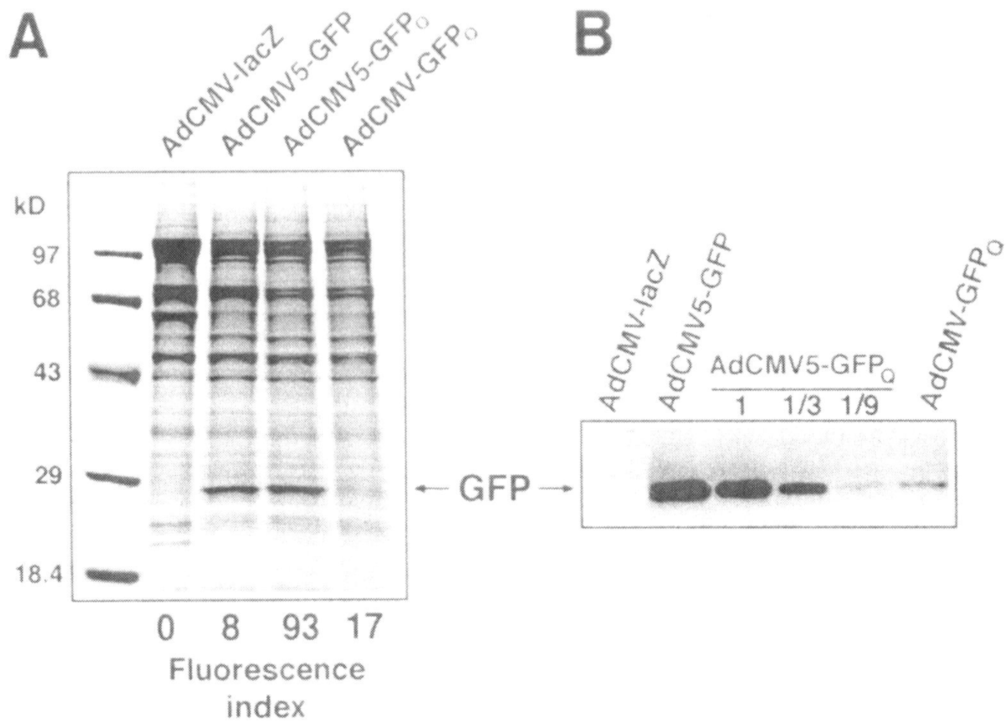

Figure 4. Comparison of GFP expression and fluorescence levels in 293 cells infected with AdV's expressing either the S65T mutant of GFP (GFP) or the red-shifted variant (GFP_Q) under the control of either the CMV-IE or CMV5 promoter. Coomassie blue stained gel (A) and western blot analysis (B) of GFP expression levels in protein extracts from 293 cells (10 μg/lane) infected with an MOI of 10 and collected 24 hr. after infection. The level of GFP fluorescence in the infected cells, as measured by flow cytometry, is indicated below the gel shown in 'A'. Dilutions (1/3 and 1/9) of the extract from the cells infected with the AdCMV5-GFP_Q virus are shown in panel 'B' in order to estimate the level of GFP expression in these cell relative to that of cells infected with the AdCMV-GFP_Q virus. Cells infected with a CMV-based promoter expressing lacZ (AdCMV-lacZ) is included as a control.

293 cells, in which the copy number of the vector can be as high as 100 000, a difference of 6-fold in favor of CMV5 as compared to CMV-IE was measured whereas in HeLa cells at lower copy number a 12-fold difference is obtained, suggest that at copy numbers greater than 5000 a saturation of GFP expression is occurring. The results of the western blot analysis, in agreement with the fluorescence index measurements, shows that although an approximate 10-fold increase in GFP_Q levels was achieved by increasing the MOI from 100 to 1000, an increase from 1000 to 5000 resulted in only less than a 2-fold increase in GFP_Q levels (Figure 5B). Thus, for functional studies gene expression can be increased proportionally up to an MOI of 1000 reaching near saturation, although more typically, studies should be conducted at lower MOIs in order to minimize any side effects associated with the viral load. Owing to the increased sensitivity of GFP_Q, we were able to established the minimal MOI (around 3 to 5) required to transfer a gene in almost 100% of HeLa cells with AdV.

Conclusions

Addition of elements for the co-expression of GFP significantly increases the utility of the AES that we have previously shown is capable of producing high levels of recombinant proteins. Use of a tetracycline-inducible expression system makes it possible to generate recombinant Ad's expressing proteins that are toxic and capable of interfering with Ad replication. Co-expression of GFP assists in the identification of recombinant Ad's since the positive plaques can be easily seen with an inverted fluorescent microscope. Furthermore, co-expression of GFP makes it possible to gauge the level of transgene expression following infection. For *in vivo* applications it allows for the identification of productively infected cells and so can be used to measure the effectiveness of the gene transfer protocol.

Expression of GFP from an independent promoter (K7 vectors) gives high levels of GFP fluorescence without expression of the gene of interest (pTR5 vec-

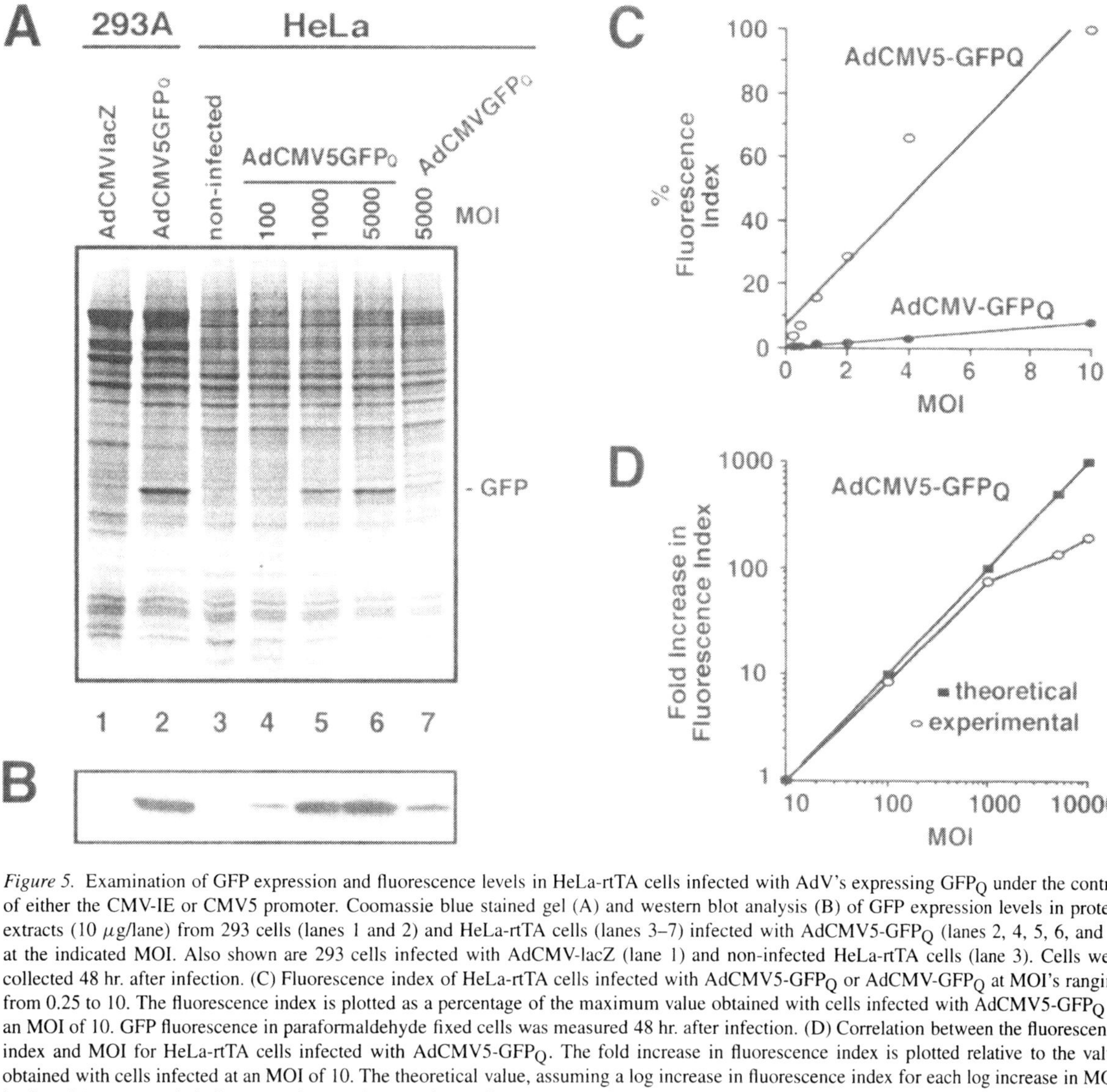

Figure 5. Examination of GFP expression and fluorescence levels in HeLa-rtTA cells infected with AdV's expressing GFP_Q under the control of either the CMV-IE or CMV5 promoter. Coomassie blue stained gel (A) and western blot analysis (B) of GFP expression levels in protein extracts (10 μg/lane) from 293 cells (lanes 1 and 2) and HeLa-rtTA cells (lanes 3–7) infected with AdCMV5-GFP_Q (lanes 2, 4, 5, 6, and 7) at the indicated MOI. Also shown are 293 cells infected with AdCMV-lacZ (lane 1) and non-infected HeLa-rtTA cells (lane 3). Cells were collected 48 hr. after infection. (C) Fluorescence index of HeLa-rtTA cells infected with AdCMV5-GFP_Q or AdCMV-GFP_Q at MOI's ranging from 0.25 to 10. The fluorescence index is plotted as a percentage of the maximum value obtained with cells infected with AdCMV5-GFP_Q at an MOI of 10. GFP fluorescence in paraformaldehyde fixed cells was measured 48 hr. after infection. (D) Correlation between the fluorescence index and MOI for HeLa-rtTA cells infected with AdCMV5-GFP_Q. The fold increase in fluorescence index is plotted relative to the value obtained with cells infected at an MOI of 10. The theoretical value, assuming a log increase in fluorescence index for each log increase in MOI, is plotted (square symbol) for comparison to the experimentally measured value (open circle).

tors) and the recombinant plaques are much easier to identify than those generated with the dicistronic GFP vector. GFP can be seen in plaques on 293 cells generated from the pTR5-DC/GFP vector due to the low level of expression from this promoter in these cells in the uninduced state. These plaques can be rapidly verified by infection of 293-tTA cells (Massie et al., 1998) with the eluted viruses. Since the red-shifted variant of GFP present in the K7 transfer vectors is much brighter than the S65T mutant GFP, we have recently constructed dicistronic transfer vectors with both GFP_Q and BFP_Q. An advantage of the dicistronic configuration is that the levels of fluorescence are directly proportional to the level of expression of the gene of interest. When using the tetracycline-inducible system, dicistronic expression of GFP identifies which cells are expressing a gene of interest following induction and the level of expression can be inferred from the level of GFP fluorescence. Also, dicistronic expression of GFP will give high levels of GFP expression in any cell that co-expresses the transactivator, while the K7 vectors may be limited to cells in which the CMV-IE promoter is strongly activated which is the case for the majority of the cells with a few exceptions (Hu et al., 1997; Xu et al., 1995).

Acknowledgements

We thank Arik Elfassy for measurement of GFP expression in AdV with the dicistronic expression cassette and Lucie Bourget for technical assistance with the cytofluorometer. This work was supported by grants from the National Research Council and the Medical Research Council of Canada. This is a NRC publication No. 41438.

References

Acsadi G, Jani A, Massie B, Simoneau M, Holland P, Blaschuk K and Karpati G (1994) A differential efficiency of adenovirus-mediated *in vivo* gene transfer into skeletal muscle cells of different maturity. Hum Mol Genetic 3: 579–584.

Acsadi G, Massie B and Jani A (1995) Adenovirus-mediated gene transfer into striated muscles. J Mol Med 73: 165–180.

Berkner KL (1988) Development of adenovirus vector for the expression of heterologous genes. BioTechniques 6: 616–629.

Berkner KL (1992) Expression of heterologous sequences in adenoviral vectors. Curr Top Microbiol Immunol 158: 39–66.

Dayan A, Bertrand R, Beauchemin M, Chahla D, Mamo A, Filion M, Skup D, Massie B and Jolivet J (1995) Cloning and characterization of the human 5,10-methenyltetrahydrofolate synthetase-encoding cDNA. Gene 165: 307–311.

Garnier A, Côté J, Nadeau I, Kamen A and Massie B (1994) Scale-up of the adenovirus expression system for the production of recombinant protein in human 293S cells. Cytotechnology 15: 145–155.

Gossen M and Bujard H (1992) Tight control of gene expression in mammalian cells by tetracycline-responsive promoters. Proc Natl Acad Sci USA 89: 5547–5551.

Gossen M, Freundlieb S, Bender G, Müller G, Hillen W and Bujard H (1995) Transcriptional activation by tetracyclines in mammalian cells. Science 268: 1766–1769.

Graham FL and Prevec L (1992) Adenovirus-based expression vectors and recombinant vaccines, In RW Ellis (Ed.) Vaccines: New Approaches to Immunological Problems. Butterworth-Heinemann, Boston, Mass, pp. 363–390.

Heim R, Cubitt AB and Tsien RY (1995) Improved green fluorescence. Nature 373: 663–664.

Hu SX, Ji W, Zhou Y, Logothetis C and Xu HJ (1997) Development of an adenovirus vector with tetracycline-regulatable human tumor necrosis factor alpha gene expression. Cancer Res 57: 3339–3343.

Jani A, Lochmüller H, Acsadi G, Simoneau M, Huard J, Garnier A, Karpati G and Massie B (1997) Generation, validation and large scale production of adenoviral recombinants with large inserts such as 6.3 kb human dystrophin cDNA. J Virolog Meth 64: 111–124.

Lamarche N, Massie B, Richer M, Paradis H and Langelier Y (1990) High level expression in 293 cells of the herpes simplex virus type 2 ribonucleotide reductase subunit 2 using an adenovirus vector. J Gen Virol 71: 1785–1792.

Massie B (1998) BMAdE1, a new complementing cell line for the large scale production of E1-deleted Adenovirus (Ad) vectors devoid of RCA (replication competent Ad). (filed on 26/04/96; Issued 09/98).

Massie B, Couture F, Lamoureux L, Mosser DD, Guilbault C, Jolicoeur P, Belanger F and Langelier Y (1998) Inducible overexpression of a toxic protein by an adenovirus vector with a tetracycline-regulatable expression cassette. J Virol 72: 2289–2296.

Massie B, Dionne J, Lamarche N, Fleurent J and Langelier Y (1995) Improved adenovirus vector provides herpes simplex virus ribonucleotide reductase R1 and R2 subunits very efficiently. Bio/Technology 13: 602–608.

McKinnon RD, Bacchetti S and Graham FL (1982) Tn5 mutagenensis of the transforming genes of human adenovirus type 5. Gene 19: 33–42.

Mittereder N, March K and Trapnell BC (1996) Evaluation of the concentration and bioactivity of adenovirus vectors for gene therapy. J Virol 70: 7498–7509.

Mosser DD, Caron AW, Bourget L, Jolicoeur P and Massie B (1997a) Use of a dicistronic expression cassette encoding the green-fluorescent protein for the screening and selection of cells expressing inducible gene products. Biotechniques 22: 150–161.

Mosser DD, Caron AW, Bourget L, Denis-Larose C and Massie B (1997b) Role of the human heat shock protein hsp70 in protection against stress–induced apoptosis. Mol Cell Biol 17: 5317–5327.

Palm GJ, Zdanov A, Gaitanaris GA, Stauber R, Pavlakis GN and Wlodawer A (1997) The structural basis for spectral variations in green fluorescent protein. Nat Struct Biol 4: 361–365.

Rao L, Modha D and White E (1997) The E1B 19K protein associates with lamins *in vivo* and its proper localization is reuired for inhibition of apoptosis. Oncogene 15: 1587–1597.

Stauber RH, Ilorie K, Carney P, Hudson EA, Tarasova NI, Galtanaris GA and Pavlakis GN (1998) Development and applications of enhanced green fluorescent protein mutants. BioTechniques 24: 462–471.

Stratford-Perricaudet L D and Perricaudet M (1994) Gene therapy: the advent of adenovirus, In John A Wolff (Ed.) Gene Therapeutics: Methods and Applications of Direct Gene Transfer. Birkhauser Boston, Cambridge, Mass, pp. 344–362.

Teodoro JG and Branton PE (1997) Regulation of apoptosis by viral gene products. J Virol 71: 1739–1746.

Trapnell BC and Gorzilia M (1994) Gene therapy using adenoviral vectors. Curr Opinion in Biotech 5: 617–625.

White E, Cipriani R, Sabbatini P and Denton A (1991) Adenovirus E1B 19-kilodalton protein overcomes the cytotoxicity of E1A proteins. Virol 65: 2968–2978.

Xu ZZ, Krougliak V, Prevec L, Graham FL and Both GW (1995) Investigation of promoter function and animal cells infected with human recombinant adenoviruses expressing rotavirus antigen VP7sc. J Gen Virol 76: 1971–1980.

Cytotechnology **28**: 65–72, 1998.
© 1998 *Kluwer Academic Publishers. Printed in the Netherlands.*

Modulation of cell cycle progression and of antibody production in mouse hybridomas by a nucleotide analogue

František Franěk[1]*, Antonín Holý[2], Ivan Votruba[2] & Tomáš Eckschlager[3]
[1] *Institute of Molecular Genetics, Academy of Sciences, Videnska 1083, CZ-14220 Praha 4, Czech Republic*
E-mail: franek@biomed.cas.cz.
[2] *Institute of Organic Chemistry and Biochemistry, Academy of Sciences, CZ-16610 Praha 6, Czech Republic*
[3] *University Hospital and 2nd Medical Faculty, CZ-15018 Praha 5, Czech Republic*

Received 8 September 1998; accepted 8 September 1998

Key words: acyclic nucleoside phosphonate, cell cycle, hybridoma, specific MAb production rate

Abstract

The nucleotide analogue 9-[2-(phosphonomethoxy)ethyl]guanine (PMEG) has been identified as a powerful antiproliferative substance when acting on hybridoma cells. In the range of 10 nM to 100 nM concentrations this agent reduces cell growth rate, while its apoptosis-inducing activity is marginal. Marked induction of apoptosis can be observed at micromolar and higher order concentrations. In PMEG-supplemented media the cell cycle progression is perturbed, the flow-cytometric DNA profile shows a higher proportion of cells in the S and G_2/M phases of the cell cycle. Concomitantly with the reduction of the growth rate, the specific monoclonal antibody production rate may rise by 20–27%. Addition of PMEG at the end of the exponential phase of a batch culture results in an enhancement of the final monoclonal antibody concentration.

Abbreviations: MAb – monoclonal antibody; PMEG – 9-[2-(phosphonomethoxy)ethyl]guanine

Introduction

The production of MAbs is evidently not associated with the growth of hybridoma cells in a unique manner. Positively-, negatively-, or non-growth associated MAb production rates have been observed in various laboratories. The explanation of diverse findings by variation in culture conditions has been provided by the extended structural model of Suzuki et al. (1992). In our study of hybridoma growth in non-traditional unbalanced media we arrived at a conclusion that determinants different from those regulating growth and death of the cells control the intensity of MAb synthesis and secretion (Franěk and Šrámková, 1997).

Persistent cell proliferation and simultaneous significant death rate, occurring commonly in the stationary and decline phases of batch cultures, as well as in cultures run in the perfusion mode, are unfavorable for the production of secreted protein. A considerable portion of the nutrients is spent on synthesis of living cell mass, that is converted to dead cell mass within a short time period. The wasting of nutrients on synthesis of structural and informational macromolecules is particularly significant with hybridoma cells. Due to their somatic-hybrid nature hybridoma cells are hypotetraploid and possess thus an excessive load of DNA. It can be expected that the synthesis of the secreted MAbs would increase if hybridoma cell growth is inhibited without impairing cell viability.

Reduction of hybridoma growth by controlled feeding had, in our experience, a negative impact on the specific MAb production rate (Franěk, 1995; Franěk and Šrámková, 1997, 1998). Therefore, we decided to pursue an alternative approach based on the introduction of an antiproliferative molecular signal in a culture system saturated with substrates. Arresting the cell cycle while maintaining the cells in viable state has been recognized as a promising way to enhancing

* Author for all correspondence.

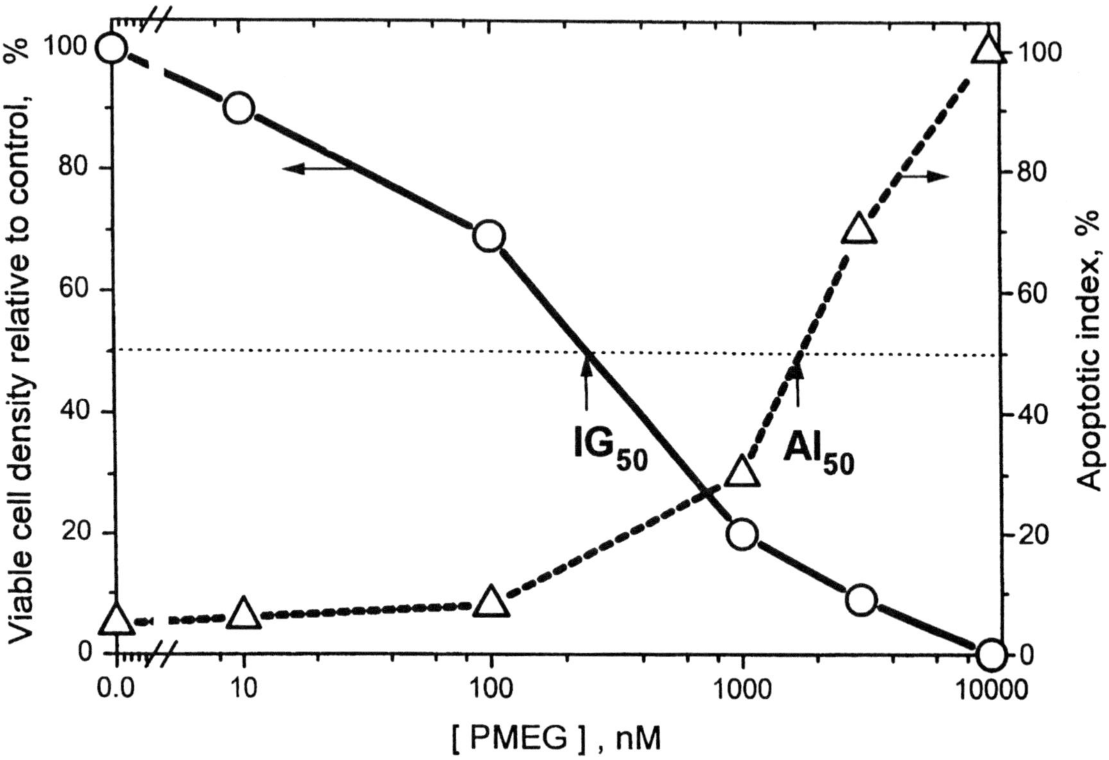

Figure 1. Inhibition of growth and induction of apoptosis in cultures of hybridoma ME-750 supplemented with PMEG at various concentrations. The data were obtained after 2 days in static culture. IG_{50}-PMEG concentration causing 50% inhibition of growth. AI_{50}-PMEG concentration at which the apoptotic index is equal to 50%. The experimental error of the data was ±10 and ±15% for cell number and for apoptotic index, respectively (Franěk, 1995).

protein production by mammalian cells (Takahashi et al., 1994).

Substances endowed with antiproliferative capacity ('cytostatics') regularly induce apoptosis, depending on the concentration applied. Two criteria for selection of a suitable antiproliferative agent are relevant. First, the value of the lowest concentration exhibiting a significant inhibition of growth. Second, the ratio of the apoptosis-inducing concentration to the growth-inhibiting concentration. Among several acyclic nucleotide analogues the acyclic nucleoside phosphonate PMEG (Holý, 1994) was found to meet the first of above indicated criteria, being active at concentrations in the range of 10 to 100 nM, and was therefore selected for exploration of its effect on the cell cycle progression and on MAb production rate.

Materials and methods

Cell culture

Mouse hybridomas ME-750 and PVA-187 were cultured in DMEM/F 12/RPMI 1640 (2:1:1) medium supplemented with BME amino acids, 2.0 mM glutamine, 0.4 mM each of alanine, serine, asparagine and proline (Franěk and Šrámková, 1996), 15 mM HEPES, 2.0 g L^{-1} sodium bicarbonate and with the iron-rich protein-free growth-promoting mixture containing 400 μM ferric citrate (Franěk et al., 1992). All cultures were kept at 37 °C in a humidified atmosphere containing 5% CO_2. The static cultures were conducted in 25 cm^2 T-flasks. For estimation of growth inhibition, apoptosis and cell cycle profiles the cultures were inoculated at viable cell density (400±50) $\times 10^3$ mL^{-1} and incubated for 2 days. Stirred cultures were carried out in spinner flasks.

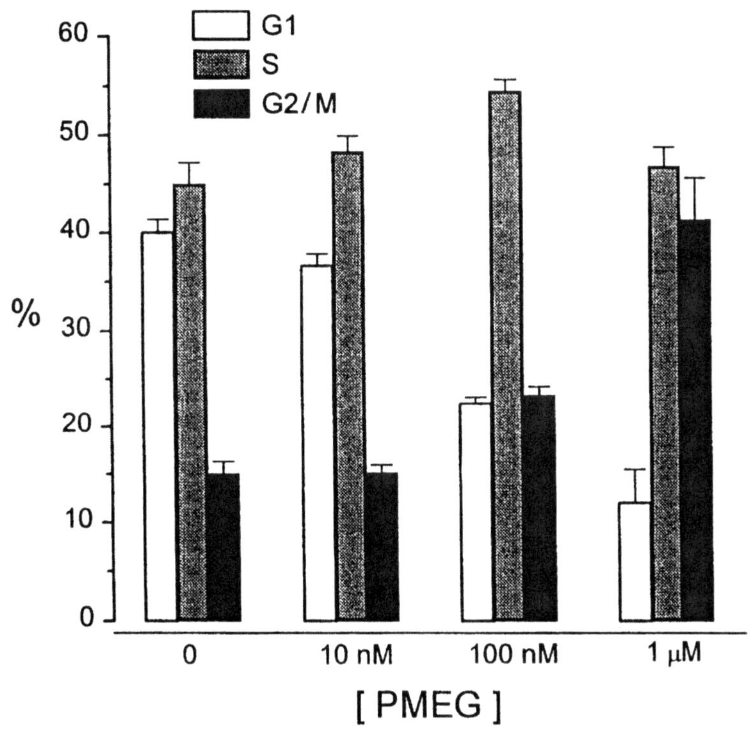

Figure 2. Distribution of cell cycle phases in cultures of hybridoma ME-750 supplemented with PMEG at various concentrations. The data were obtained by DNA flow cytometry of cells harvested after 2 days in static culture.

Sample analyses

Viable cells and dead cells were counted in a hemocytometer using trypan blue exclusion test.The experimental error involved in the estimation of cell count and viability was ±10%. The MAb concentration was determined by immunoturbidimetry (Fenge et al., 1991). Two parallels were set up for each sample evaluation. The experimental error associated with the estimation of MAb concentration was ±5%.

The apoptotic index, i.e., the percentage of apoptotic cells relative to total cell density, was determined by microscopic counting of morphologically distinct shrunken cells with ruffled membrane, as described in detail before (Franěk, 1995). The concentration of glutamine was determined using TC L-glutamic acid kit with asparaginase (Boehringer Mannheim, Germany). For the determination of glucose the kit Oxochrom glucose (Lachema, Brno, Czech Republic) was employed.

Cell cycle profiles

For determination of the cell cycle profiles the cells ($5–10 \times 10^6$ mL^{-1}) were permeabilized and stained with propidium iodide using DNA Prep Reagent Kit (Coulter Immunology, Hialeah, FL) according to manufacturer's protocol. Twenty to 120 min after completing the staining procedure the samples were measured with EPICS Profile II flow cytometer (Coulter Immunology, Hialeah, FL) and data were analyzed by Multicycle software (Phoenix Flow Software Products, San Diego, CA). Percentage of cells in G_1, S, and in G_2/M phases were evaluated.

Antiproliferative agent

PMEG, i.e. 9-[2-(phosphonomethoxy)ethyl]guanine, was synthesized as described before (Holý et al., 1989). Stock solution at 1 mM concentration was prepared in PBS and stored at +4 °C. The solution was stable for several months.

Results

Relation of growth inhibition and apoptosis

With the aim at evaluating quantitatively the relationship between growth-inhibiting and apoptosis-inducing activities of PMEG an experiment was set

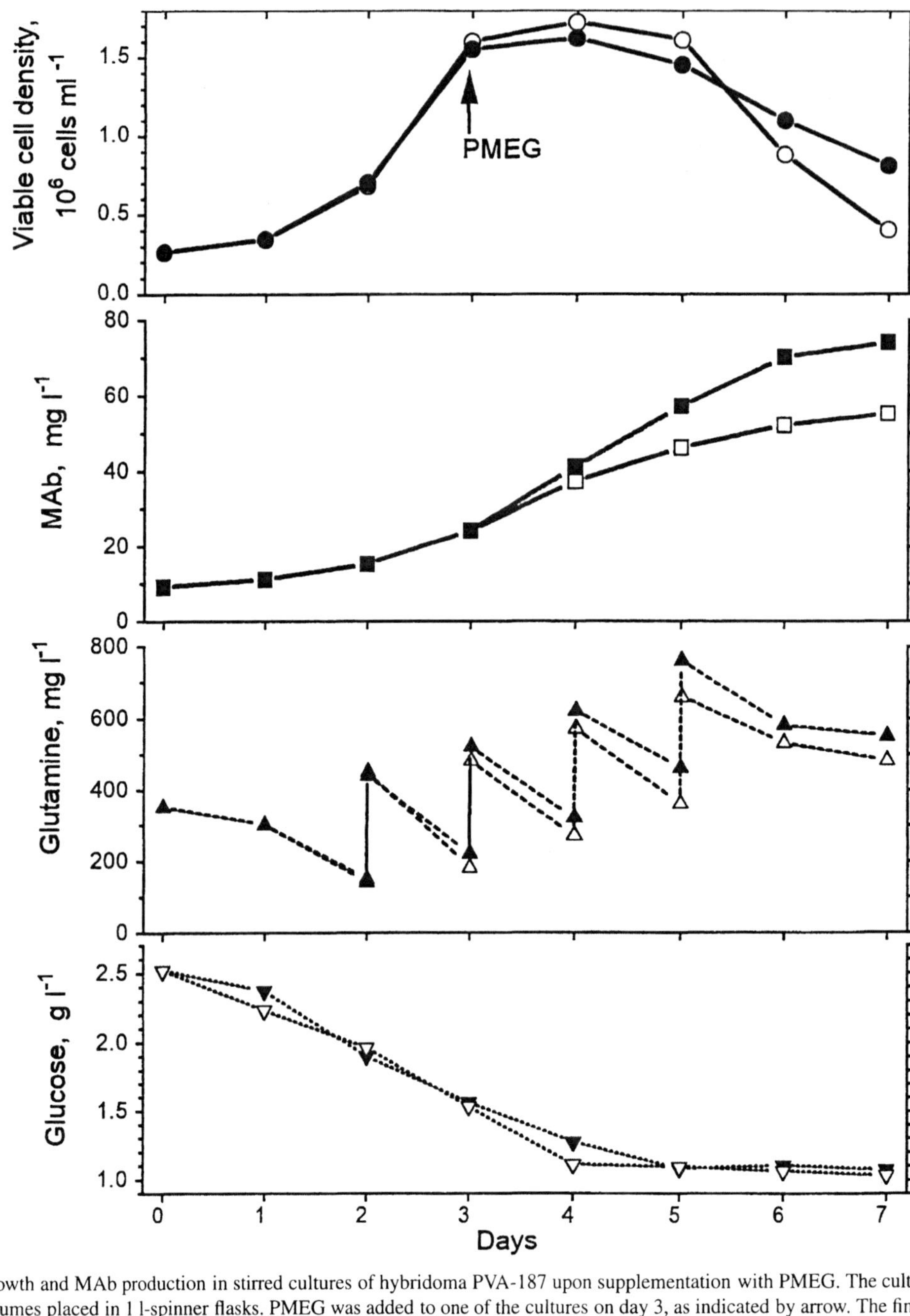

Figure 3. Growth and MAb production in stirred cultures of hybridoma PVA-187 upon supplementation with PMEG. The cultures were set up in 300-ml volumes placed in 1 l-spinner flasks. PMEG was added to one of the cultures on day 3, as indicated by arrow. The final concentration of PMEG was 50 nM. The culture was fed on days 2 to 5 by doses of 300 mg L^{-1} glutamine. Open symbols – control; closed symbols – culture supplemented with PMEG.

up, in which ME-750 hybridoma cells were incubated for 2 days in the presence of various concentrations of PMEG. The inhibition of growth was expressed as the viable cell density relative to the viable cell density in the control without PMEG. The degree of apoptosis was represented by the apoptotic index (Figure 1). The two sigmoid curves intersected at a value of approximately 30%. The concentrations representing the 50% changes of the parameters, i.e. IG_{50} (PMEG concentration causing 50% inhibition of growth), and AI_{50} (PMEG concentration at which the apoptotic index is equal to 50%), differed by almost one order of magnitude. The data in Figure 1 show that in the concentration range up to 100 nM the growth-inhibiting activity of PMEG is not accompanied with any substantial increase of apoptosis.

The changes in the cell cycle phases under influence of PMEG were examined by DNA flow cytometry (Figure 2). The inhibition of growth by PMEG, observable at PMEG concentrations >10 nM (Figure 1), was accompanied with a shift in the flow-cytometric DNA profile towards a higher proportion of the S and/or G_2/M phases of the cell cycle, and, consequently, with a reduction of the G_1 phase fraction. The degree of the shift progressed with increasing PMEG concentration. While at 10 nM PMEG the G_1 phase fraction represented 90% of the value observed in the control culture, at 100 nM PMEG the G_1 phase fraction dropped to 55% of the value in the control culture (Figure 2). The flow-cytometric analysis of DNA content thus showed that any significant reduction of hybridoma growth rate by application of PMEG would be accompanied with an altered proportion of cells in individual cell cycle phases.

MAb production

The influence of growth inhibition by PMEG on the specific MAb production rate was first examined in cultures to which PMEG was added at inoculation and was present during the whole culture period. The specific MAb production rate was found to be higher by 27%, relative to control, in the presence of 100 nM PMEG (Table 1). However, the final MAb concentration was lower than in the control, due to lower final viable cell density. Therefore, in further experiments PMEG was added at concentrations in the range of 10 to 50 nM at later stages of culture progression, and, in one case, in two portions.

A pair of stirred batch cultures of ME-750 hybridoma were set up starting from a single inoculation suspension. PMEG was added to one of the spinner cultures in two portions (Table 2). The final MAb concentration was 58 mg L^{-1} in the control, and 63 mg L^{-1} with PMEG, respectively. Thus, in this simple mode of batch culture the enhancement of the final MAb concentration upon addition of PMEG was found to be marginal. On the other hand, notable difference between the specific MAb production rates was found at the early decline period, i.e., between days 3 and 5. While in the control the average q_{MAb} dropped in this period to 6.95 pg cell^{-1} d^{-1}, in the culture with PMEG the average q_{MAb} value 8.70 compared well with the highest q_{MAb} values attained in the exponential phase. The longer persistence of high specific MAb production rate in the PMEG-supplemented culture, and the fact that the pool of substrates was identical in both the control and in the PMEG-supplemented culture, indicated that available substrates were preferentially utilized for the synthesis of the secreted antibody.

The drawback of the above reported experiment was the decline of the specific MAb production rate on days 5 to 6 (Table 2). We explain this fact by exhaustion of critical substrates, likely to occur in cultures run in the trivial batch mode. Therefore, in similar experiments carried out with the hybridoma PVA-187 the spinner cultures were fed with glutamine to ensure maximum energy supply for MAb synthesis (Franěk, 1995; Franěk and Šrámková, 1998). The data of a representative experiment of this series are illustrated in Figure 3. Upon addition of PMEG to one of the spinner cultures on day 3 the MAb concentration began to rise above the level of MAb concentration in the control culture. An analysis of the cell cycle profile of samples withdrawn at day 4 showed 58 and 71% of S + G_2/M phase fractions in the control and in the PMEG-supplemented culture, respectively. These data are in agreement with the general trend of increase of the S + G_2/M phase proportion accompanying the addition of PMEG.

The experiment revealed a delayed decline of viable cell density in the PMEG-supplemented culture on days 6 and 7 (see also Table 2, days 5 and 6). The reason for this favorable persistence of viability is not obvious. More data and more detailed analyses are needed to understand this feature of PMEG activity.

Discussion

In the search for antiproliferative drugs analogues of nucleic acid components belong to one of the most promising classes of substances. Efficient cytostatics, such as 6-thioguanine, 6-mercaptopurine or 5-fluorouracil are known to be incorporated both to DNA and to RNA, after intracellular activation, and, in addition, might inhibit activities of some enzymes (Pizzorno et al., 1997).Their complex mode of action, that is advantageous for therapeutic purposes, represents a disadvantage in production cultures, because damage of RNA inevitably impairs the synthesis of secreted proteins.

Table 1. Specific MAb production rate in hybridoma ME-750 cultures carried out in the presence of PMEG. Parameters on day 3 in static cultures set up in 25 cm^2 flasks are given. Inoculation density was 410×10^3 mL^{-1}

PMEG nM	Viable cell density $\times 10^{-3}$ mL^{-1}	Viability (%)	MAb mg L^{-1}	q_{MAb} pg $cell^{-1}$ d^{-1}	q_{MAb} rel. %
0 (Control)	1590	83	43	10.7	100
10	1560	84	43	10.9	102
100	960	77	39	13.6	127

q_{MAb} – Specific MAb production rate.

Table 2. Specific MAb production rate upon addition of PMEG to a grown culture of hybridoma ME-750. Batch cultures were set up in 120-ml volumes placed in 250-ml spinner flasks

Day of culture	Spinner I PMEG nM	$N_v \times 10^{-3}$ mL^{-1}	Viability (%)	MAb mg L^{-1}	q_{MAb} pg $cell^{-1}$ d^{-1}	Spinner II PMEG nM	$N_v \times 10^{-3}$ mL^{-1}	Viability (%)	MAb mg L^{-1}	q_{MAb} pg $cell^{-1}$ d^{-1}
0	0	470	90	12		0	470	90	12	
					8.2					8.2
1	0	750	95	17		0	750	95	17	
					8.7					8.8
2	0	1310	94	26		10	1300	93	26	
					8.3					8.6
3	0	1570	86	38		30	1480	86	38	
					7.1					8.9
4	0	1260	70	48		30	1210	68	50	
					6.8					8.5
5	0	810	45	55		30	920	52	59	
					4.3					4.8
6	0	580	31	58		30	740	42	63	

N_v – Viable cell density; q_{MAb} – Specific MAb production rate.

The acyclic nucleotide analogs have been designed with a specific aim to obtain substances resistant to most catabolic reactions in the target system (Holý, 1994). Among several N-(2-phosphonomethoxyethyl) derivatives, the guanine derivative PMEG was found superior to the adenine derivative, and to other derivatives, in the cytostatic efficiency (Veselý et al., 1990). PMEG belongs to agents that damage DNA during replication, i.e. it perturbs cellular processes in the S phase of the cell cycle. Its putative diphosphate derivative, PMEGpp, exhibits inhibitory activity toward cellular DNA polymerases α, δ and ε, and terminates DNA chain (Štětina et al., 1993; Kramata et al., 1996; Pisarev et al., 1997). *In vivo* antitumour activity of PMEG has been reported (Rose et al., 1990). Because the mode of action of PMEG is perturbation of DNA synthesis, slow-down in the rate of progression through S and G_2/M phases can be expected. Our observation of the progressive increase of the fraction of S and G_2/M phases with increasing PMEG concentration are in agreement with the above assumption. It is necessary to have in mind, that the distinction between the S and G_2/M phases might be impaired by the fact that the DNA content of cells in the late S phase and in the G_2 phase is hardly distinguishable. Therefore, in our present interpretations of the data we preferred to adhere to the rough division into G_1 and S + G_2/M phases.

It has been assumed that transient inhibition of DNA synthesis, caused by low concentrations of PMEG, will not be accompanied by inhibition of protein synthesis, and that transient unbalanced growth (Kung et al., 1990) will result in enhanced specific MAb production rate. The guanine derivative PMEG has attracted our interest because its antiproliferative activity is apparent at concentrations ~1000-fold lower than are the concentrations at which traditional antiproliferative agents, butyrate (Kruh, 1982), thymidine (Terada et al., 1997), or mimosine (Orren et al., 1997), are applied.

The favorable ratio of growth-inhibiting activity to apoptosis-inducing activity of this compound, as shown in Figure 1, pointed to the possibility of a harmless application of PMEG in the concentration range of 10 to 100 nM. The shift in the flow-cytometric DNA profile towards a higher proportion of the S and G_2/M phases is consistent with the assumed perturbation of DNA replication.

The expected enhancement of the specific MAb production rate could be shown both in static cultures and in stirred cultures (Tables 1 and 2). However, for the application in the field of animal cell technology the final yield of MAb is of decisive importance. The batch experiments (Table 2 and Figure 3) have shown that the addition of small doses of PMEG to a highly grown hybridoma culture is a promising mode of practical utilization of this substance in MAb production and, possibly, for production of recombinant proteins by cultured animal cells.

Providing an antiproliferative signal to a production culture through addition of a small synthetic molecule is an alternative to the genetic approach that utilizes the capacities of recombinant technology. Controlled overexpression of physiological growth regulators, such as IRF-1 (Kirchhoff et al., 1996) or p21, p27 and P53175P (Fussenegger et al., 1997) led to elevation of model protein production. Eventhough the spectrum of capacities of the recombinant techniques for improvement of production lines is higher than the reduction of growth rate, we believe that further research on growth control by synthetic substances of antimetabolite character may prove to be a short and inexpensive way to increasing efficiency of production processes.

The present pilot study has answered some basic questions. Further investigation is needed to accumulate more data and to explore the possible favorable effects of PMEG application in the perfusion mode of culture, or in systems with immobilized cells.

Acknowledgements

This work was supported by the Grants No. 303/96/0519 and No. 203/96/K001 of the Grant Agency of the Czech Republic, the Grant No. 2066-5 of the Ministry of Health Care of the Czech Republic, by the Contract No. BIO4 CT95-0207 of the Commission of the EC, DGXII/Contract No. OK 120 of the Czech State Budget, and by the Grant No. I/68957 of the Volkswagen Foundation. The authors also thank M. Fálová and I. Fismolová for excellent technical assistence.

References

Fenge C, Fraune E, Freitag R, Scheper T and Schügerl K (1991) On-line monitoring of monoclonal antibody formation in high-density perfusion culture using FIA.Cytotechnology 6: 55–63.

Franěk F (1995) Starvation-induced programmed death of hybridoma cells: Prevention by amino acid mixtures. Biotech. Bioeng 45: 86–90.

Franěk F and Šrámková K (1996) Cell suicide in starving hybridoma culture: survival-signal effect of some amino acids. Cytotechnology 21: 81–89.

Franěk F and Šrámková K (1997) Unbalanced media for hybridoma cell culture – An alternative reality. In: Carrondo MJT et al. (Eds.) Animal Cell Technology, Kluwer Academic Publishers, Dordrecht, pp. 675–680.

Franěk F and Šrámková K (1998) Increasing monoclonal antibody productivity by semicontinuous substitution of production medium for growth medium. In: Merten O-W et al. (Eds.) Animal Cell Technology, Kluwer Academic Publishers, Dordrecht, pp. 55–57.

Franěk F, Vomastek T and Dolníková J (1992) Fragmented DNA and apoptotic bodies document the programmed way of cell death in hybridoma cultures. Cytotechnology 9: 117–123.

Fussenegger M, Mazur X and Bailey JE (1997) A novel cytostatic process enhances the productivity of Chinese hamster ovary cells. Biotech Bioeng 55: 927–939.

Holý A (1994) Isopolar phosphorus-modified nucleotide analogues. In: De Clercq E (Ed.) Advances in Antiviral Drug Design (pp. 179-232) JAI Press, Inc., Greenwich, CT (U.S.A.).

Holý A, Rosenberg I and Dvořáková H (1989) Synthesis of N-(2-phosphonylmethoxyethyl)derivatives of heterocyclic bases. Collect Czech Chem Commun 54: 2190–2210.

Kirchhoff S, Kröger A, Cruz H, Tümmler M, Schaper F, Köster M and Hauser H (1996) Regulation of cell growth by IRF-1 in BHK-21 cells. Cytotechnology 22: 147–156.

Kramata P, Votruba I, Otová B and Holý A (1996) Different inhibitory potencies of acyclic phosphonomethoxyalkyl nucleotide analogs toward DNA polymerases α, δ and ε. Mol Pharmacol 49: 1005–1011.

Kruh J (1982) Effects of sodium butyrate, a new pharmacological agent, on cells in culture. Mol Cell Biochem 42: 65–82.

Kung AL, Zetterberg A, Sherwood SW and Schimke RT (1990) Cytotoxic effects of cell cycle phase specific agents: Result of cell cycle perturbation. Cancer Res 50: 7307–7317.

Orren DK, Petersen LN and Bohr VA (1997) Persistent DNA damage inhibits S-phase and G_2 progression, and results in apoptosis. Mol Biol Cell 8: 1129–1142.

Pizzorno G, Cheng YC and Handshumacher RE (1997) Pyrimidine and purine antimetabolites. In: Holland JF, Bast RC Jr, Morton DL, Frei E III, Kufe DW and Weichselbaum RR (Eds.) Cancer medicine, Williams and Wilkins, Baltimore, pp. 923–948.

Pisarev VM, Lee SH, Connely MC and Fridland A (1997) Intracellular metabolism and action of acyclic nucleoside phosphonates on DNA replication. Mol Pharmacol 52: 63–68.

Rose WC, Crosswell AR, Bronson JJ and Martin JC (1990) *In vivo* antitumor activity of 9-[(2-phosphonylmethoxy)ethyl]-guanine and related phosphonate nucleotide analogues. J Natl Cancer Inst 82: 510–512.

Štětina R, Votruba I, Holý A and Merta A (1993) The effect of purine phosphonomethoxyalkyl derivatives on DNA synthesis in CHO Chinese hamster cells. Neoplasma 40: 373–378.

Suzuki E, Takahashi K and Ollis DF (1992) A simple structured model predicted positively-, negatively-, or non-growth associated antibody production rate depending on culture conditions.

In: Murakami H et al. (Eds.) Animal Cell Technology: Basic and Applied Aspects. Vol. 4, Kluwer Academic Publishers, Dordrecht, pp. 375–381.

Terada S, Fukuoka K, Fujita T, Komatsu T, Takayama S. Reed JC and Suzuki E (1997) Anti-apoptotic genes, bag-1 and bcl-2, enabled hybridoma cells to survive under treatment for arresting cell cycle. Cytotechnology 25: 17–23.

Takahashi K, Terada S, Ueda H., Makishima F and Suzuki E (1994) Growth rate suppression of cultured animal cells enhances protein productivity. Cytotechnology 15: 54–64.

Veselý J, Merta A, Votruba I, Holý A and Rosenberg I (1990) The cytostatic effects and mechanism of action of antiviral acyclic nucleotide analogues in L1210 mouse leukemia cells. Neoplasma 37: 105–110.

Cytotechnology **28:** 73–80, 1998.
© 1998 *Kluwer Academic Publishers. Printed in the Netherlands.*

Engineering Chinese hamster ovary (CHO) cells to achieve an inverse growth – associated production of a foreign protein, β-galactosidase

Frank W. F. Lee*, Cynthia B. Elias†, Paul Todd & Dhinakar S. Kompala‡
Department of Chemical Engineering, University of Colorado, Boulder, CO 80309-0424, U.S.A.

Received 26 August 1998; accepted 26 August 1998

Key words: adenovirus major late promoter, β-galactosidase, Chinese hamster ovary cells, continuous culture, G1 phase expression, *inverse*-growth associated production

Abstract

Protein synthesis in mammalian cells can be observed in two strikingly different patterns: 1) production of monoclonal antibodies in hybridoma cultures is typically *inverse* growth associated and 2) production of most therapeutic glycoproteins in recombinant mammalian cell cultures is found to be growth associated. Production of monoclonal antibodies has been easily maximized by culturing hybridoma cells at very low growth rates in high cell density fed-batch or perfusion bioreactors. Applying the same bioreactor techniques to recombinant mammalian cell cultures results in drastically reduced production rates due to their growth associated production kinetics. Optimization of such growth associated production requires high cell growth conditions, such as in repeated batch cultures or chemostat cultures with attendant excess biomass synthesis. Our recent research has demonstrated that this growth associated production in recombinant Chinese hamster ovary (CHO) cells is related to the S (DNA synthesis)-phase specific production due to the SV40 early promoter commonly used for driving the foreign gene expression. Using the stably transfected CHO cell lines synthesizing an intracellular reporter protein under the control of SV40 early promoter, we have recently demonstrated in batch and continuous cultures that the product synthesis is growth associated. We have now replaced this S-phase specific promoter in new expression vectors with the adenovirus major late promoter which was found to be active primarily in the G1-phase and is expected to yield the desirable *inverse* growth associated production behavior. Our results in repeated batch cultures show that the protein synthesis kinetics in this resulting CHO cell line is indeed *inverse* growth associated. Results from continuous and high cell density perfusion culture experiments also indicate a strong *inverse* growth associated protein synthesis. The bioreactor optimization with this desirable *inverse* growth associated production behavior would be much simpler than bioreactor operation for cells with growth associated production.

Introduction

Two strikingly different patterns of the specific glycoprotein production rate versus mammalian cell growth rates are commonly found in the literature. The first pattern, observed primarily in monoclonal antibody production from murine hybridoma cells, shows an *inverse* relationship between the specific (per cell) antibody production rate and the cell growth rate. This essentially means that the specific production rate increases several fold as the specific growth rate decreases in continuous chemostat or perfusion cultures (Miller et al., 1988; Batt et al., 1990; Suzuki and Ollis, 1990). Repeated observations of higher specific monoclonal antibody production rate in many growth limiting culture conditions, such as increased osmolality (Ozturk and Palsson, 1991a), high ammonia concentrations, low and high pH (Miller et al., 1988; Ozturk and Palsson, 1991b) suggest the hypothesis that reduced cell growth rate under these different sub-optimal conditions may be the common mechanistic

* Current address: Regeneron Pharmaceuticals, Inc., Tarrytown, NY, U.S.A.

† Current address: Biotechnology Research Institute, National Research Council, Montreal, Canada

‡ Author for all correspondence.

factor causing higher antibody production rates. The second pattern of glycoprotein production rate versus the cell growth rate is observed increasingly in recombinant mammalian cell cultures. In an early example of this pattern, the specific antibody production rate in transfected myeloma cell lines is found to be strongly proportional to growth rate (Robinson and Memmert, 1991). Further examples of this directly proportional relationship between specific production rate and cell growth rate have been observed during the production of recombinant glycoproteins in CHO cells at elevated pCO_2 levels (Aunins and Henzler, 1993; Kimura and Miller, 1996) and at varying serum concentrations (Leelavatcharamas et al., 1994).

These two different patterns of protein synthesis have been reconciled by hypothesizing different cell cycle phase specific expression in a mathematical model for an intracellular reporter protein (Gu et al., 1994). The *inverse* relationship between the specific monoclonal antibody production rate and the cell growth rate has been generated by a previous cell cycle model (Suzuki and Ollis, 1989), which assumes that the cells secrete the antibody only during the G1 phase of the mammalian cell cycle. Experimental confirmations of the G1 phase expression of antibody by hybridoma cells have now been obtained by several researchers (Suzuki and Ollis, 1990; Ramirez and Mutharasan, 1990; Kromenaker and Srienc, 1991). Prediction of the *proportional* relationship between the intracellular reporter protein content and the cell growth rate is based on the assumption that the reporter protein is synthesized only during the S (DNA synthesis) phase of the cell cycle (Gu et al., 1994).

Several experimental studies suggest that commonly used promoters and enhancer elements driving the expression of foreign proteins are active maximally in the S phase. Mariani et al. (1981) found S phase specific synthesis of dihydrofolate reductase (DHFR), presumably due to its native promoter, in CHO cells. Kubbies and Stockinger (1990) found that the expression of tissue plasminogen activator (tPA) controlled by the SV40 early promoter in CHO cells was again S phase specific. The expression of β-galactosidase under the control of the cytomegalovirus (CMV) promoter in CHO cells was also shown to be S phase specific (Gu et al., 1993). Kubbies and Stockinger (1990) also found that the expression of DHFR controlled by the adenovirus major late (AML) promoter in CHO cells occurred primarily during G1 phase.

These literature reports suggest that the cell cycle phase in which the foreign genes are expressed is not controlled by the structural gene or by the site of integration on the chromosome, but by the promoter and enhancer elements used in the vector for their expression. For example, changing the promoter and enhancer elements driving the expression of the same structural gene *dhfr* (from its native promoter to the AML promoter) appears to alter the cell cycle phase in which DHFR is expressed from S to G1 phase. Further, the two distinct cell cycle phase expression results (Kubbies and Stockinger, 1990) from the two expression cassettes (SV40 early promoter driving the tPA expression and AML promoter driving the expression of DHFR) on a single plasmid stably transfected and presumably integrated together on the chromosome indicate that cell cycle phase expression characteristics of the two different promoters override any possible cell cycle effects prevalent at the random site of integration on the chromosome. Based on such observations we have hypothesized that the upstream promoter and enhancer elements determine the cell cycle phase in which the downstream structural genes are expressed. The expression of β-galactosidase driven by the S–phase specific SV40 promoter in stably transformed CHO 202 cells was found to be strongly growth associated (Banik et al., 1996), as predicted by Gu et al. (1994). In this paper, we determine the intracellular levels of the same reporter protein, β-galactosidase, expressed from the G1 phase specific AML promoter in stably transfected CHO 202 cells at different growth rates in batch and continuous suspension cultures to investigate whether the predicted *inverse* growth associated production pattern is obtained.

Materials and methods

Cell line construction and maintenance

Escherichia coli XL1-Blue (F′ [*proA*$^+$ *proB*$^+$ *lacI*^{q} *lacZ*DM15 Tn*10* (*Tet*^{R})]/*supE*44 *hsdR*17 *recA*1 *gyrA*46 *thi relA*1 *lac*$^-$) (Stratagene, CA) was used for plasmid maintenance and construction. Plasmid (*Amp*^{r}) containing cells were selected using Luria broth (LB) medium containing 100 mg L^{-1} ampicillin sodium salt (Sigma), as described in Sambrook et al. (1989).

Restriction endonucleases, alkaline phosphatase, and T4 DNA ligase were obtained from New England Biolabs, MA and Gibco, NY. Plasmid minipreparation

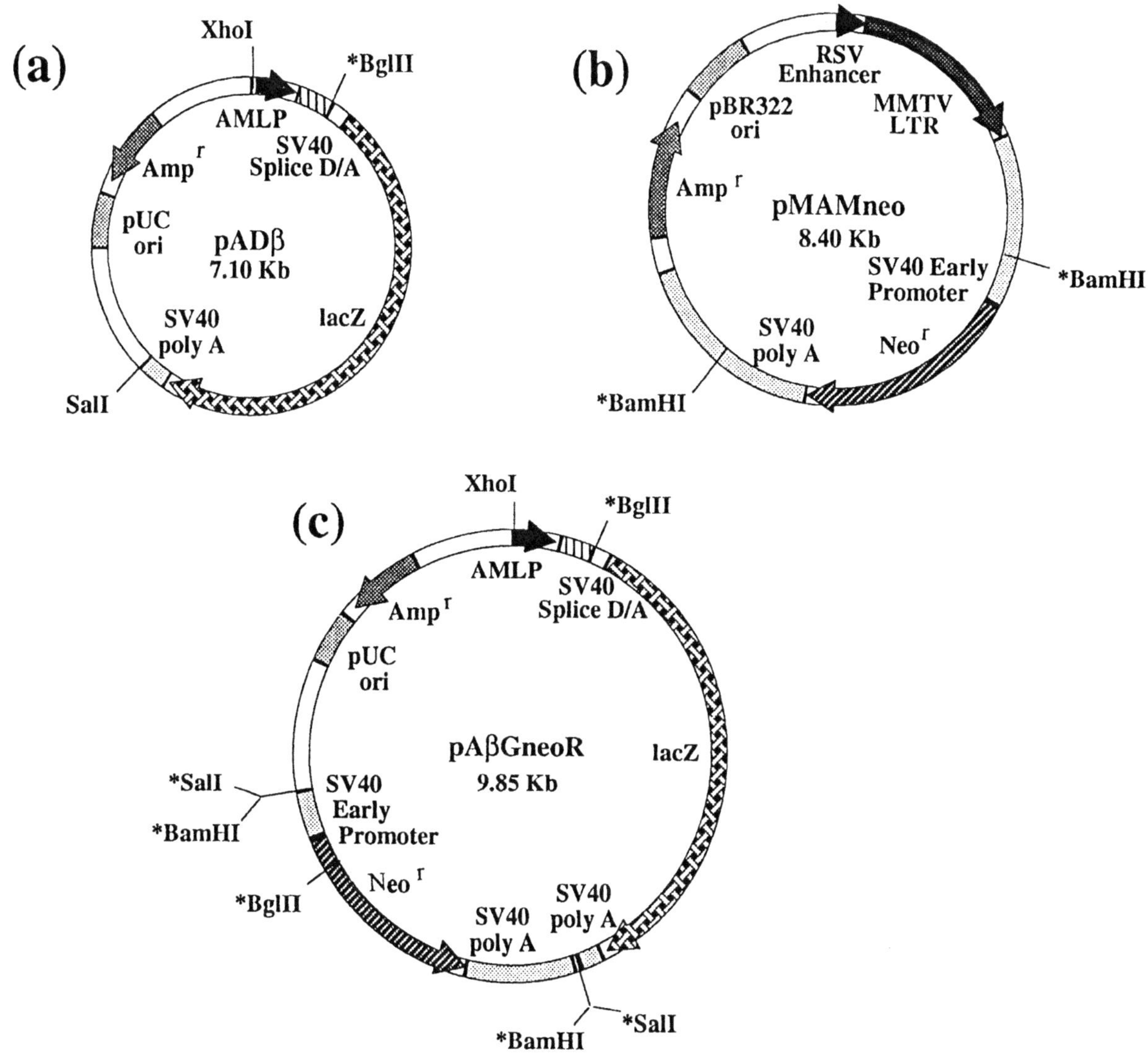

Figure 1. Plasmid maps of (a) pADβ, (b) pMAMneo, and (c) pAβGneoR showing relative size, restriction sites, and location of inserted DNA.

and DNA electrophoresis were conducted as described in Sambrook et al., (1989). DNA fragments were purified from agarose gels using the GENECLEAN kit (Bio 101, CA). Large-scale plasmid DNA preparation from *E. coli* was carried out using the Qiagen Plasmid Maxi kit (Qiagen, CA). *E. coli* transformations were performed using the calcium chloride method as described earlier by Sambrook et al. (1989).

Plasmid pADβ (Clontech, CA; Figure 1a) carries the bacterial *lacZ* gene under the control of the adenovirus major late (AML) promoter. Transcription of the *lacZ* gene is terminated by employing the polyadenylation (poly A) sequence from the simian virus SV40. The unique SalI site downstream to the SV40 poly A sequence in the plasmid pADβ was used as a cloning site for the neomycin resistance (*neo*^{r}) selection marker. Plasmid pMAMneo (Clontech, CA; Figure 1b) contains the *neo*^{r} gene driven by the SV40 early promoter, and its transcription is also terminated by the SV40 poly A sequence. The *neo*^{r} gene confers resistance to the antibiotic G418, a neomycin analog, and provides a selection marker in screening plasmid-transfected mammalian cells. This *neo*^{r} expression cassette (2.9 Kb) was isolated by a BamHI restriction enzyme digestion; both ends of this BamHI fragment was subsequently adapted to SalI. After ligation of this fragment into the unique SalI site of the plasmid pADβ, an insertion with transcription in the reverse

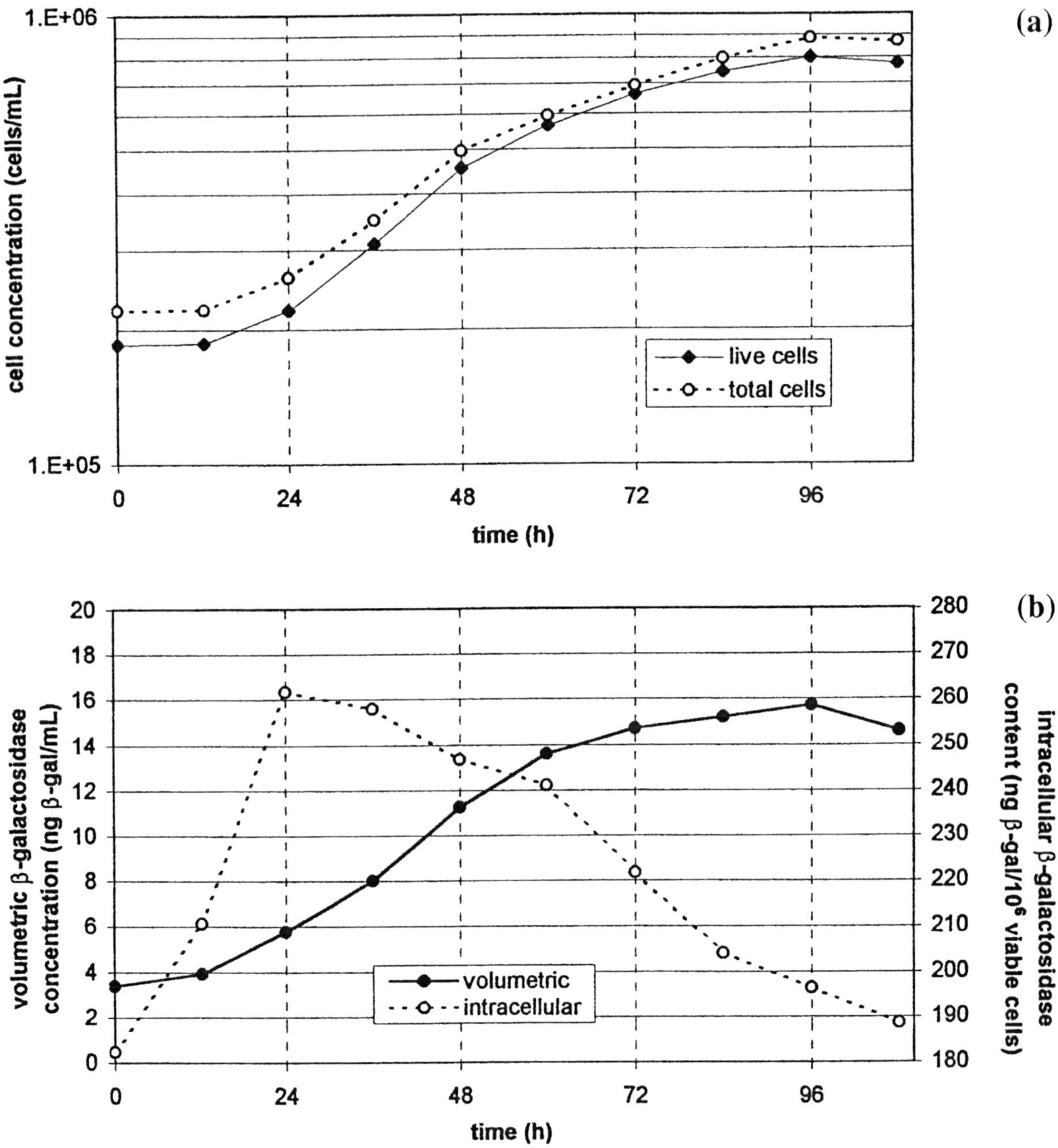

Figure 2. a) Batch culture kinetics of viable and total cell concentration. CHO cells were stably transfected with the plasmid pAβGneoR and grown in suspension culture in spinner flasks. b) The volumetric β-galactosidase concentrations and the intracellular B-gal content versus batch culture time. While the volumetric concentration appears to follow the increase in cell concentration shown on the semilog graph in Figure 2, the intracellular content clearly shows a large increase in the pre-exponential growth phase and gradual decline during mid and post exponential growth phases.

direction as the *lacZ* gene were desired to prevent the potential structural instability due to the homologous recombination between two copies of the SV40 poly A sequence. The construct, including direction of this insertion was confirmed by a double digestion with BglII and XhoI; the final plasmid was designated pAβGneoR (Figure 1c).

Host cell line CHO-202 was stably transfected with plasmid pAβGneoR constructed as described above (containing the *lac Z* gene under the influence of the AML promoter and the gene for neomycin resistance driven by the SV40 early promoter) using Lipofectin^{TM} as recommended by the manufacturer (Life Technologies NY.). The transfected cells were cultured and cloned under a selection pressure of G418 (400 μg mL^{-1}), and a high producing cell line (2GAR-24) was used in all the experiments described below.

Medium and supplements

The stably transfected clones were routinely maintained in α-MEM (Life Technologies, NY) medium supplemented with 10% fetal calf serum (FCS, Life Technologies). The selected clone was adapted for growth in serum free IS-CHO medium (Irvine Scientific, CA), containing 8 g L^{-1} glucose and 8 mM glutamine. Antibiotics, penicillin (100 units mL^{-1}) and streptomycin (10^{-4}g mL^{-1}), were added to the medium. The medium was further supplemented with the neomycin analog G418 (400 μg mL^{-1}) for selection pressure.

Cell culture and analytical methods

The recombinant 2GAR-24 cells were adapted to suspension culture in spinner flasks (Techne, Cambridge, U.K.) agitated on a magnetic stirrer at a speed of 45 rpm placed inside a controlled environmental chamber at 37 °C with 5% CO_2. All the data shown in this paper were obtained from cells growing in suspension. Batch culture experiments were carried out in 350 mL spinner flasks with a working volume of 150 mL. Cells were seeded at a density of 1–2 × 10^5 cells mL^{-1}.

Continuous culture experiments were carried out in a 1.5 L Celligen bioreactor (New Brunswick, NJ) with a working volume of 1.1 L. The dissolved oxygen was controlled at 40% air saturation and the pH was maintained at 7.2±0.1 by sparging a mixture of air, nitrogen, carbon dioxide and oxygen into the culture medium.

Samples from the spinner flasks and the bioreactor were collected at fixed time intervals, and the cell numbers (both viable and total) were determined by counting on a haemacytometer slide after staining with 0.4% trypan blue solution. The activity of intracellular β-galactosidase was measured using the standard o-nitrophenyl β-D-galactoside (ONPG) assay described in more detail by Banik et al. (1996). This kinetic assay is highly reproducible and accurate with a standard error of about 5%.

Results

Batch culture

Stably transfected CHO cells (2GAR 24) were grown in spinner flasks for eight sequential batch cultures, to confirm reproducibility of the results. Data from a single representative batch culture are shown in Figures 2a and b. The cells were seeded at an initial density of 1.8 × 10^5 viable cells mL^{-1}. The cells show an initial lag period and a slow growth rate during the first 24 h. Cells grew exponentially at their maximum specific growth rate till about 48 h, after which the cell growth slowed and cells reach their maximum density at 96 h. The volumetric β-galactosidase concentration (ng mL^{-1}) was found to increase with increase in cell number and reached a maximum value at the end of the exponential phase (Figure 2b). The intracellular β-galactosidase content expressed as ng 10^{-6} cells increased rapidly during the initial lag/slow-growth period of 24 h after which a gradual decrease in the intracellular content was observed (Figure 2b).

Continuous culture studies

The continuous culture experiments with the 2GAR 24 cells were carried out in a 1.5 L Celligen bioreactor with a working volume of 1.1 L. The cells were seeded at an inoculum density of 1.8 × 10^5 cells mL^{-1} and allowed to grow in batch culture initially. Continuous feeding and removal of medium were started during mid-exponential phase. The dilution rates were varied stepwise from 0.3 to 0.65 day^{-1} after a steady state was well established at each setting. At the dilution rate of 0.65 day^{-1} steady state concentration of cells could not be reached and wash-out conditions were observed, consistent with $\mu_{max} \approx 0.76$ day^{-1} from Figure 2a.

The viable and total cell number concentrations during the continuous culture are shown in Figure 3a.

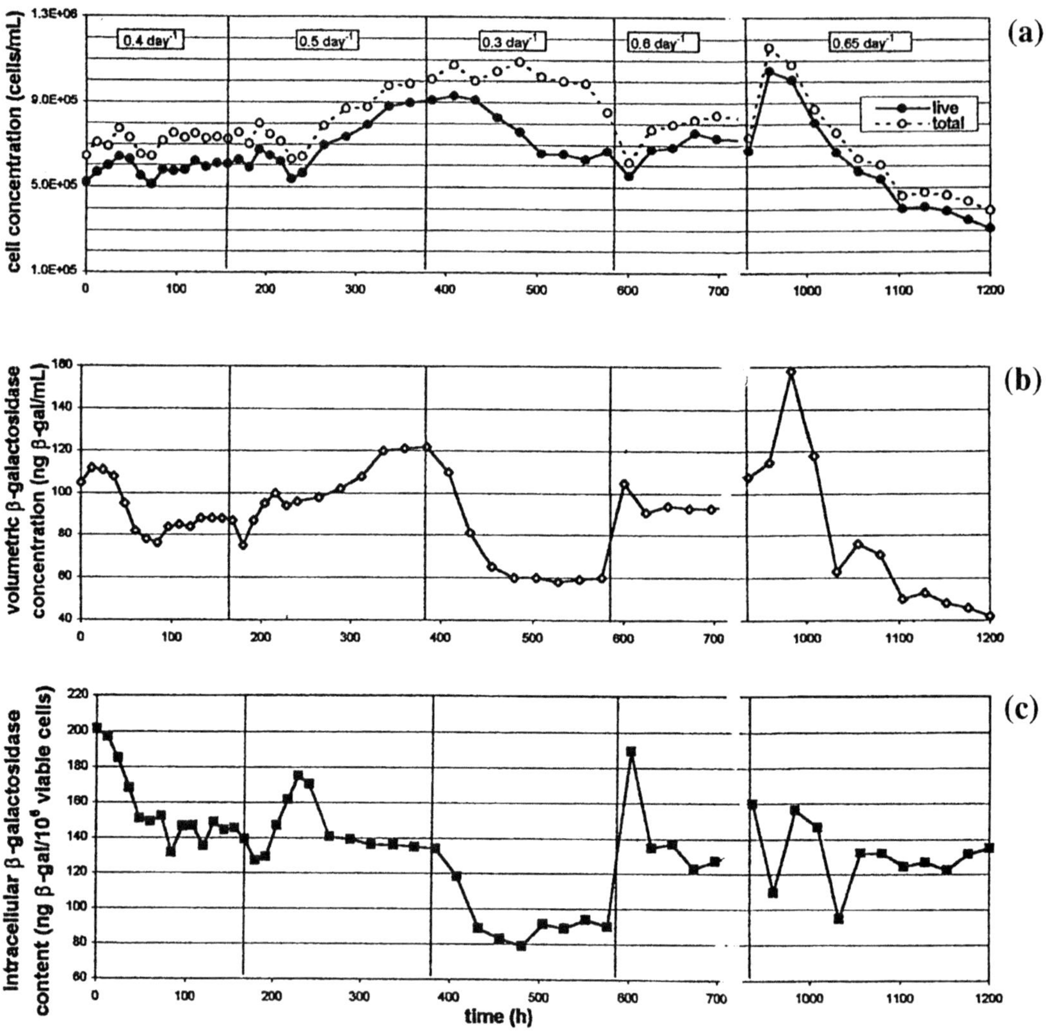

Figure 3. a) Viable and total cell concentrations of stably transfected CHO cells in continuous bioreactor culture. The dilution rates are changed stepwise after a steady state is well established. Volumetric β-galactosidase concentration; b) and intracellular β-galactosidase content; c) in continuous suspension cultures at different dilution rates.

From these results it can be seen that the steady state concentration of cells was higher at the higher dilution rates. The profiles of the volumetric β-galactosidase concentrations and specific intracellular β-galactosidase content are shown in Figures 3b and c respectively. Although the volumetric β-galactosidase concentration increased along with cell number with an increase in dilution rate, the specific intracellular β-galactosidase content showed the opposite trend and was higher at the lower dilution rates. This trend is clearly seen in Figure 4 where the steady state values of the intracellular β-galactosidase content are plotted as a function of the specific cell growth rate μ. The only exception to this trend occurred at the lowest dilution rate of 0.3 day^{-1}, where the cell viabilities were very low.

The specific cell growth rate μ is typically equal to the dilution rate when the cell viability is close to 100%, as in continuous cultures of bacterial cells. In chemostat cultures of mammalian cells, the cell viability is typically in the range of 90 to 95%. Hence the specific cell growth rate is calculated using the equation ($\mu = DX_T/X_V$) obtained from steady state cell balances. During the transient washout state observed at the highest dilution rate shown in Figure 3, the intracellular β-galactosidase content reaches a steady state after the initial transients during the first 200 hr, since the cells maintain their highest specific growth rate of

0.76 day^{-1}, as calculated from the transient balance and also confirmed by the batch culture data shown in Figure 2.

Discussion

Several investigators (Gu et al., 1994; Suzuki and Ollis, 1989) have suggested that foreign protein production can be maximized by utilizing different bioreactor operating strategies appropriate for the particular pattern in which the foreign production varies with the cell growth rate. For example, the *inverse*-growth associated relationship between the production rates and growth rate, may be exploited by growing cells at reduced growth rates in high cell density fed-batch or perfusion cultures to maximize the production of foreign proteins. A directly proportional relationship between the two will require maintaining the cells at high cell growth rates in repeated batch cultures or cause severely reduced production during the low growth fed-batch and perfusion cultures. For hybridoma cultures, which primarily exhibit the inverse relationship between the specific antibody production rate and cell growth rate (Miller et al., 1988; Suzuki and Ollis, 1990), the volumetric monoclonal antibody production rate can be increased several fold by growing cells at low specific growth rates in fed-batch (Bibila and Robinson, 1995) and continuous perfusion cultures (Batt et al., 1990).

In contrast, glycoprotein production in recombinant mammalian cells is typically found to be strongly growth associated with respect to their specific productivity (Aunins and Henzler, 1993; Leelavaatcharamas et al., 1994; Robinson and Memmert, 1991). This directly proportional pattern of specific production rate with growth rate requires constant high cell growth rates such as in repeated batch cultures for the maximization of production rate. However, as the CHO cells are merely the catalysts or mini-factories for the synthesis of glycoproteins it is undesirable to keep on producing these cells, when the real objective is to maximize the synthesis of glycoproteins. High cell density fed-batch or perfusion cultures, which represent the optimized production systems for hybridoma cells, however, cause severely reduced glycoprotein synthesis in cells possessing a directly proportional (relationship) between the production rate and growth rate.

The intracellular content of foreign protein has been predicted to either increase (for S-phase synthesis) or decrease (for G1-phase synthesis) with growth rate, depending on the cell cycle specific promoters that control the expression of foreign genes. Recent studies in our laboratory (Banik et al., 1996) have shown that the expression of an intracellular reporter protein (β-galactosidase) under the control of the S-phase specific SV40 promoter was growth associated in recombinant CHO cells, i.e., the steady state intracellular β-galactosidase content was found to increase almost linearly with the increasing specific cell growth rate in chemostat cultures. The same host cell line (CHO 202) stably transfected with a similar expression plasmid, but containing the G1-phase specific AML promoter for driving the expression of the same reporter protein, reverses the variation of intracellular β-galactosidase content with cell growth rate. The chemostat results summarized in Figure 4 show that the intracellular content of the reporter protein increases as the dilution rate and thereby the specific growth rate is decreased. Thus by simply switching to the G1 phase specific AML promoter, we have demonstrated that the intracellular foreign protein production can be switched from the undesirable growth associated pattern to the more desirable *inverse*-growth associated pattern.

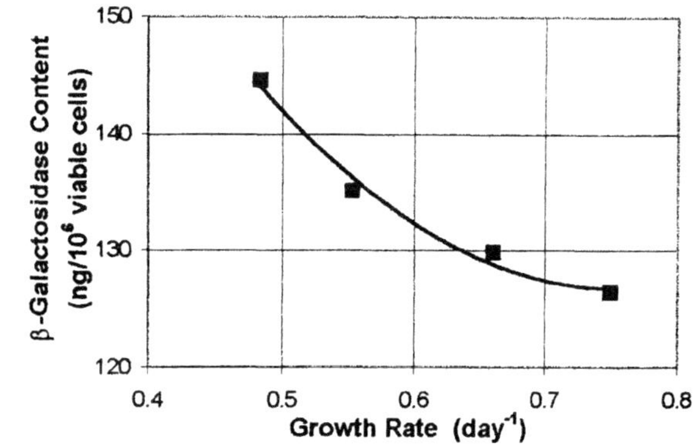

Figure 4. The *inverse*-growth-associated pattern of intracellular β-galactosidase content versus the specific growth rate, obtained in chemostat cultures of CHO cells expressing β-galactosidase from the G1 phase specific AML promoter.

The batch results shown in Figures 2a and b are consistent with these *inverse*-growth associated chemostat results in Figure 4 during the early pre-exponential slow growth phase, but not during the late-exponential slow growth period. Further in chemostat culture at the lowest dilution rate studied, the intracellular β-galactosidase content and the cell viability were significantly lower than the trend shown in Figure 4. It is suspected that the accumulated metabolic by-products and dead cells in these conditions may have some bearing on these results. We are now re-investigating them through the selective removal of these components in a continuous perfusion bioreactor with an attached inclined settler (Searles et al., 1993; Batt et al., 1990). Nevertheless, the *inverse*-growth associated pattern of intracellular β-galactosidase reported here has been found to be reproducible for these stably transfected CHO cells in subsequent chemostat cultures at these medium to high cell growth rates. If this *inverse*-growth associated production pattern with the G1 phase specific AML promoter is reproduced with secreted reporter proteins in future experiments, such results will have significant impact on the maximizing of glycoprotein production rates from the slow growing CHO cells in high cell density fed-batch or perfusion bioreactors.

Conclusions

Switching from the S-phase specific SV40 promoter to the G1 phase specific AML promoter for driving the expression of the intracellular reporter protein β-galactosidase in a stably transfected CHO cell line, modifies intracellular β-galactosidase production from the growth associated to the more desirable inverse-growth associated pattern. These findings confirm our earlier predictions (Gu et al., 1994) concerning the effects of cell cycle phase specific expression characteristics. If these findings are reproduced with secreted reporter proteins in future chemostat cultures of recombinant CHO cells, they will have a large impact on maximizing glycoprotein synthesis in production scale bioreactors.

Acknowledgements

This research was funded by NSF Grant # BES-9504840. We thank Irvine Scientific, CA for the IS-CHO medium. Technical assistance from Stephanie Bryant, Jennifer Shull, Jennifer Plant and Bill Perry is appreciated.

References

Aunins JG and Henzler HL (1993) Aeration in cell culture bioreactors. In: Rehm HJ, Reed G, Puhler A and Stadler PJW (eds.) and Stephanopoulos G (vol. ed.) Biotechnology: Bioprocessing (reaction and process engineering) Vol. 3 VCH Verlag, Weinheim, Germany.

Banik GG, Todd P and Kompala DS (1996) Foreign protein production from S-phase specific promoters in continuous cultures of recombinant CHO cells. Cytotechnology 22: 179–184.

Batt BC, Davis RH and Kompala DS (1990) Inclined sedimentation for selective retention of viable hybridomas in a continuous suspension bioreactor. Biotechnol Prog 6: 458–464.

Bibila TA and Robinson DK (1995) In pursuit of the optimal fed-batch process for monoclonal antibody production. Biotechnol Prog 11: 1–13.

Gu MB, Todd P and Kompala DS (1993) Foreign gene expression (β-galactosidase) during the cell cycle phases in recombinant CHO cells. Biotechnol Bioeng 42: 1113–1123.

Gu MB, Todd P and Kompala DS (1994) Analysis of foreign protein overproduction in recombinant CHO cells: Effect of growth kinetics and cell cycle traverse. Ann NY Acad Sci (Recombinant DNA technology II) 721: 194–207.

Kimura R and Miller WM (1996) Effects of elevated pCO_2 and/or osmolality on the growth and recombinant tPA production in CHO cells. Biotechnol Bioeng 52: 152–160.

Kromenaker SJ and Srienc F (1991) Cell cycle dependent protein accumulation by producer and non producer murine hybridoma cell lines: A population analysis. Biotechnol Bioeng 38: 665–677.

Kubbies H and Stockinger H (1990) Cell cycle dependent DHFR and tPA production in cotransfected, MTX amplified CHO cells revealed by dual-laser flow cytometry. Exptl Cell Res 188: 665–677.

Leelavaatcharamas V, Emery AN and Al-Rubeai M (1994) Growth and interferon-γ production in batch culture of CHO cells. Cytotechnology 15: 65–71.

Mariani BD, Slate DL and Schimke RT (1981) S phase-specific synthesis of dihydrofolate reductase in Chinese hamster ovary cells. Proc Natl Acad Sci USA 78: 4985–4989.

Miller WM, Blanch HW and Wilke CR (1988) A kinetic analysis of hybridoma growth and metabolism in batch and continuous suspension culture: Effect of nutrient concentration, dilution rate and pH. Biotechnol Bioeng 32: 947–965.

Ozturk SS and Palsson BO (1991a) Effect of medium osmolality on hybridoma growth, metabolism and antibody production. Biotechnol Bioeng 37: 989–993.

Ozturk SS and Palsson BO (1991b) Growth, metabolic and antibody production kinetics of hybridoma cell culture: Effects of serum concentration, dissolved oxygen concentration and medium pH in a batch reactor. Biotechnol Prog 7:481–494.

Ramirez OT and Mutharasan R (1990) Cell cycle and growth phase dependent variations in size distributions, antibody productivity and oxygen demand in hybridoma cultures. Biotechnol Bioeng 36: 839–848.

Robinson DK and Memmert KW (1991) Kinetics of recombinant immunoglobulin production by mammalian cells in continuous culture. Biotechnol Bioeng 38: 972–976.

Sambrook J, Fritsch J and Maniatis T (1989) Molecular Cloning, a laboratory manual, 2nd edition. Cold Spring Harbor Laboratory, Cold Spring Harbor, New York.

Searles JA, Todd P and Kompala DS (1994) Viable cell recycle with an inclined settler in the perfusion culture of suspended recombinant Chinese hamster ovary cells. Biotechnol Prog 10: 198–206.

Suzuki E and Ollis DF (1989) Cell cycle model for antibody production kinetics. Biotechnol Bioeng 34: 1398–1402.

Suzuki E and Ollis DF (1990) Enhanced antibody production at slowed growth rate: Experimental determination and a simple structured model. Biotechnol Prog 6: 231–236.

Cytotechnology **28:** 81–89, 1998.
© 1998 *Kluwer Academic Publishers. Printed in the Netherlands.*

A high-yielding serum-free, suspension cell culture process to manufacture recombinant adenoviral vectors for gene therapy

G. Schoofs, T. J. Monica, J. Ayala, J. Horwitz, T. Montgomery, G. Roth & F. J. Castillo*
Fermentation and Cell Culture Development, Biotechnology R&D (CMC), Berlex Biosciences, 15409 San Pablo Avenue, Richmond, CA 94804, U.S.A.
E-mail: Frank_Castillo@Berlex.com

Received 21 September 1998; accepted 21 September 1998

Key words: adenovirus, HEK 293, perfusion, serum free media, suspension perfusion

Abstract

We have developed an efficient, reproducible, and scaleable cell culture process for a recombinant adenoviral vector expressing therapeutic transgenes for clinical trials. HEK 293 cells – which support the propagation of E1 deficient adenovirus – were first adapted to serum free media and suspension growth. Subsequent studies focused on the infection, virus production and harvest from suspension culture bioreactors. Future studies are planned to address the kinetics of adenovirus production in HEK 293 as well as in other cell lines.

Abbreviations: Ad5 – adenovirus serotype 5; cGMP – current good manufacturing practice; CPE – cytopathic effect; DMEM – Dulbecco's modified Eagle's medium; EPD – end point dilution assay; FBS – fetal bovine serum; HEK 293 – human embryonic kidney cells transformed with Ad5 E1 DNA; HPLC – high pressure liquid chromatography; ip – number of infectious virus particles; MOI – multiplicity of infection; p – total number of virus particles (infectious and non-infectious); PBS – phosphate buffered saline; SFM – serum free medium

Introduction

The number of gene therapy protocols using adenoviral vectors has increased steadily and reached ten percent of the two hundred and seventy three worldwide protocols by the end of 1996 (Marcel and Grausz, 1997). These vectors are preferable to retroviral vectors when transient gene expression is required (Crystal, 1995). The adenoviral vectors for gene therapy programs are currently produced by infection of a suitable permissive cell line with a recombinant E1 deficient adenovirus containing a transgene insert. To meet commercial and regulatory requirements, this process must be high yielding, scaleable and reproducible. This process should also preferably avoid the use of animal-derived medium constituents, such as serum.

Some researchers have described the use of suspension adapted HEK 293 cells in serum-containing media for protein production via adenoviruses (Garnier et al., 1994; Nadeau et al., 1996). Other workers have discussed their use of attached or clumped HEK 293 cells in serum-containing media to produce adenovirus for gene therapy applications (Shabram et al., 1997; Hehir et al., 1998). There is no literature, however, on adenovirus production in serum-free mediumg with suspension adapted HEK 293 cells. In addition, little data is available on optimal multiplicity of infection or in-process virus stability (Graham and Prevec, 1994).

Our goal was to develop an efficient, reproducible, and scaleable cGMP manufacturing process for the production of an adenoviral vector expressing therapeutic transgenes using HEK 293 cells. These requirements were met by establishing a cell culture process employing stirred tank fermentors, suspension culture adapted HEK 293 cells, and serum free cell culture medium. All process steps, from preparation of cell and virus banks to collection of recombinant virus

* Author for all correspondence.

from the cell culture harvest, were studied. This article specifically addresses development work on the adaptation of cells to suspension culture, the selection of serum-free medium and adaptation of cells to that medium. The characterization of adapted cells, the production of virus in adapted cells, and the cryopreservation of cells in serum free medium are also discussed, as well as performance aspects of our long-term, large-scale cultures. This article further describes clonal purification of virus, studies to determine optimal multiplicity of infection; the infection of cells at large scale, and yields and stability of raw virus.

Materials and methods

Cell culture

HEK 293 cells were obtained from Magenta Corporation (Rockville, MD). These HEK 293 cells were maintained and passed in DMEM (Gibco BRL, Life Technologies, Grand Island, NY) with 10% FBS (Hyclone Laboratories Inc., Logan, UT). Serum free media were provided by JRH Scientific (Lenexa, KS), Irvine Scientific (Santa Ana, CA), and BioWhittaker (Walkersville, MD). Cell counts were done using standard hemocytometers and a Coulter counter (Coulter Electronics LTD, Luton, Beds., England). Stock cells were passed when at $>1 \times 10^6$ viable cells mL^{-1}, and reseeded at $>4 \times 10^5$ viable cells mL^{-1}. Cells were photographed using a Sony CCD-IRIS Color Video Camera (Sony Corporation, Japan).

Shake flasks

Small scale experiments were conducted in 125 to 2800 mL (total volume) shake flasks, that were either disposable flasks (Corning, Corning, NY), reusable plastic flasks (Nalgene, Rochester, NY) or reusable glass flasks (Bellco, Vineland, NJ). The reusable shake flasks were sterilized by autoclaving. The flasks were maintained in 5% CO_2 incubators with temperature control at 37 °C (VWR Scientific, So. Plainfield, NJ). Shaker tables were operated at 140 rpm (New Brunswick Scientific, Edison, NJ). The pH was checked by off-line measurements with a pH meter (Corning Inc., Corning, NY). Dedicated instruments (YSI Instruments, Yellow Springs, OH) measured glucose and lactate off-line.

Bioreactors

Long term perfusion cell cultures and batch cell infections were conducted in 3 and 10 L Applikon glass bioreactors (Applikon, Schiedam, the Netherlands). The reactors were all sterilized by autoclaving, and all sterile connections were made using a sterile tubing welder (Terumo Medical Corporation, Elkton, MD). Temperature was maintained at 37 °C by a single loop controller (Fenwall, Ashland, MA), using a platinum RTD to sense temperature, and a heating blanket (Applikon, Schiedam, the Netherlands) to heat the reactor. Dissolved oxygen was maintained at 50% of air saturation by single loop controller (B. Braun Biotech, Allentown PA), with a oxygen sensing probe (Mettler-Toledo Process Analytical Inc., Wilmington, MA) and pulsed sparging by time-proportioning on/off control of a solenoid valve. The pH was maintained at 7.1 by a single loop controller (B. Braun Biotech, Allentown PA) with a pH sensing probe (probe (Mettler-Toledo Process Analytical Inc., Wilmington, MA) and time-proportioning on/off control of either a pump feeding 0.2 N NaOH, or control of a solenoid valve adding CO_2 to the sparge system. The pH was checked by off-line measurements with a pH meter (Corning Inc., Corning, NY). Liquid level was controlled at 3 or 10 L working volume by a single loop controller (B. Braun Biotech, Allentown PA) using a conductance probe to sense level, and time-proportioning on/off control of a media feed pump (Cole-Parmer, Vernon Hills, IL). Cells were retained in perfusion culture using an external vortex flow filtration device (Membrex Inc., Fairfield, NJ) adapted for this purpose, as described earlier (Roth, et al., 1997). The rate of recirculation through and the rate of harvest from the vortex flow filtration device were set manually at the recirculation and harvest pumps (Cole-Parmer, Vernon Hills, IL). Typical recirculation rates were 50–200 mL min^{-1} and harvest rates were 1–2 culture volumes day^{-1}. A cell purge rate was set manually at the purge pump (Cole-Parmer, Vernon Hills, IL), and used as needed to maintain cell density in long term perfusion culture. Dedicated instruments (YSI Instruments, Yellow Springs, OH) measured glucose and lactate off-line.

Virus plaque purification

Plaque purification was done following an adaptation of a prior method (Green and Wold, 1979). Non-confluent HEK 293 cells were used to seed uncoated 60 mm plates (Corning Glass Works, Corning, NY) at 2.5×10^6 cells/plate in DMEM with 10% FBS. After

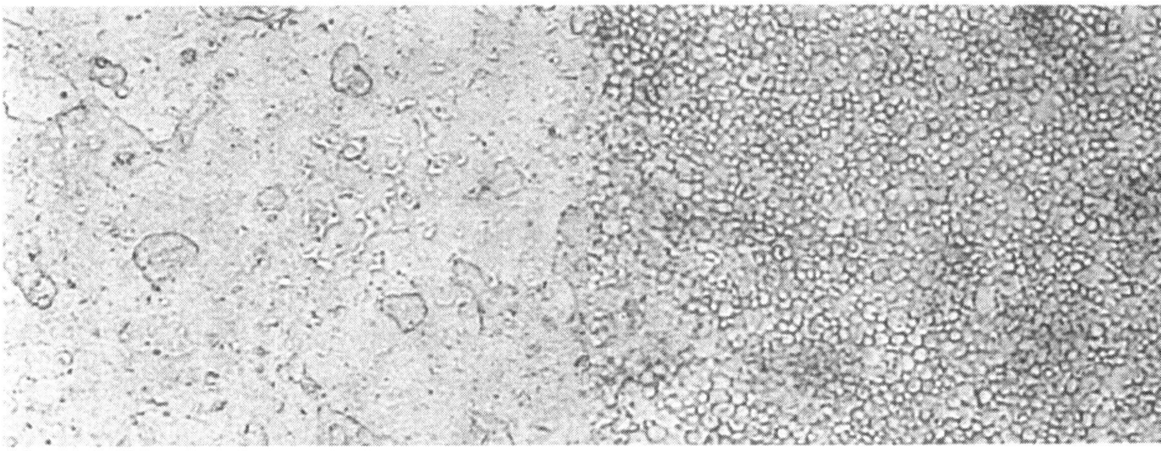

Figure 1. HEK 293 monolayers from a plaque purification dish, showing a non-plaque region on the left, and an early plaque formation (pre-lytic) region on the right. As described in the text, round pre-lytic regions such as this could be detected visually and confirmed microscopically prior to harvest from the dish, without requiring a second staining overlay.

48 hr, media was aspirated, and each plate was infected with 100–200 μL of virus diluted in DMEM without FBS. These plates were incubated at 37 °C in a humidified CO_2 incubator for 90 min, and then each plate was covered with 5 mL of a standard overlay containing 0.5% agarose and 6% FBS in DMEM without phenol red (the infection medium was not removed prior to adding the overlay). The plates were again incubated at 37 °C in a humidified CO_2 incubator. Plaques formed after five to seven days, and the associated areas of the agarose gel, were individually removed with a sterile wire loop or Pasteur pipette and diluted in 1 mL PBS with 2 mM $MgCl_2$. These collected plaques were stored at –70 °C.

Harvest

Adenovirus preparations were harvested from the infected HEK 293 cells by adaptation of a prior method (Shabram et al., 1997). Infected cultures were centrifuged at 700 x g for 5 min. Cell pellets were re-suspended in PBS containing 2 mM $MgCl_2$. The suspensions were frozen in dry ice and isopropanol, then thawed in a 37 °C water bath. These suspensions were freeze-thawed for two to three cycles, and then centrifuged at 700 x g for 5 min The pellets were discarded and the supernatants stored at –70 °C.

Assay for total virus particles

Determination of total viral particles was by analytical HPLC following the method of Shabram et al., 1997.

Assays for infectious virus particles

Infectious virus particles were determined using a variation of the end-point dilution assay (Nielsen et al., 1992), and by a simple cytopathic effect (CPE) assay, as follows.

In the end-point dilution (EPD) assay, confluent HEK 293 cells from a T-75 at >90% viability were used to seed a 96 well plate at 5×10^3 cells well^{-1}. The virus samples were first diluted in DMEM by an estimated three orders of magnitude, based on expected infectious titer, then were serially two-fold diluted in DMEM up to twelve times, and finally 25 μL of each viral dilution in DMEM was added to each of 8 wells. Negative controls (DMEM only) and positive controls (Virus standards diluted to known infectious virus particles mL^{-1}) were used on other plates run in parallel. The plates were incubated at 37 °C in a 5% CO_2 incubator, with media supplementation at day 1 (60 μL well^{-1}) and day 4 to 6 (100 μL well^{-1}). Wells were checked under a microscope for cytopathic effect on day 10 to 12, and scored either positive or negative for infection (infected cells are rounded and detached). A Microsoft Excel (Microsoft Corp., Redmond WA) spreadsheet was used to compute the infectious titer, based on the assumption that a dilution that produces 50% positive wells implies 0.5 infectious virus particles well^{-1} at that viral dilution.

In the CPE assay, HEK 293 cells were plated in 6 well plates at $\sim 2.5 \times 10^6$ cells well^{-1} and grown to confluence. The virus samples were initially diluted three-fold in PBS (Gibco BRL, Life Technologies, Grand Island, NY). Five wells were then infected with

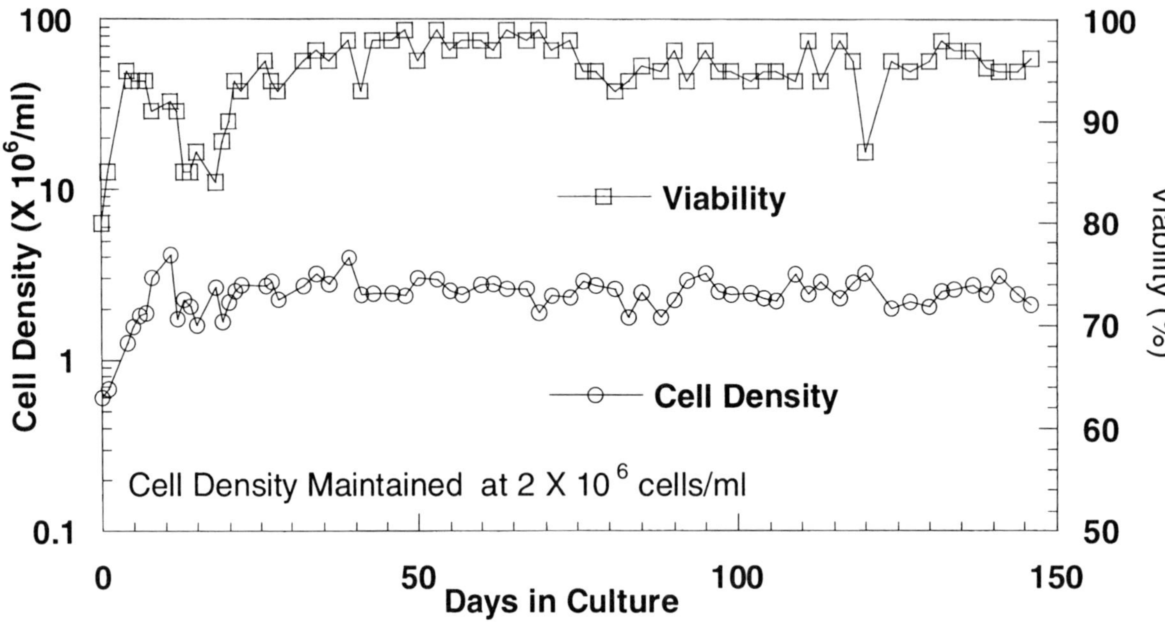

Figure 2. Long-term culture of HEK 293 cells in a 10 L perfused bioreactor using serum-containing medium. The reactor controls and conditions are detailed in the Section Materials and Methods. As described in the text, cells were removed from this reactor throughout the run and infected to produce virus. Over this time period, there were no detectable changes in the quality or amount of virus produced from these cells.

100 μL of serial 10-fold dilutions of the viral samples in PBS, and one well was left un-infected as a negative control. Infectivity was scored at 48 hr where the well that had approximately 50% total cell rounding and detachment (CPE) was scored as half-infected. For each well, it was assumed that there were 4×10^5 cells per cm^2 at confluence and 21 cm^2 per plate, giving approximately 8.4×10^6 cells/plate. The infectious titer per mL was estimated for each well assuming one infectious virus per infected cell. If half the cells were infected, there were 4.2×10^6 infectious virus particles well^{-1}, or 4.2×10^7 infectious virus particles mL^{-1} in the diluted sample. This rapid assay usually agrees with the EPD assay within an order of magnitude and allows for a rapid estimate of infectious titer.

Results

Plaque purification of clonal virus

A method was developed for plaque purification of virus, based on previously reported techniques (Green and Wold, 1979). In contrast to earlier methods, however, early plaque formation (pre-lytic) was identified by microscopy which is a faster approach than the second overlay of neutral red staining typically executed for Ad5 plaques (see Figure 1). These plaques could be detected visually at an early stage, and they grew radially as expected for slowly migrating plaques. Three sequential plaque purifications were executed using this technique to produce three or more clonal virus seeds for a master viral bank.

Long-term culture in bioreactors in serum-containing medium

Initially, HEK 293 cells were suspension adapted, and cultured in stirred tanks using serum-containing medium. These cultures were continuously perfused and cells were retained using a vortex flow filtration device. High-density growth at the 3 and 10 L scales were sustained for 99 and 148 days respectively (see Figure 2, data from a 10 L reactor). Cells were removed periodically for batch infection and proved to be capable of virus production over the life of these reactors. No loss in cell productivity or change in infectivity of the viral harvest was detected (data not shown). This study demonstrated that a long-term perfusion culture could provide cells for batch infections in reactors. Subsequently, we attempted to develop a similar process using cells adapted to SFM.

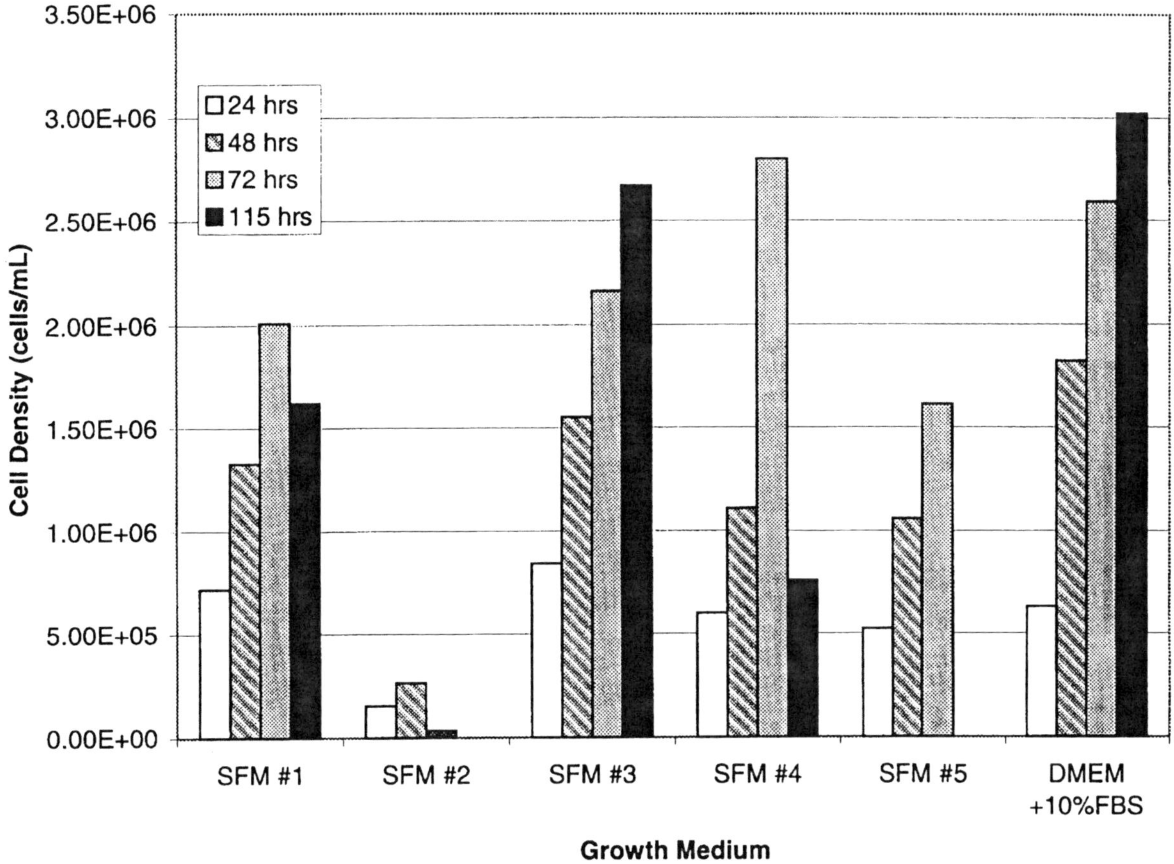

Figure 3. Evaluation of commercial serum-free media with HEK 293 cells using non-weaned cells. The graph shows cells densities for five SFM over 115 hr, compared to DMEM with 10% FBS. Attachment dependent cells were seeded at 5×10^5 cells mL^{-1} in 6-well plates, and a plate was harvested and counted at each time point. The DMEM w 10% FBS culture and the SFM#3 culture had viabilities >90% over the time of the experiment. All other cultures were less viable.

Serum-free medium

Five serum free media (SFM) from three different vendors, Irvine Scientific, BioWhittaker, and JRH Scientific, were screened for use with the HEK 293 cells. These media were first evaluated in 6-well culture plates with adherent cells (see Figure 3). One culture medium (SFM#3) supported growth and viability comparable to serum-containing medium. This medium (SFM#3) also yielded the best growth characteristics in suspension culture (shake flasks). Growth rate and viability of cells in this medium were comparable to that in serum-supplemented medium (see Table 1). Adenovirus production and infectivity of the product were found to be equivalent in both SFM#3 and the serum-supplemented medium, with yields of ~2–5 × 10^4 viral particles per cell.

Table 1. Growth and viability in serum-containing vs serum-free medium. HEK 293 growth rates and viabilities in a serum-containing formulation compared with a commercial serum free media. Cells were seeded in multiple shake flasks at ~5 × 10^5 cells mL^{-1}, and cultured for ~72 hr. Specific growth rates and viabilities were calculated from the first cell doubling

		WEC 5.2.1 + 2% calf serum	SFM#3
Mean specific growth rate	(1 day^{-1})	0.47	0.42
Mean viability	(%)	96	91

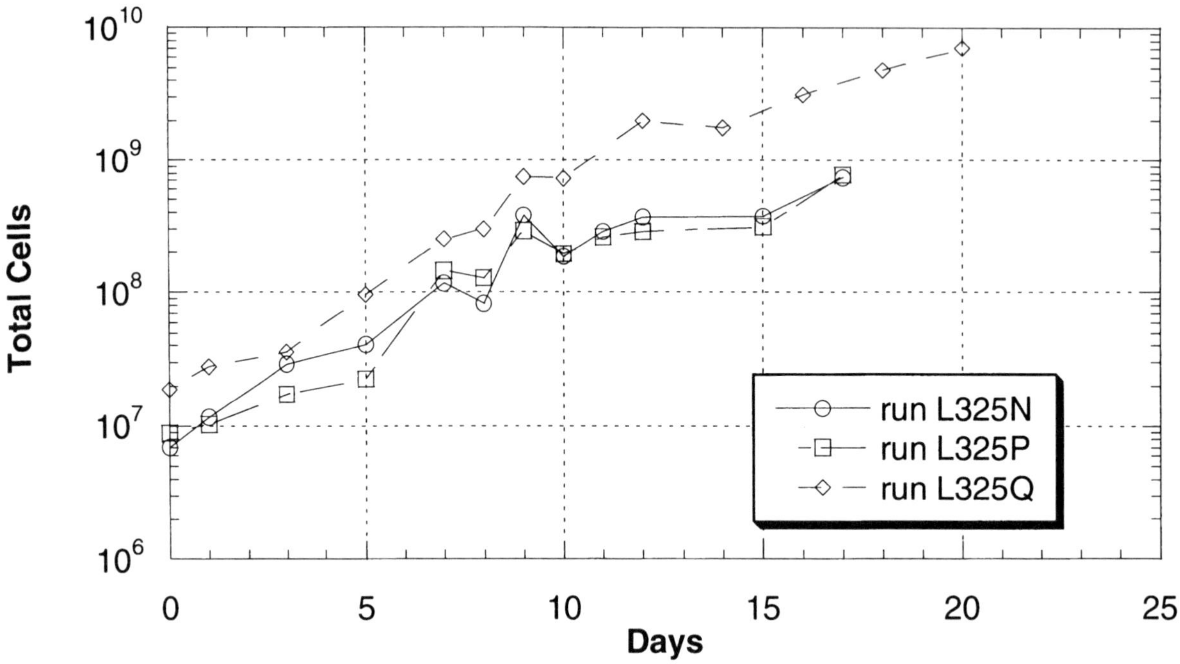

Figure 4. Cell growth qualification for serum-free medium and suspension adapted HEK 293 cells in shake flasks. All three cell cultures graphed on the figure were passaged into new flasks approximately every other day (as described in the materials and methods section). For that reason, the total cells are calculated to provide a clearer depiction of the cell growth. Viability was greater than 90% for all of these cultures.

Adaptation of HEK 293 cells to suspension and serum-free medium for cGMP

After selecting a SFM, HEK 293 cells were adapted to suspension and SFM under cGMP conditions. Suspension adaptation in shake flasks in DMEM + 10% FBS took 9 passages over 20 days. These adapted cells were used for long-term perfusion bioreactor runs using serum-containing medium.

Once the cells were fully adapted to suspension, with similar doubling times and viabilities as achieved in serum-containing cultures, adaptation to SFM#3 was begun. The cells were first passaged multiple times in SFM#3 + 5% FBS until doubling times of no more than 2 days and viability above 90% were achieved. At that point, the cells were assumed to be fully adapted to the reduced level of serum and were transferred into and passaged in fresh medium containing further sequential step-reductions in serum concentration. The cells were finally passaged in SFM#3 without added serum after 30 days and 15 passages starting from the initial suspension adapted, cultures. The final suspension and serum-free medium adapted cells were frozen in medium with 10% DMSO. They were subsequently thawed and expanded in shake flasks. The cultures exhibited doubling times from 1.8 to 2.4 days (see Figure 4). In separate studies, samples of the adapted cells were infected in suspension, and virus production averaged 7×10^4 particles per cell and $1–3 \times 10^3$ infectious particles per cell, which was comparable to production in serum-containing medium.

Small scale cultures

Viral production requires thawing of cells from the cell bank and expansion in small shake flasks. Our best conditions for freezing HEK 293 cells in SFM was 5% DMSO and freezing at –70 °C, with later transfer to liquid nitrogen. These cells were best thawed by direct recovery into SFM, without centrifugation to remove DMSO.

HEK 293 cells are prone to aggregation in SFM. The aggregation of HEK 293 cells in shake flasks was found to be dependent on the flask size. Cell clumping was prevalent in 250 mL shake flasks containing 50 mL of culture (agitation set at 140 RPM), while mono-disperse cultures could be obtained in larger flasks (e.g. Fernbach's). Since clumping of cells was reversed by transfer of cells to a larger flask, the seed cell process included transfer of cells as soon as they

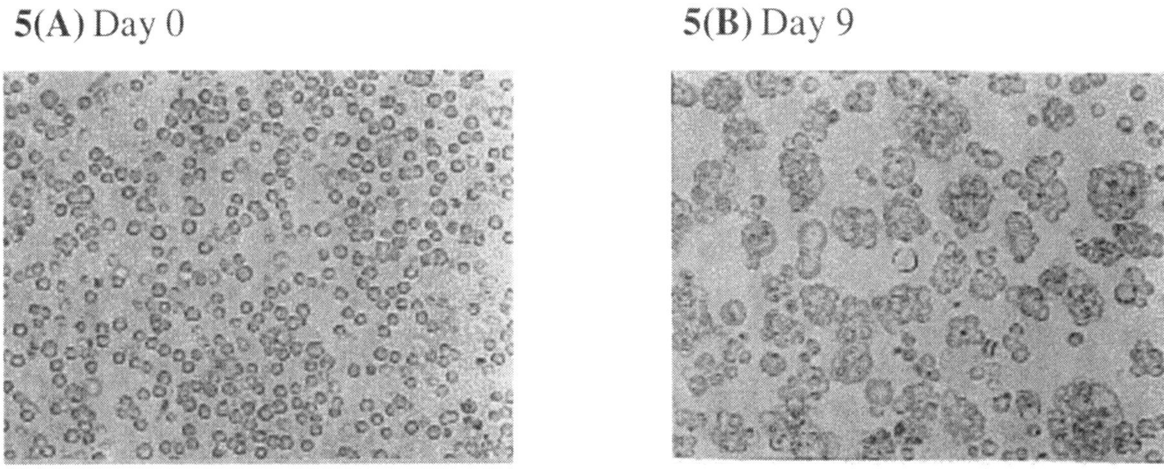

Figure 5. (A) HEK 293 cells in serum-free medium sampled from a bioreactor run at day zero, and (B) the same bioreactor sampled on day nine. The photographs indicate the progression of clumping of the cells over this time frame.

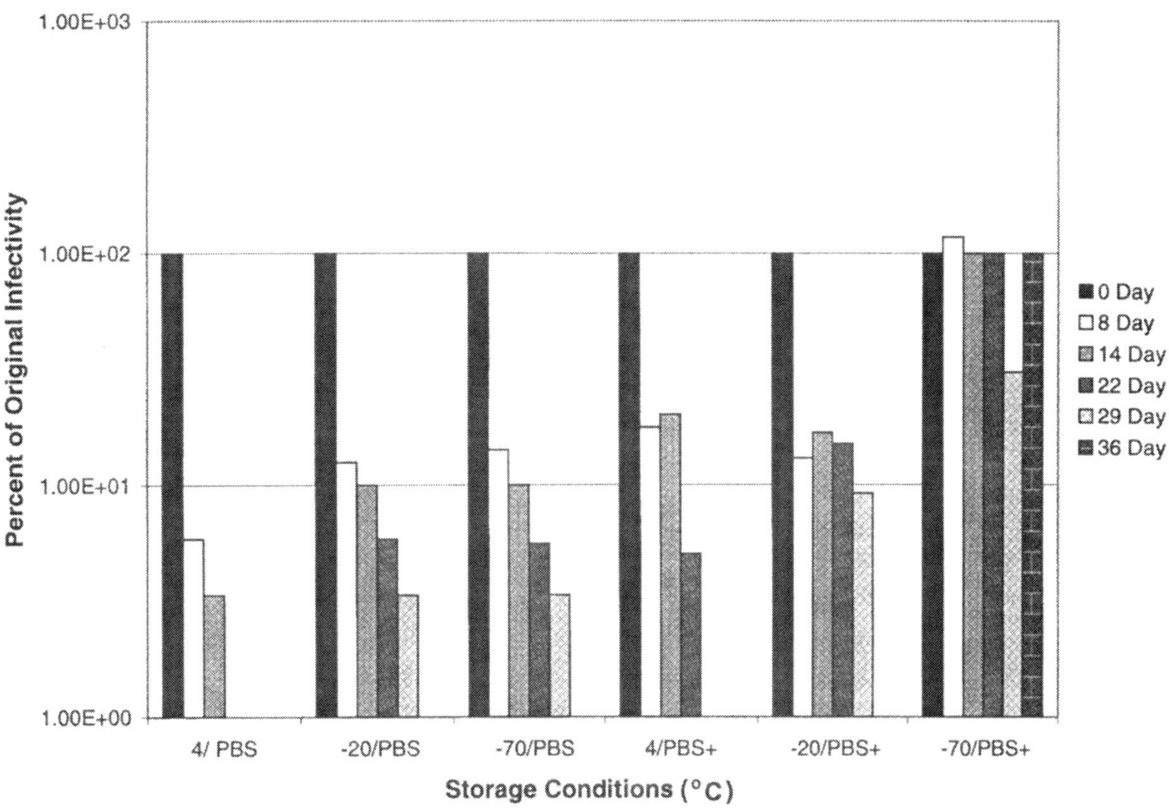

Figure 6. The stability of adenovirus, measured in percent infectivity of the original samples as assayed by CPE (see the Section Materials and Methods), when stored at various temperatures and in various diluents (PBS + is PBS with sucrose and $MgCl_2$). This data clearly shows that under the conditions of PBS+ at –70 °C, the raw virus is stable for at least one month.

Table 2. Determination of optimal MOI. Yields of total and infectious particles from 100 mL culture volume shake flask cultures at various MOI. Particles were measured by HPLC, and infectious particles were measured by EPD

MOI	Particles/cell	Infectious particles/cell
125	6.4×10^4	2.8×10^3
25	8.9×10^4	3.9×10^3
5	5.9×10^4	3.0×10^3
1	4.1×10^4	1.9×10^3
0	0	0

can seed the next largest flask (i.e., at $>4 \times 10^5$ cells mL^{-1} in 20% of the total volume of the flask).

Growth in shake flask cultures in SFM was shown to be dependent on gassing conditions. The SFM used with these cells does not require CO_2 buffering, thus we have traditionally used airtight flasks rather than vented flasks in our 5% CO_2 incubators. Doubling times were lower (1.8 days) in vented flasks exposed to the incubator's atmosphere, compared to sealed flasks with doubling times of 2.4 days. This effect was shown not to be due to pH, but maybe due to either oxygen limitations or accumulation of CO_2 in the sealed flasks (we have observed that both occur, data not shown). This issue was not studied using serum-containing media. To maintain higher growth rates we now routinely culture cells in shake flasks with vented caps.

Long-term bioreactor cultures in serum-free medium

To develop long-term perfusion cultures, ten bioreactor runs were initiated in serum free medium. In all cases, the cells formed large aggregates (see Figure 5) and stopped growing after several days. A variety of conditions were tested, including increased agitation and decreased calcium to reduce the aggregation, and increased surfactants and decreased temperatures to reduce cell shear sensitivity. None of these conditions, however, resulted in a successful culture. While it might be possible to prevent aggregation with additives that inhibit cell/cell interaction, that is probably not a satisfactory approach due to difficulties in removal of such additives.

Effect of multiplicity of infection (MOI)

Table 2 lists viral yields from a wide range of MOIs. Cells were grown in small scale shake flask cultures up to densities of $\sim 1 \times 10^6$ cells mL^{-1} prior to infection. The flasks were harvested at 48 hr post infection, and assayed for total and infectious viral particles. The results indicate that the best MOI for these conditions was greater than or equal to one. A similar experiment in spinner flasks indicated that an MOI of 0.1 significantly reduced the yield of virus. There was no significant increase or loss of per cell productivity with MOIs from 1 to 125.

Adenovirus production in bioreactors in serum-free medium

Characterization of adenovirus production in bioreactors was done in nineteen 3 L runs using serum-free medium. HEK 293 cells were expanded in large shake flasks, introduced into a batch reactor, and infected at MOIs of ~ 1 to 100. At approximately 48 hr post infection, cells were harvested by centrifugation, resuspended, and lysed by repeated freeze-thaws. Lysates were assayed for total and infectious viral particles. Data from these nineteen runs is presented in Table 3. The data indicates process reproducibility, with reasonably consistent viral yields (both total and infectious).

Storage conditions for stability of the raw virus

Conditions for long-term storage of the raw, unpurified adenovirus, liberated from cell lysates were studied to determine maximum hold times prior to further purification. Two diluents and three storage temperatures were examined. Storage at −70 °C in PBS with sucrose and $MgCl_2$ provided the greatest viral stability (see Figure 6), and was chosen as the standard condition for storage of raw virus.

Conclusions

We have developed an efficient, reproducible, and scaleable manufacturing process for an adenoviral vector expressing therapeutic transgenes. HEK 293 cells were adapted to suspension cultures and serum-free medium. These adapted cells exhibited similar growth, viability, and virus production to serum dependent cells. Although perfusion cultures of suspension adapted cells were maintained for up to 148 days

Table 3. Summary of adenoviral production runs in 3 L bioreactors (Suspension, SFM). Results from fourteen 3 L batch bioreactor runs producing adenovirus, infected at an MOI of 1 to 100, and harvested at ~48 hr. Viral particles were measured by HPLC and infectious particles were measured by EPD

Run result		Mean (n=14)	Standard deviation
Cell density at harvest	$(\times 10^{-6})$	0.82	0.38
Viral particles per run	$(\log_{10})$	14.2	0.22
Infectious particles per run	$(\log_{10})$	12.5	0.24
Viral particles per cell	$(\times 10^{-4})$	8.6	4.6
Infectious particles per cell	$(\times 10^{-3})$	1.8	1.0
Infectivity	(%)	2.2	1.0

in serum, these long-term cultures could not be established for SFM. The long-term SFM cultures appeared to fail due to cell clumping, and attempts to reduce this clumping were so far unsuccessful. It appears that in shake flasks, clumping is less pronounced in larger flasks.

Plaque purification steps for isolating clonal virus were simplified and accelerated by replacing the staining overlay with microscopic inspection. The best multiplicity of infection (MOI) in shake flasks was found to be in the range of 1 to 125, as confirmed by viral production in serum-free suspension bioreactors. The raw virus was shown to have good long-term stability when stored in a PBS buffer at –70 °C.

Development work continues, to establish long-term SFM suspension cultures, and to better characterize the kinetics of virus production in serum-free, suspension cultures.

Acknowledgments

We would like to thank Cindy Clements, Paul Farragher, Gregory Ja, Mike McCaman, Mei Tan and Spencer Tse for their contributions to this work.

References

Crystal, RG (1995) Transfer of genes to humans: early lessons and obstacles to success. Science 270: 404–410.

Garnier A, Cote J, Nadeau I, Kamen A and Massie B (1994) Scale-up of the adenovirus expression system for the production of recombinant protein in human 293S cells. Cytotechnology 15: 145–155.

Graham FL and Prevec L (1991) Manipulation of adenovirus vectors. In: Murray EJ (ed.) Methods in Molecular Biology. Vol. 7, The Humana Press, Clifton, NJ, pp. 109–128.

Green M and Wold WSM (1979) Human adenoviruses: Growth, purification, and transfection assay. Meth Enzym 58: 425–435.

Hehir K, Keegan J, Martin J, Pratt D, Auger C, Davis M, Everton M, Narayana R and Karey K (1998) Production issues for adenoviral gene therapy vectors. Presented at Cell Culture Engineering VI, February 7–12, 1998, San Diego, CA.

Marcel T and Grausz JD (1997) The TMC worldwide gene therapy enrollment report, end 1996. Human Gene Therapy 8: 775–800.

Nadeau I, Garnier A, Cote J, Massie B, Chavarie C and Kamen A (1996) Improvement of recombinant protein production with the human adenovirus/293S expression system using fed-batch strategies. Biotech Bioeng 51: 613–623.

Nielsen LK, Smyth GK and Greenfield PF (1992) Accuracy of the endpoint assay for virus titration. Cytotechnology 8: 231–236.

Roth G, Smith C, Schoofs GM, Montgomery TJ, Ayala JL and Horwitz JI (1997) Using an external vortex flow filtration device for perfusion cell culture. BioPharm 10: 30–35.

Shabram PW, Giroux DD, Goudreau AM, Gregory RJ, Horn MT, Huyghe BG, Liu X, Nunnally MH, Sugarman BJ and Sutjipto S (1997) Analytical anion-exchange HPLC of recombinant type-5 adenoviral particles. Human Gene Therapy 8: 453–465.

Cytotechnology **28**: 91–99, 1998.
© 1998 *Kluwer Academic Publishers. Printed in the Netherlands.*

Recombinant insulin-like growth factor-I (IGF-I) production in *Super*-CHO results in the expression of IGF-I receptor and IGF binding protein 3

Noelle-Anne Sunstrom[1]*, Masood Baig[1], Louise Cheng, Derick Payet Sugyiono & Peter Gray
CRC for Biopharmaceutical Research[1], *Department of Biotechnology, University of New South Wales, Kensington, NSW 2052, Australia*
E-mail: n.sunstrom@unsw.edu.au

Received 26 August 1998; accepted 26 August 1998

Key words: autocrine growth, CHO, IGF-I, IGFBP, IGF-I receptor

Abstract

Previously, we described the genetic construction *Super*-CHO, a cell line capable of autocrine growth under fully defined protein-free conditions. *Super*-CHO cells constitutively express insulin growth factor-I (IGF-I) and transferrin in sufficient amounts to support long-term, stable growth without the addition of exogenous growth factors, thus making it an ideal host for the production of recombinant biopharmaceuticals. although IGF-I has been successfully expressed in Chinese Hamster Ovary (CHO) cells, the long term effects of recombinant IGF-I expression have not been explored. In particular, the expression of the endogenous IGF-I receptor in response to IGF-I production has not been reported. We report here the transcriptional induction of the type I IGF receptor gene in *Super*-CHO. In addition, we examined the conditioned medium for the presence of IGF-I binding proteins. Ligand blot analysis reveals the presence of IGF binding proteins present in the medium conditioned by *Super*-CHO cells as well as CHO cells incubated in the presence of IGF-I. Furthermore, immunoaffinity reveals that *Super*-CHO expresses IGF binding protein-3 in response to IGF-I production. These results suggest the autocrine growth of *Super*-CHO involves a complex interaction of cell type specific factors which regulate its utility of IGF-I.

Introduction

The *Super*-CHO cell line expressing recombinant IGF-I and transferrin grows indefinitely in fully defined protein free medium. Growth rate and final cell densities are comparable to CHO-K1 cells growing in medium supplemented with insulin, transferrin and selenium. *Super*-CHO is an ideal host cell line for the production of recombinant biopharmaceutical products as it does not require foetal bovine serum (FBS) or other sources of mitogens, thus providing considerable regulatory and cost advantages. IGF-I and transferrin are secreted into protein free medium where they are presumed to bind to cellular surface receptors and sustain an autocrine fashion of cellular growth.

IGF-I is a member of a family of structurally related peptides that also includes insulin and IGF-II. Insulin-like growth factors (IGF-I and IGF-II), are found in a variety of cell types where they function in an autocrine/paracrine fashion to regulate cell growth (Baxter, 1986). IGFs are potent mitogens which exert their effect principally through binding to the type-I IGF receptor (Dufourny et al., 1997; LeRoith et al., 1995). IGF-I receptor binds IGF-I with high affinity and binds insulin with 500–1000 times lower affinity, explaining the well known role of insulin as a growth factor at high concentrations.

The interaction between the IGF and the IGF-I receptor is regulated by a family of IGF-binding proteins(Camacho-Hubner et al., 1992) of which six have been identified and cloned, designated IGFBP-1 to IGFBP-6 (Kelley et al., 1996). IGFBPs bind

* Author for all correspondence.

IGFs with very high affinities and confer upon the IGF regulatory system both functional and tissue specificity which influences their bioavailability and distribution. IGFBPs are produced by many cell types, which can modify them through differential post-translational modifications including phosphorylation, glycosylation, proteolysis and multimerization alter their biological activity (Coverley and Baxter, 1997; Koedam et al., 1997). It is believed that IGFBPs inhibit IGF-I activity by sequestering the peptide and consequently preventing interaction with its receptor. However, IGFBP-3 is known to both inhibit and potentiate IGF-I action depending on whether it is soluble or cell surface bound (Conover, 1992). The majority of IGFs in the adult circulation are tightly bound to IGFBP-3 in a 150 kDa ternary complex composed of 38–42 kDa IGFBP-3, IGF peptide and an 85 kDa acid-labile protein subunit (ALS)(Baxter and Martin, 1989; Hashimoto et al., 1997; Young Lee et al., 1996)

CHO cells do not normally produce IGF-I. Although IGF-I has been successfully expressed in mammalian cells, the long term consequences of recombinant IGF-I expression have not been well defined. We investigated the dynamics of IGF-I production in *Super*-CHO with respect to expression of the type I IGF-I receptor as well as to the implication of IGF binding proteins. As a consequence of IGF-I production in *Super*-CHO cells, we observed an induction of IGF-I receptor expression. In addition, it appears that IGFBP-3 is involved in the modulation of cell growth of *Super*-CHO. Therefore, the autocrine growth of *Super*-CHO involves a complex interaction of cellular proteins that regulate IGF-I utility resembling the dynamics of an IGF autocrine system. This report aims to extend characterisation of *Super*-CHO, a host cell line constructed for the production of biopharmaceutical products. In particular, we characterised *Super*-CHO with respect to its production of the growth factor, IGF-I.

Materials and methods

Cell lines and cell culture

The *Super*-CHO cell line was derived from the Chinese hamster ovary cell line CHO-K1 (ATCC CCL 61) cotransfected with plasmids expressing the human IGF-I under the control of the CMV promoter (pCMVIGF) and the human transferrin under the control of the SV40 early promoter (pSVLTf) (Pak. et al., 1996).

Cells were grown in Dulbecco's Modified Eagle's Medium (DMEM) and Coons (1:1 mixture) medium referred to as protein-free medium (PFM). PFM was supplemented with either 10% foetal bovine serum (FBS) or IGF-I (Sigma) where indicated.

Cell labelling and immunoprecipitation

Cell monolayers in PFM were washed twice in PBS and incubated for 2 hr in labelling medium (methionine-free DMEM (GIBCO BRL). The medium was then replaced with fresh labelling medium containing 50 μCi mL^{-1}[^{35}S] methionine, 1175 Ci mmol^{-1} (ICN Pharmaceuticals, Inc. Irvine, CA). After labelling for 6 hr, the medium was removed, and cells were washed twice in PBS before solubilizing in RIPA buffer (150 mM NaCl, 1.0% NP-40, 1% deoxycholate, 0.1% SDS, 50 mM Tris pH 8.0) containing protease inhibitors (100 μm leupeptin and 1 mM PMSF) for one hour, and then centrifuged at 14 000 rpm for 15 min. Immunoprecipitation was performed using 100 μL labelled lysate with 0.4 μg of M19 antibody (Santa Cruz Biotechnology, Santa Cruz, CA), a rodent-specific anti IGFBP-3 antibody and protein G-Sepharose (Pharmacia Biotech Inc.). The immunoprecipitates were washed twice in RIPA and once in buffer consisting of 25 mM Tris-HCl, pH 7.4, 150 mM NaCl. Immunoprecipitates were solubilized in SDS-sample buffer containing 2-mercaptoethanol and denatured by heating at 95 °C for 5 min. Samples were electrophoresed on 4–20% acrylamide Tris-glycine gels (Novex, San Diego, CA). Gels were fixed in 10% (w/v) TCA and 2% (w/v) PEG-6000, and dried gels were exposed to a phosphoimaging screen (Bio-Rad, CS-imaging screen), and scanned using a Phosphoimager (Bio-Rad, model GS 363).

RNA extraction and northern blot analysis

Monolayer cultures of cells were grown in T150 flasks (Corning, New York) in PFM or PFM supplemented with 10% foetal bovine serum (GIBCO BRL). Cells were harvested at 70–80% confluence, using 0.2% EDTA in PBS. Total RNA was isolated from different cell lines by the acid-guanidine thiocyanate-phenol-chloroform method (Chomczynski and Sacchi, 1987). For Northern analysis, total RNA (20 μg) was separated on a 1.2% agarose gel containing 2.2 M formaldehyde and MOPS (a stock solution of 10X contained 0.2 M 3-(-N-morpholine) propanesulfonic acid, 5 mM sodium acetate, 5 mM EDTA, pH 7.0).

The RNA was transferred to Hybond-N nylon membrane (Amersham Life Science) by vacuum blotting with 20X SSC for 2–3 hr and RNA was bound to nylon membrane by UV cross-linking (Amersham Life Science). To confirm the integrity and the loading uniformity of the RNA on the membrane, methylene blue staining was employed (Herrin and Schmidt 1988). Complementary DNA (cDNA) for IGF-IR (a kind gift from Dr. R. Baserga, Jefferson Cancer Institute, Pennsylvania) and for IGF-I were used in this experiment. The probes were prepared using Gene Images Random Prime Labelling Module (Amersham, Life Science). Prehybridization of the nylon blots was carried out for at least 10–12 hr at 42 °C in a solution containing 50% deionised formamide, 5X Denhardt's solution (1X, 0.02% ficoll, 0.02% polyvinylpyrrolidone and 0.02% BSA), 1% SDS, 5X SSC and 200 μg mL^{-1} herring sperm DNA. After prehybridization, DNA probes, previously heated to 100 °C were applied to the blots. Hybridization was conducted overnight at 42 °C. The membranes were washed at room temperature in 1X SSC buffer containing 0.1% SDS(3×15 min) and (2×20 min) at 42 °C in 0.1% SSC-0.1% SDS. RNA detection was carried out using Gene Images CDP-star detection module (Amersham Life Science) according to the manufacturer's protocol.

Western blotting and chemiluminescent dot blot

Conditioned medium containing IGF-I and IGF-I complexes were assessed by Western blotting. In some cases, cross-linking of proteins was performed using disuccinimidyl suberate (DSS, Pierce, Rockford, IL) according to the manufacturer's protocol. Prior to electrophoresis, the samples were mixed with an equal volume of 2X Tris-glycine SDS sample buffer (Novex, San Diego, CA) and heated at 95 °C for 5 min. Samples were electrophoresed on 4–20% acrylamide Tris-glycine gels (Novex, San Diego, CA). Resolved proteins were electroblotted onto nitrocellulose membrane (Sartorius, Germany) using a semi-dry system at 2–3 mA cm^{-2} for 90 min. The transfer buffer consisted of 25 mM Tris-HCl (pH 8.5), 192 mM glycine and 20% (v/v) methanol. Following transfer, the blotted membranes were blocked by incubating for 1 hr at room temperature in a blocking solution composed of PBS containing 5% nonfat dry milk and 0.1% Tween-20. For the detection of IGF-I, the membranes were incubated with a rabbit anti-human IGF-I primary antibody (Peninsula Laboratories, Inc.), diluted 1:1000 and incubated for 2 hr at room temperature. The membranes were washed (4×10 min) with PBS/0.1% Tween-20 and then incubated with an alkaline phosphatase-conjugated mouse anti-rabbit conjugated IgG (Boehringer Mannheim) diluted 1:5000 in PBS containing 0.1% Tween-20. Membranes were washed and IGF-I and IGF-I complexes were visualised using Western Blue stabilised substrate for alkaline phosphatase (Promega Corp. Madison, WI). IGF binding protein-3 was detected using the antibody M-19 (Santa Cruz Biotechnology, Santa Cruz, CA), a rodent specific goat anti-IGFBP-3 followed by an alkaline phosphatase-conjugated rabbit anti-goat antibody (Pierce, Rockford, Il.).

A chemiluminescent dot blot assay was performed to detect IGFBP-3 in conditioned media as described by De Leon and Asmeron, 1997). Briefly, conditioned media were applied to ECL-hybond nitrocellulose (Amersham) membrane using a vacuum blotter (Hoeffer Scientific Instruments). Following adsorption, the membrane was dried and blocked in PBS containing 0.1% Tween-20 and BSA. Three successive 10 min washes in PBS containing 0.1% Tween-20 preceded incubation with primary antibody, M-19 (Santa Cruz Biotechnology, Santa Cruz, CA) diluted 1:5000 in PBS containing 0.1% Tween-20. The membrane was washed three times in wash buffer before incubation in Supersignal reagent (Pierce, Rockford, Il.) followed by exposure to the Fluor-S MultiImager (Bio-Rad) system with maximum aperture without the 660 SP filter.

[^{125}I]IGF western ligand blot

This was performed according to the method described by Hossenlopp et al., 1986. Briefly, conditioned media were subjected to 12.5% SDS PAGE under nonreducing conditions followed by electrotransfer onto nitrocellulose (Sartorius, Germany). IGFBPs were detected by incubation with 100 000 cpm ^{125}I-IGF-I (ICN Pharmaceuticals, Inc. Irvine, CA) followed by autoradiography at –70 °C for 5 days using Kodak Autoradiographic film.

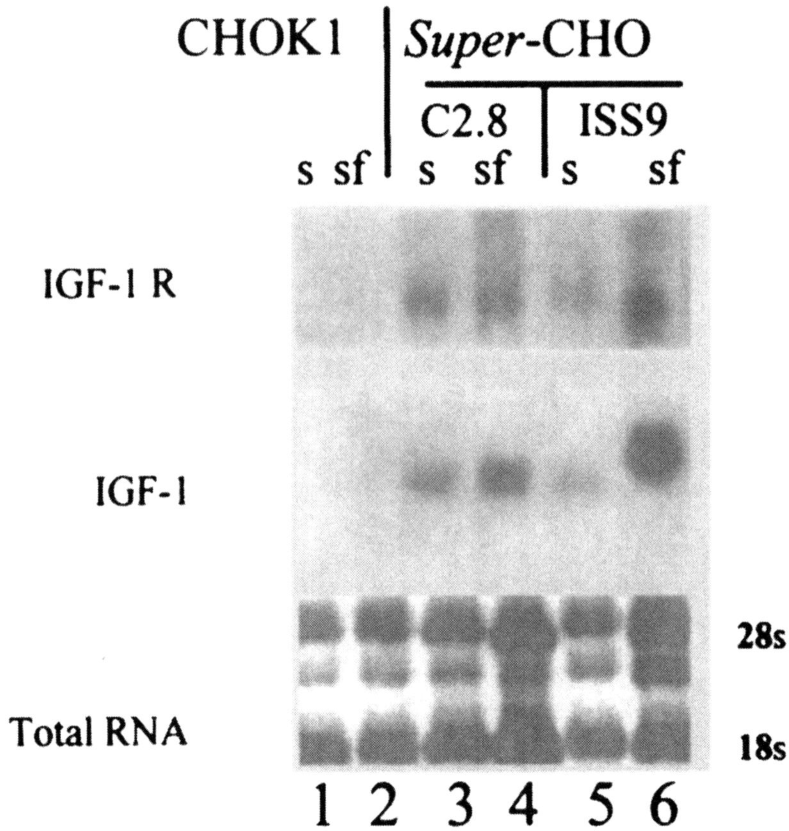

Figure 1. Northern blot analysis of IGF-I and IGF-I receptor. Expression of IGF type I receptor and IGF-I mRNA from *Super*-CHO clones C2.8 and ISS9 as well as the parental cell line, CHOK1 was analysed by Northern blot. IGF-I receptor-specific transcripts as well as transcripts of IGF-I were detected in *Super*-CHO clones but not in the control CHOK1 cells. Cells in serum-containing medium (s) or in serum-free medium (sf) were harvested for RNA isolation. Total RNA was visualised by methylene blue staining to confirm the integrity and loading uniformity of the RNA.

Results and discussion

IGF-I receptor expression

We examined the expression of the type I IGF receptor in the parent CHOK1 cell line compared to that of *Super*-CHO. We chose to investigate two clonal isolates of *Super*-CHO, named C2.8 and ISS9. Although both clones produce similar amounts of IGF-I (100 ng mL^{-1}) as assayed by RIA analysis (data not shown), they differ in their growth rates (doubling times are 24 and >35 hr, respectively). RNA was analysed for IGF-I and IGF-I receptor expression by Northern blot (Figure 1). RNA extractions were done using cells reaching confluence either in serum-containing (s) or serum-free medium (sf). IGF-I receptor-specific transcripts were detected in both clones of *Super*-CHO, C2.8 and ISS9 under serum and serum-free conditions, but not in the parent cell line, CHOK1 (top panel, Figure 1). Similarly, IGF-I-specific transcripts were detected in *Super*-CHO clones but not in CHOK1 (middle panel). Total RNA (bottom panel) with predominant 28S and 18S species is included to indicate comparable gel loading strategy.

RNA transcripts corresponding to IGF-I receptor in parental CHO cells were below our detection limits. Normally, CHO cells have between 5000 and 25 000 IGF-I receptors per cell (Kato et al., 1993; Steele-Perkins et al., 1988). The number of endogenous IGF-I receptors in CHO cells is known to fluctuate due to many factors. An upregulation of receptor by the respective ligand is not without precedent. Transcription of *Xenopus* thyroid hormone receptor genes is strongly up-regulated by their own ligand (Machuca et al., 1995). Expression of type I IGF receptors in keratinocytes were activated as a result of IGF-I exposure (Tavakkol et al., 1992; Ando et al., 1993).

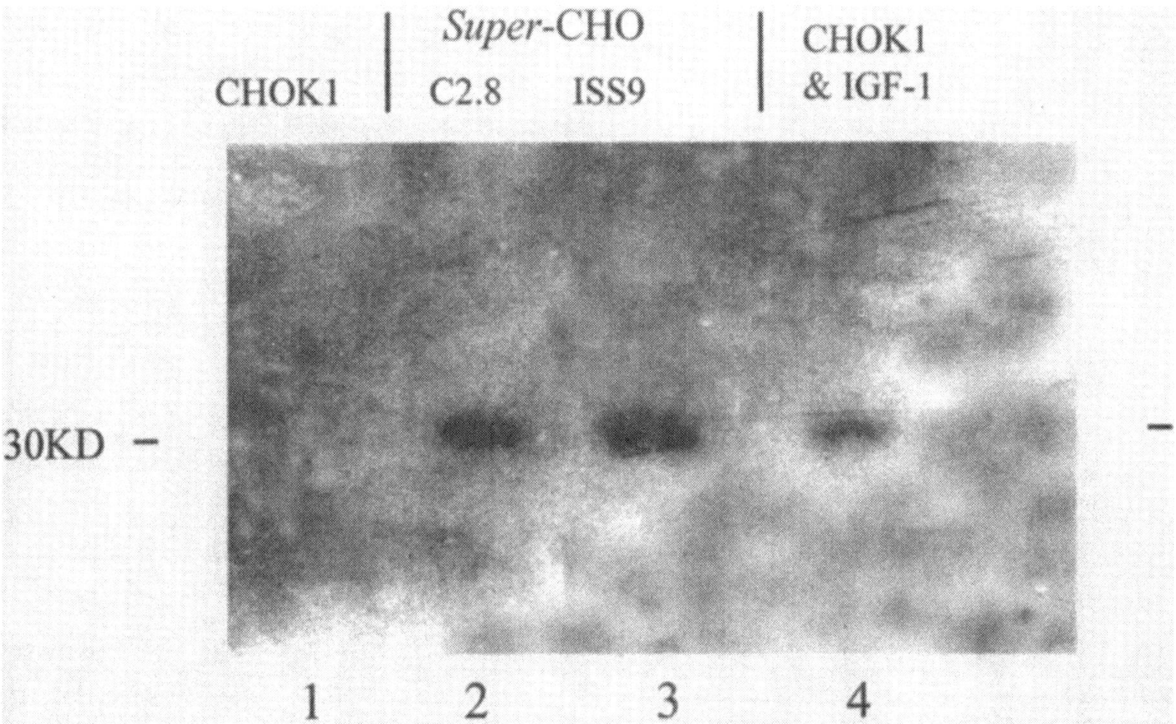

Figure 2. Western ligand blotting of IGF-I. Samples were prepared and processed for ligand blotting as described under Materials and Methods. Samples are media conditioned by CHOK1 cells (lane 1), in the presence of 50 ng mL^{-1} IGF-I (lane 4), and *Super*-CHO clones C2.8 and ISS9 (lanes 2 and 3). The ligand used was ^{125}I-IGF-I.

An up-regulation of IGF-I receptor expression was demonstrated in rat adult myocytes by IGF-I stimulation (Reiss et al., 1997). IGF-I binding is increased by platelet-derived growth factor (PDGF) treatment which activates the IGF-I receptor promoter (Baserga et al., 1997). In fact, an overexpression of the receptors for PDGF, epidermal growth factor (EGF) and IGF-I is observed when cells are grown in serum-free medium supplemented with their respective ligands (Rosenthal et al., 1991; Stewart et al., 1990). An upregulation of IGF-I receptor expression in *Super*-CHO may be due to the production and secretion of recombinant IGF-I in these cells.

Detection of IGF binding proteins in conditioned medium

The production of binding proteins in response to IGF-I was assessed by [^{125}I]IGF Western ligand blotting and shown in Figure 2 (Hossenlopp et al., 1986). Binding assays were performed on conditioned supernatents from confluent cell monolayers in PFM. Cells were seeded and grown overnight in serum-containing medium, washed twice with PFM, followed by a subsequent 48 hr incubation in PFM. A band of approximately 30 kDa is present in conditioned medium from *Super*-CHO, clones C2.8 and ISS9, but not in the that of the parent cell line, CHOK1. However, an IGF binding protein with similar migration is apparent in the conditioned medium of CHOK1 cells incubated in the presence of exogenously added IGF-I (50 ng mL^{-1}).

The production of binding proteins in response to IGF-I has been demonstrated in several studies (Conover, 1992; Donnelly and Holly, 1996), and it has been shown that expression and secretion of IGFBPs fluctuate under varying growth conditions (Birnbaum et al., 1994; Mouhieddine et al., 1994).

Western blot of IGF-I binding complexes in Super-CHO conditioned medium

Further evidence for the presence of IGF binding proteins was obtained in the conditioned medium of *Super*-CHO cells. Proteins present in PFM conditioned by *Super*-CHO cells were treated with the chemical cross linker, disuccinimidyl suberate (DSS) before separation on SDS-PAGE and transfer onto a

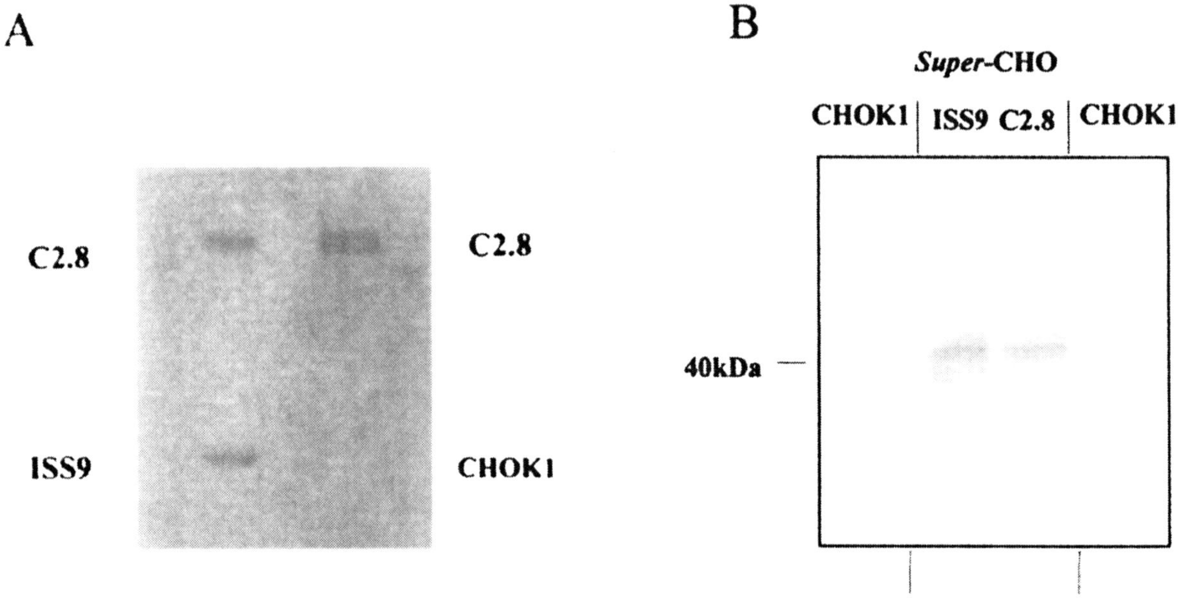

Figure 3. IGF-I and IGF-I binding proteins in conditioned media were detected using Western blot analysis. Cells were seeded in media containing 10% FBS to allow for attachment overnight, after which they were washed in PBS and incubated in PFM. Conditioned media was collected after 48 hr and incubated in the presence (+) or absence (–) of the cross-linker disuccinimidyl suberate (DSS). Following TCA precipitation, samples were electrophoresed, transferred to nitrocellulose and probed using antibody against IGF-I (see Section Material and Methods).

nitrocellulose membrane. The membrane was probed with anti-IGF-I antibody and IGF-I complexes were visualised as described in materials and methods and shown in Figure 3. IGF-I and complexes of IGF-I are present in the conditioned medium of *Super*-CHO, clone C2.8 in the presence of cross-linker (lane 1). A prominent IGF-I complex migrates at approximately 150 kDa and a doublet at 30–36 kDa . Minor complexes appear to migrate at 125 and 78 kDa. A 150 kDa complex consisting of IGF-I, IGFBP-3 and an acid-labile subunit (ALS) has been demonstrated in human plasma. This complex is considered as a reservoir of IGFs (Young Lee and Rechler, 1996; Ooi et al., 1997). IGFBP-3, a protein of approximately 35 kDa, has been demonstrated to form dimers in solution. Recently, it has been shown that the multimerization of IGFBP-3 may be responsible for generating distinctive conformations of the protein which in turn can regulate its ability to bind to the cell surface.(Koedam et al., 1997). IGFBP-3 is also subject to proteolysis, and it is possible that the 125- and 78 kDa protein complexes may be cross-linked dimers of IGFBP-3, IGF-I and their proteolytic products (Lalou and Binoux, 1993; Mohseni-Zadeh and Binoux, 1997). In the absence of cross-linker (lane 2), IGF-I is visualised as a complex migrating at 36 and 150 kDa, suggesting a strong association with these proteins. The absence of a 30 kDa complex may represent a weak association with IGF-I. The 150 kDa complex appears less intense in the absence of cross-linker, probably due to disassociation during electrophoresis. The IGF-I binding protein profile of clone ISS9 differs from that of clone C2.8 both in the presence and in the absence of cross-linker (lanes 3 and 4). Cross-linking proteins in the conditioned medium of *Super*-CHO clone ISS9 results in the appearance of the 150 kDa IGF-I complex while the 30- or 36 kDa complex is not visible (lane3). This suggests that under normal circumstances, IGF-I is bound tightly in the 150 kDa complex, but disassociates during electrophoresis if not chemically cross-linked (lane 4). The difference in profiles of clones C2.8 and ISS9 may help to explain the differences in growth capabilities of both *Super*-CHO clones. Although similar quantities of IGF-I are produced in both C2.8 and ISS9, their growth may depend on the nature of the interaction of IGF-I and IGF binding proteins, regulating the utility of IGF-I by the cells. IGF-I and/or IGF-I complexes are not present in PFM conditioned by CHO-K1 cells in the presence (lane 5) or in the absence (data not shown) of

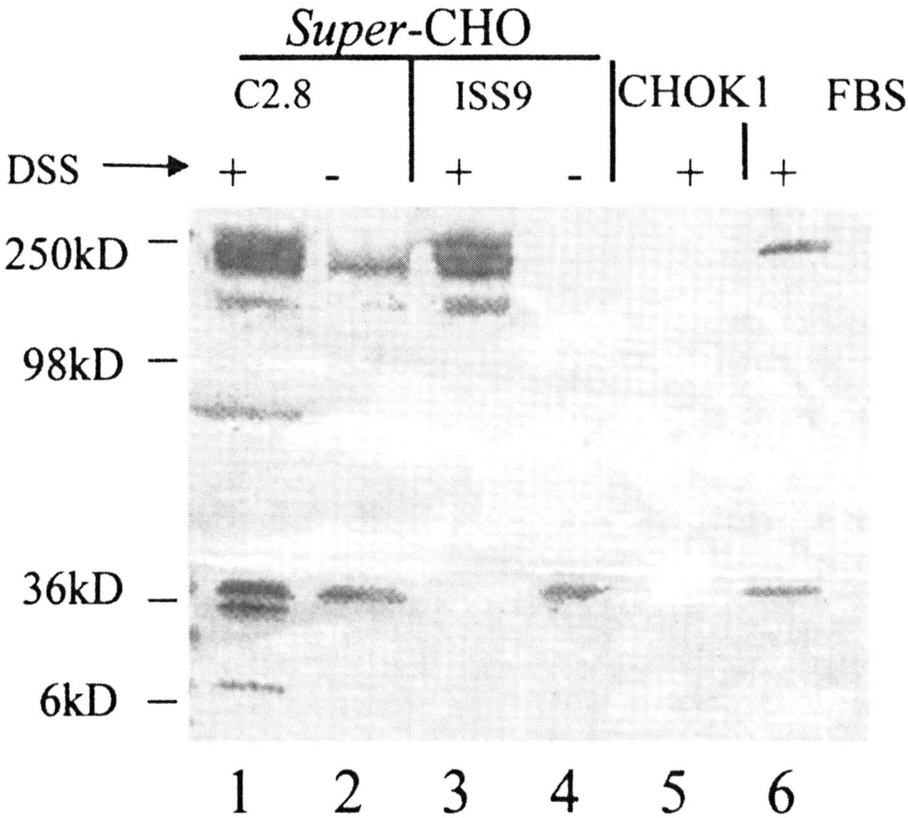

Figure 4. Detection of IGFBP-3 in conditioned media and in cellular lysates. A) IGFBP-3 dot blot. Conditioned media was obtained from cells maintained in PFM for 48 hr and applied to a nitrocellulose membrane using a vacuum blotting apparatus (see Materials and Methods). IGFBP-3 was detected using the rodent-specific anti-IGFBP-3 antibody, M-19 (Santa Cruz Biotechnology). B) Immunoprecipitation of ^{35}S Methionine-labelled cellular lysates using the M-19 antibody. Estimation of molecular weight was determined by co-migration of See Blue size markers (Novex, San Diego, CA).

cross-linker. Unconditioned medium containing 10% FBS (lane 6) includes IGF-I complexes and are visible when cross-linked to the molecule.

Detection of IGFBP-3

The M-19 antibody (Santa Cruz Biotechnology) which is specific for rodent IGFBP-3 was used to detect IGFBP-3 both in conditioned media and cellular lysate of *Super*-CHO and CHOK1 (Figure 4). IGFBP-3 can be detected in the conditioned media from *Super*-CHO clones C2.8 and ISS9, but not from that of CHOK1 cells (Figure 4A). In addition, a protein of approximately 40 kDa corresponding to CHO IGFBP-3 was detected in ^{35}S-methionine labelled cellular lysates of *Super*-CHO clones, but not from that of CHOK1 when immunoprecipated using the same antibody (Figure 4B). Cell surface associated IGFBP-3 has been demonstrated to either inhibit or stimulate IGF action depending on the cell type (Conover et al., 1996). Our results suggests that *Super*-CHO expresses its own IGFBP-3, but it remains to be determined whether IGFBP-3 is membrane bound and/or intracellular.

Conclusion

The engineered *Super*-CHO cell line is capable of stable growth under fully defined protein-free conditions making it an ideal host for the production of biopharmaceutical products. The constitutive expression of IGF-I and transferrin eliminates the need for exogenous growth factors in the culture medium. The growth of *Super*-CHO has been shown to be due to the autocrine effect of the endogenous IGF-I (Pak. et al., 1996). We detected IGF-I receptor-specific RNA transcripts in two different clonal isolates of *Super*-CHO which differ in their growth rates whereas RNA transcripts corresponding to IGF-I receptor in parental CHO cells were below our limits of detection.

Although the biological effects of IGFs are mediated by cell surface receptors, their bioavailability is regulated by a family of at least six homologous proteins that have co-evolved with the IGFs called IGF binding proteins (IGFBPs). The IGFBPs modulate the actions of IGF in endocrine, paracrine and autocrine settings. They influence IGF bioavailability and can either inhibit or potentiate IGF action. Inhibition is achieved by sequestration of IGFs from their receptors (Angervo et al., 1991; Cohen. et al., 1993) and the mechanism of potentiation involves a cellular association by IGFBPs bound to IGFBP receptors or otherwise (Conover, 1992). Potentiation of IGFs can also be achieved by the proteolysis of IGFBPs into fragments with lower affinity for IGFs, thereby increasing free IGF bioavailability (Cohen et al., 1994).

Since IGFBPs play an important role in directly regulating cell growth we investigated the IGF autocrine system of *Super*-CHO to determine the implication of IGFBPs. We report that *Super*-CHO cells synthesise IGFBP-3 in response to IGF-I production. We noted an apparent difference in the size of soluble IGFBP (about 30 kDa) in conditioned media and that of endogenous IGFBP-3 (40–45 kDa). Proteolysis of IGFBP-3 may be occurring extracellularly since it is known that specific IGFBP proteases can release IGFs from reservoir complexes. The activation of these proteases might be mediated by binding to IGF (Mohseni-Zadeh and Binoux, 1997; Zumbrunn and Trueb, 1996). Proteolytic enzymes specific to the host CHO cell may be operative in the dynamics of IGF-I regulated growth. In conclusion, the IGF-I-mediated growth characteristics of *Super*-CHO is governed by a complex interaction of host cell specific factors which regulate its utility of the growth factor and are coordinated by altering the expression of IGFBPs, IGF-I receptor and perhaps other relevant genes.

References

Ando Y and Jensen PJ (1993) Epidermal Growth Factor and Insulin-Like Growth Factor I Enhance Keratinocyte Migration. J Invest Dermatol 100, 633–639.

Angervo M, Koistinen R, Suikkari AM and Seppala M (1991) Insulin-like Growth Factor Binding Protein Inhibits the DNA Amplification induced by Insulin-like Growth Factor-I in Human Granulosaluteal cells. Hum Reprod 6, 770–773.

Baserga R, Hongo A, Rubini M, Prisco M and Valentinis B (1997) The IGF-I Receptor in Cell Growth, Transformation and Apoptosis. Biochimica et Biophysica Acta 1332, F105–F126.

Baxter RC (1986) The Somatomedins: Insulin-like Growth Factors. Adv Clin Chem 25, 49–115.

Baxter RC and Martin, JL (1989) High Molecular Weight Insulin-like Growth Factor Binding Protein Complex. J Biol Chem 264, 11843–11848.

Birnbaum RS and Wiren KM (1994) Changes in Insulin-Like Growth Factor-Binding Protein Expression and Secretion during the Proliferation, Differentiation, and Mineralization of Primary Cultures of Rat Osteoblasts. Endocrinology 135, 223–230.

Camacho-Hubner C, Busby WH, Mccusker RH, Wright G and Clemmons DR (1992) Identification of the Forms of Insulin-like Growth Factor-binding Proteins Produced by Human Fibroblasts and the Mechanisms that Regulate their Secretion. J Biol Chem 267, 11949–11956.

Chomczynski P and Sacchi W (1987) Single-step method of RNA isolation by acid guanidium thiocyanate-phenol-chloroform extraction. Anal Biochem 162, 165–169.

Cohen P, Lamson G, Okajima T and Rosenfeld RG (1993) Transfection of the Human Insulin-Like Growth Factor Binding Protein-3 gene into Balb/c Fibroblasts Inbhibits Cellular Growth. Mol Endcrinol 7, 380–386.

Cohen P, Peehl DMBB, Liu FHRL and Rosenfeld RG (1994) Insulin-like Growth Factor Axis Abnormalities in Prostatic Stromal Cells from Patients with Benign Prostatic Hyperplasia. J Clin Endocrinol Metabol 79, 1410–1414.

Conover CA (1992) Potentiation of Insulin-Like Growth Factor (IGF) Action by IGF-Binding Protein-3: Studies of Underlying Mechanism. Endocrinology 130(6): 3191–3199.

Conover CA, Clarkson JT and Bale LK (1996) Factors Regulating Insulin-Like Growth Factor-Binding Protein 3 Binding, Processing, and Potentiation of Insulin-Like Growth Factor Action. Endocrinology 137, 2286–2292.

Coverley JA and Baxter RC (1997) Phosphorylation of Insulin-Like Growth Factor Binding Proteins. Mol and Cellular Endocrinology 128, 1–5.

De Leon DD and Asmerom Y (1997) Quantification of Insulin-Like Growth Factor I (IGF-I) without Interference by IGF Binding Proteins. Endocrinology 138, 2199–2202.

Donnelly MJ and Holly JMP (1996) The Role of IGFBP-3 in the Regulatlion of IGFBP-4 Proteolysis. J Endocrinology 149, R1–R7.

Dufourny B, Alblas J, van Teeffelen HAAM, Van Schaik FMA, Van der Burg B, Steenbergh PH and Sussenbach JS (1997) Mitogenic Signaling of Insulin-like Growth Factor I in MCF-7 Human Breast Cancer Cells Requires Phosphatidylinositol 3-Kinase and is Independent of Mitogen-activated Protein Kinase. J Biol Chem 272, 31163–31171.

Hashimoto R, Ono M, Fujiwara H, Higashihashi N, Yoshida M, Enjoh-Kimura T and Sakano K (1997) Finding Sites and Binding Properties of Binary and Ternary Complexes of Insulin-like Growth Factor-II (IGF-II), IGF-binding Protein-3, and Acid-labile Subunit. J Biol Chem 272, 27936–27942.

Herrin DL and Schmidt GW (1988) Rapid, Reversible Staining of Northern Blots Prior to Hybridization. Biotechniques 6, 196–200.

Hossenlopp P, Seurin D, Segovia-Quinson B, Hardouin S, and Binoux M (1986) Analysis of Serum Insulin-like Growth Factor Binding Proteins Using Western Blotting: Use of the Method for Titration of the Binding Proteins and Competitive Binding Studies. Anal Biochem 154, 138–143.

Kato H, Faria TN, Stannard B, Levy-Toledano R, Taylor SI, Roberts CTJ and LeRoith D (1993) Paradoxical Biological Effects of Overexpressed Insulin-like Growth Factor-I Receptors in Chinese Hamster Ovary Cells. J Cell Physiol 156, 145–152.

Kelley KM, Youngman OH, Gargosky SE and Gucev ZS (1996) Insulin-like Growth Factor-binding Proteins (IGFBPs) and their Regulatory Dynamics. Int J Biochem Cell Biol 28, 619–637.

Koedam JA, Hoogerbrugge CM and Van Buul-Offers SC (1997) Insulin-like Growth Factor Binding Proteins-3 and -5 Form Sodium Dodecyl Sulfate-Stable Multimers. Biochem Biophys Res Comm 240, 707–714.

Lalou C and Binoux M (1993) Evidence that Limited Proteolysis of Insulin-like Growth Factor Binding Protein-3 (IGFBP-3) occurs in the normal state outside of the bloodstream. Regul Pept 48, 179–188.

Lee CY and Rechler MM (1995) A Major Portion of the 150-kilodalton Insulin-Like Growth Factor Binding-Protein (IGFBP) Complex in Adult Rat Serum Contains Unoccupied, Proteolytically Nicked IGFPB-3 that Bind IGF-II Preferentially. Endocrinology 136, 668–678.

LeRoith D, Werner H, Beitner-Johnson D and Roberts Jr, CT (1995) Molecular and Cellular Aspects of the Insulin-like Growth Factor I Receptor. Endocrine Reviews 16, 143–163.

Machuca I, Esslemont G, Fairclough Lynne and Tata JR (1995) Analysis of Structure and Expression of the *Xenopus* Thyroid Hormone Receptor-B Gene to Explain its Autoinduction. Mol Endcrinol 9, 96–107.

Mohseni-Zadeh S and Binoux M (1997) Insulin-like Growth Factor (IGF) Binding Protein-3 Interacts with the Type I Receptor, Reducing the Affinity of the Receptor for its Ligand: an Alternative Mechanism in the Regulation of IGF Action. Endocrinology 138, 5645–5648.

Mouhieddine OB, Cazals V, Maiatre B, Le Bouc Y, Chadelat K and Clement A (1994) Insulin-Like Growth Factor-II (IGF-II) Type 2 IGF Receptor, and IGF-Binding Protein-2 Gene Expression in Rat Lung Alveolar Epithelial Cells: Relation to Proliferation. Endocrinology 135, 83–91.

Ooi GT, Cohen FJ, Tseng LYH, Rechler MM and Boisclair YR (1997) Growth Hormone Stimulates Transcription of the Gene Encoding the Acid-Labile Subunit (ALS) of the Circulating Insulin-Like Growth Factor-Binding Protein Complex and ALS Promoter Activity in Rat Liver. Mol Endcrinol 11, 997–1007.

Pak SCO, Hunt SMN, Bridges MW, Sleigh MJ and Gray PP (1996) Super-CHO – A cell line capable of autocrine growth under fully defined protein-free conditions. Cytotech 22, 139–146.

Reiss K, Cheng W, Pierzchalski P, Kodali S, Li B, Wang S, Liu Y and Anversa P (1997) Insulin-like Growth Factor-I Receptor and its Ligand Regulate the Reentry of Adult Ventricular Myocytes in the Cell Cycle. Exp Cell Res 235, 198–209.

Rosenthal SM, Brown EJ, Brunetti A and Goldfine ID (1991) Fibroblast Growth Factor Inhibits Insulin-Like Growth Factor-II (IGF-II) Gene Expression and Increases IGF-I Receptor Abundance in BC3H-1 Muscle Cells. Mol Endcrinol 5(678): 684.

Steele-Perkins G, Turner J, Edman JC, Hari J, Pierce SB, Stover C, Rutter WJ and Roth RA (1988) Expression and Characterization of a functional Human Insulin-like Growth Factor I receptor. J Biol Chem 263, 11486–11492.

Stewart AJ, Johnson MD, May FEB and Westley BR (1990) Role of Insulin-Like Growth Factors and the Type I Insulin-Like Growth Factor Receptor in the Estrogen-stimulated Proliferation of Human Breast Cancer Cells. J.Biol.Chem 265, 21172-21178.

Tavakkol A, Elder JT, Griffiths CEM, Cooper KD, Talwar H, Fisher GJ, Keane KM, Foltin SK and Voorhees JJ (1992) Expression of Growth Hormone Receptor, Insulin-Like Growth Factor I (IGF-I) and IGF-I Receptor mRNA and Proteins in Human Skin. J Invest Dermatol 99, 343–349.

Young Lee C and Rechler MM (1996) Proteolysis of Insulin-Like Growth Factor (IGF)-Binding Protein-3 (IGFBP-3) in 150-Kilodalton IGFBP Complexes by a Cation-Dependent Protease Activity in Adult Rat Serum Promotes the Release of Bound IGF-I. Endocrinology 137, 2051–2058.

Zumbrunn J and Trueb B (1996) Primary Structure of a Putative Serine Protease Specific for IGF-Binding Proteins. FEBS Letters 398, 187–192.

Cytotechnology **28:** 101–109, 1998.
© 1998 *Kluwer Academic Publishers. Printed in the Netherlands.*

Attachment and growth of anchorage-dependent cells on a novel, charged-surface microcarrier under serum-free conditions

James Varani*, Felicia Piel, Sean Josephs, Ted F. Beals & William J. Hillegas
[1] *Department of Pathology, The University of Michigan, 1301 Catherine Road, P.O. Box 0602, Ann Arbor, MI 48109 U.S.A.*
E-mail: varani@umich.edu
[2] *SoloHill Engineering, Inc., Ann Arbor, MI, U.S.A.*

Received 25 August 1998; accepted 25 August 1998

Key words: cell lines, diploid fibroblasts, microcarriers, serum-free culture

Abstract

The present study describes a novel microcarrier substrate consisting of a swellable, copolymer of styrene and divinylbenzene, derivatized with trimethylamine. The co-polymer trimethylamine microcarriers support the growth of a number of different cell lines – Madin Darby Bovine Kidney, Madin-Darby Canine Kidney, Vero and Cos-7 – under serum-free conditions, and human diploid fibroblasts in serum-containing medium. Cells attach to the co-polymer trimethylamine microcarriers as rapidly as they attach to other charged-surface microcarriers (faster than they attach to collagen-coated polystyrene microcarriers) and spread rapidly after attachment. All of the cells examined grow to high density on the co-polymer trimethylamine microcarriers. Furthermore, cells are readily released from the surface after exposure to a solution of trypsin/EDTA. In this respect, the co-polymer trimethylamine microcarriers are different from other charged-surface microcarriers. Madin-Darby Bovine Kidney cells grown on this substrate support production of vaccine strain infectious bovine rhinotracheitis virus as readily as on other charged-surface or collagen-coated microcarriers. Thus, the co-polymer trimethylamine microcarriers combine the positive characteristics of the currently available charged-surface and adhesion-peptide coated microcarriers in a single product. The viral vaccine production industry is undergoing considerable change as manufacturers move toward complete, animal product-free culture systems. This novel substrate should find application in the industry, especially in processes which depend on viable cell recovery.

Abbreviations: MDBK – Madin Darby Bovine Kidney Cells; MDCK – Madin Darby Canine Kidney Cells; CP-TMA – Co-polymer trimethylamine; DMEM-FBS – Dulbecco's Modified Medium Essential Medium of Eagle with 10% fetal bovine serum; IBR – Infectious bovine rhinotracheitis virus.

Introduction

The ideal microcarrier should i) support rapid cell attachment and spreading; ii) support high-density cell growth; iii) not interfere with the elaboration or secretion of metabolic products; and iv) allow cells to be easily-detached. These characteristics should be expressed under serum-free conditions as well as in the presence of low serum-containing medium. Any product that combines these features along with low manufacturing costs, long shelf-life, and ease of handling will find acceptance in the biologicals industry for the large-scale cultivation of anchorage-dependent cells.

There are two basic strategies for achieving the ideal microcarrier. One involves covalently linking a strongly charged surface moiety such as diethyl amino ethyl (DEAE) to a porous or semi-porous neutral substrate such as dextran or polysytrene plastic (Van Wezel, 1967). The other involves coating the substrate surface with an adhesion-supporting extracellu-

* Author for all correspondence.

lar matrix component such as collagen or fibronectin – or with synthetic peptides from the attachment-supporting domains of these matrix components (Ruoslahti and Pierschbacher, 1986; Varani et al., 1989; Varani et al., 1993). Cellular attachment to charged surfaces involves electostatic interactions (Ginsburg, 1987) while interactions with adhesion peptides occurs through specific cell surface receptors (Swartz, 1993). In general, charged surfaces support better cell attachment under the high shear-stress conditions present in bioreactor cultures, but once cells have attached, they often spread and begin proliferating more slowly than on collagen-coated microcarriers (Varani et al., 1989; Varani et al., 1995). Furthermore, it is often difficult to harvest viable cells from charged surfaces, while cells are readily released (and, therefore, retain high viability) from matrix-coated materials (Varani et al., 1986). In the present report we describe a novel substrate consisting of a swellable copolymer of styrene and divinylbenzene, derivatized with trimethylamine (CP-TMA). The CP-TMA microcarriers combine the desirable properties of both types of currently-available microcarriers in a single product. This novel substrate should find application in the viral vaccine-manufacturing industry (especially in processes which depend on viable cell recovery) as the industry moves toward complete animal product-free culture systems.

Materials and methods

Microcarriers

CP-TMA microcarriers and collagen-coated polystyrene plastic microcarriers were produced by Solohill Labs (Ann Arbor, MI). DEAE-dextran microcarriers (Cytodex I) were obtained from Amersham – Pharmacia Biotechnology (London, UK). All three microcarriers products were prepared for use as indicated by their respective manufacturers. Briefly, the dried collagen-coated polystyrene plastic microcarriers were resuspended in PBS and autoclaved. The CP-TMA microcarriers and DEAE-dextran microcarriers were swelled in PBS and washed several times. Following this, these substrates were sterilized by autoclaving. After sterilization and cooling, all three preparations were rinsed in PBS and brought to a final bead concentration of 50 mg mL^{-1}.

Experiments were conducted in 2-mL plastic culture dishes and in 100-mL, bulb-stirred suspension culture flasks. When plastic culture dishes were used, the surface was first coated with a polyhydroxymethacrylate solution to prevent cell-sticking to the polystyrene plastic. CP-TMA microcarriers were added to the cultures at 25 mg mL^{-1}, providing a surface area of approximately 18 cm^2 mL^{-1}. DEAE-dextran and collagen-coated polystyrene microcarriers were used at comparable amounts of surface area.

Cells

Four different cell lines were used in these studies. These were Madin-Darby Bovine Kidney (MDBK), Vero (African Green Monkey Kidney), Madin-Darby Canine Kidney (MDCK) and Cos-7. All were obtained from the American Type Culture Collection. In addition to the cell lines, low-passage human diploid fibroblasts obtained from neonatal foreskin as described previously (Tamm et al., 1984) were also used. The cells were maintained in monolayer culture using Dulbecco's Modified Minimal Essential Medium of Eagle with Earles Salts, nonessential amino acids and 10% fetal bovine serum (DMEM-FBS) as culture medium. Incubation was at 37 °C in an atmosphere containing 95% air and 5% CO_2. Cell lines were subcultured as required, while human neonatal fibroblasts were isolated from fresh tissue weekly.

Cell attachment and spreading on CP-TMA microcarriers

Attachment assays were carried out in 2-mL culture dishes. Cells (2.5×10^5 per mL) were added to cultures containing 25 mg per mL of CP-TMA microcarriers (or an equivalent amount of DEAE-dextran or collagen-coated polystyrene microcarriers) and incubated at 37 °C. After 0.5, 1, 2 and 4 hr, microcarriers with cells attached were removed. Cells were separated from the microcarriers with trypsin/EDTA and counted. From these counts, the percentage of cells that were attached at each time point was determined. Additionally, microcarriers with attached cells were examined microscopically. Cells that were fully rounded were scored as non-spread, while fully- or partially-flattened cells were scored as spread. The percentage of spread cells was determined from this.

Cell growth

Growth studies were conducted in 100-mL suspension cultures. Cells ($4\text{-}5 \times 10^5$ per mL) were added to cultures containing 25 mg per mL of CP-TMA at time-zero and incubated. Fifty percent of the culture medium (serum-free Aim V culture medium; [GIBCO, Grand Island, NY] unless otherwise indicated) was replaced every other day. At days 2, 4 and 6, microcarriers with cells attached were removed. Cells were separated from the microcarriers with trypsin/EDTA and counted to give a total cell count. A fraction of the harvested cells was washed two times, resuspended in DMEM-FBS and plated in plastic cell culture dishes. Four hours later, the cells that had reattached were harvested and counted. Cells that had successfully reattached were considered viable, and from this the percentage of viable cells was determined.

Scanning electron microscopy

Cells and microcarriers were fixed *in situ* with 4% Sorenson's buffered glutaraldehyde and processed for scanning electron microscopy exactly as described previously (Varani et al., 1983). Briefly, this involved post-fixation with 1% osmium tetroxide buffered s-collidine followed by staining with uranyl acetate, subsequent dehydration in graded ethanol – liquid carbon dioxide and finally gold sputter-coated. Specimens were examined using an ISI Super IIIA scanning electron microscope.

Cell detachment assay

Cells were grown for two days on CP-TMA microcarriers or on the control substrates in serum-free Aim V medium. At the end of the incubation period, the cells were exposed to a solution containing 0.5% trypsin (w/v) and 0.5 mM EDTA in Ca^{2+} – and Mg^{2+} – free PBS. The cells were incubated in this solution for 5, 10 or 15 min. At each time point, the cells and microcarriers were gently triturated to facilitate cell release. The released cells were counted and a percentage of released cells at each time point calculated.

Production of infectious bovine rhinotracheitis (IBR) virus

IBR virus (ATCC VR-864) was obtained from the American Type Culture Collection. Virus stocks were prepared in MDBK cells, titered and frozen in liquid nitrogen. To assess virus growth, MDBK cells were grown for one day in 100-mL suspension cultures containing either CP-TMA microcarriers or one of the two control types of microcarriers. Virus was seeded at 0.5 infectious units per cell on day zero. After 1, 2 and 3 days, samples of culture fluids were obtained. The culture fluids were clarified by low-speed centrifugation and the virus quantified in MDBK cell monolayers.

Table 1. Attachment and spreading of four different cell lines and human diploid fiboblasts on CP-TMA microcarriers

Cell type	Percent attached (mean ± s.d.)	Percent spread (mean ± s.d.)
MDCK	89±10	71±5
MDBK	75±6	92±7
Vero	94±8	89±16
Cos-7	90±2	55±6
Diploid fibroblasts	70±5	60±6

Cells were added to the microcarriers at time-zero and allowed to attach for a period of one hour. At the end of the incubation period, the percent attached was assessed by direct cell counts made following removal of the cells from the substrate with trypsin/EDTA. The percent spread was determined by assessing cells on individual beads under phase-contrast microscopy and determining the percentage of cells that were flattened out on the bead (spread) and the percentage that were still round and refractile to light (not spread). MDBK = Madin-Darby Bovine Kidney; MDCK = Madin-Darby Canine Kidney; Vero = African Green Monkey Kidney; Diploid fibroblasts are derived from human neonatal foreskin.

Results

Attachment of cells to CP-TMA microcarriers: Comparison with attachment to collagen-coated polystyrene microcarriers and to DEAE-dextran microcarriers

In the first series of experiments, MDBK cells were examined for attachment to the experimental microcarriers under serum-free conditions. Attachment to collagen-coated polystyrene microcarriers and DEAE-dextran microcarriers was assessed in parallel as controls. As expected, the cells attached rapidly to the DEAE-dextran microcarriers and more slowly to the collagen-coated polystyrene microcarriers (Figure 1). When the CP-TMA microcarriers were examined, MDBK cells attached as rapidly to this substrate as to the DEAE-dextran microcarriers (Figure 1). By one hour, 75–90% of the cells were attached to both the CP-TMA and DEAE-dextran microcarriers as compared to approximately 50% to the collagen-coated polystyrene microcarriers (Figure 1). Table 1 summarizes attachment and spreading data for the four different cell lines and the low-passage human diploid fibroblasts on CP-TMA microcarriers. All of the cells

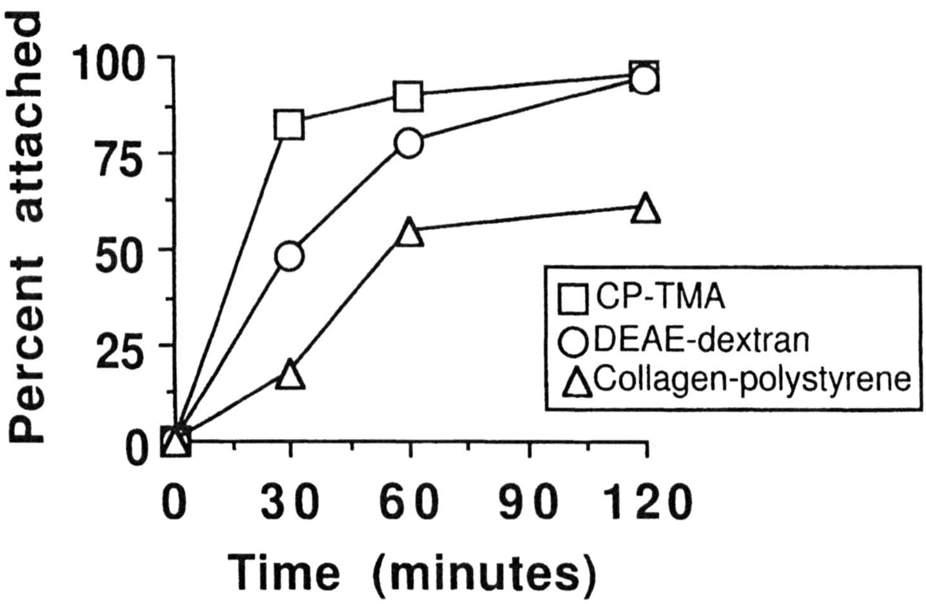

Figure 1. Attachment of MDBK cells to CP-TMA microcarriers: Comparison with DEAE-dextran and collagen-coated polystyrene microcarriers. The percent attached was determined by direct cell counts made following removal of the cells from the substrate with trypsin/EDTA. The data represent average cell counts where standard deviations were within 10% of mean values. The experiment was conducted three times with similar results.

rapidly attached to this substrate under serum-free conditions, and all of the cells rapidly spread after attachment. In previous studies it had been shown that while cells attached more rapidly to charged surface microcarriers than to collagen-coated microcarriers, cells spread more rapidly on collagen-coated polystyrene microcarriers after attachment than they did on the charged surface (Varani et al., 1995). Since the cells rapidly spread after attaching to the CP-TMA microcarriers, this suggests that the CP-TMA microcarriers exhibit optimal adhesive characteristics of both the DEAE-dextran and collagen-coated polystyrene microcarriers.

Cell growth on CP-TMA microcarriers

In the next experiments, the four cell lines and the human diploid fibroblasts were examined for growth on the CP-TMA microcarriers under serum-free conditions. All of the cell lines grew to greater than 1×10^6 cells per mL within six days (Figure 2). Viable cell counts indicated greater than 90% viability in all cases (not shown). Phase-contrast and scanning electron microscopy showed that the microcarriers became completely covered with cells as the cell counts increased. Figure 3 shows a scanning electron micrograph of CP-TMA microcarriers completely covered with Vero cells.

In contrast to results with the cell lines, low passage human diploid fibroblasts did not proliferate well on the CP-TMA microcarriers under serum-free conditions (Figure 2, middle-right). However, human diploid fibroblasts do not grow well in the absence of serum under any conditions and the same lack of growth was seen when these cells were grown in plastic cell culture dishes under the same conditions (not shown). To confirm that human diploid fibroblasts would grow on the CP-TMA microcarriers, the following experiment was peformed. Diploid fibroblasts were allowed to attach to CP-TMA microcarriers under serum-free conditions. One day later, the serum-free Aim V medium in half the culture dishes was replaced with DMEM-FBS. The other half received fresh Aim V medium. Cell counts made over the subsequent three day period indicated that proliferation did occur in the cultures to which serum-containing medium was added (Figure 2, lower-right). Thus, the failure of the human diploid fibroblasts to grow under serum-free conditions on the CP-TMA microcarriers was attributed to their serum-requirement rather than to incompatability with the substrate.

In a final set of experiments, small CP-TMA microcarriers (average diameter; 38–75 μm) and

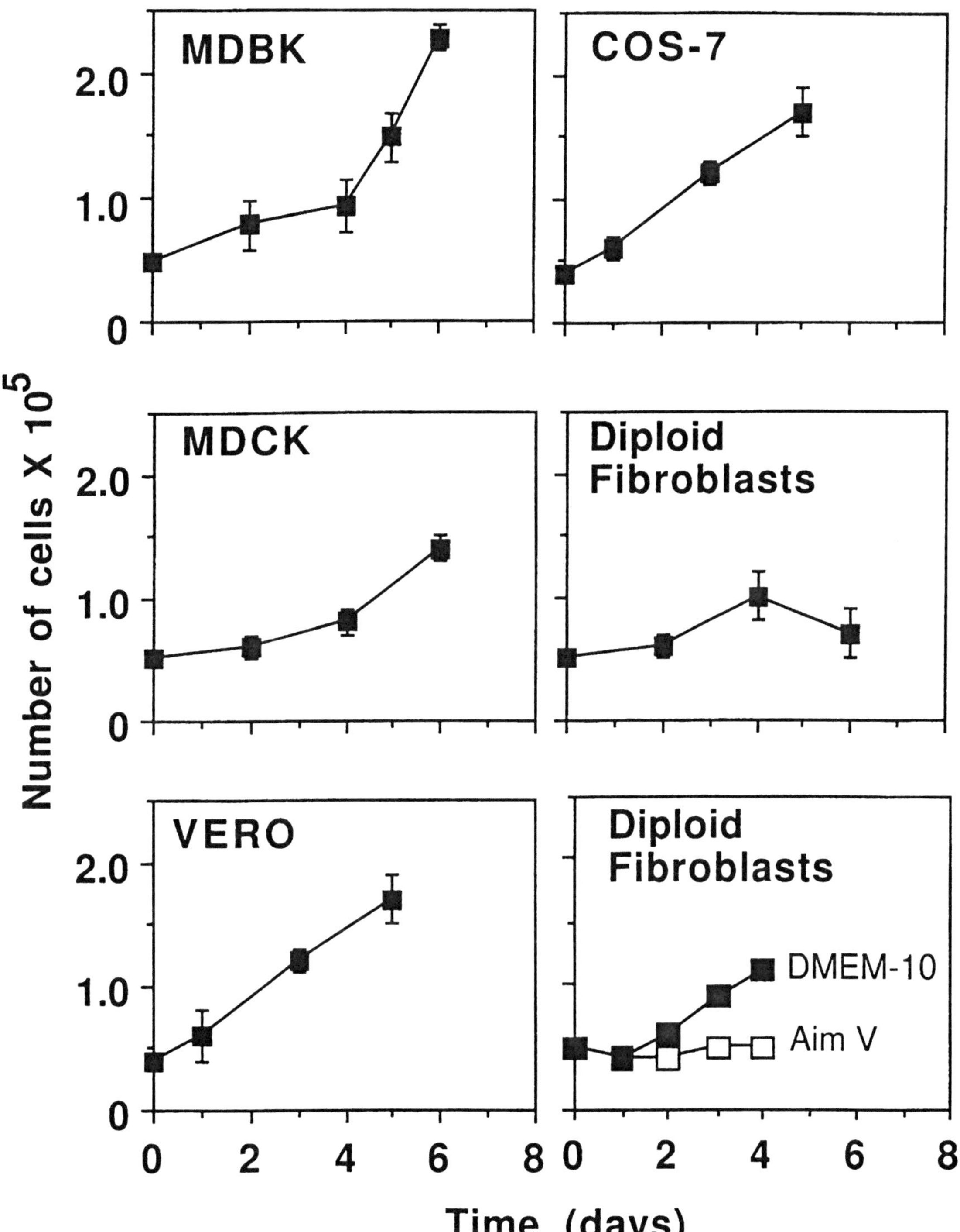

Figure 2. Cell Growth on CP-TMA microcarriers. Cells were added to 100-mL suspension cultures at 4–5 × 10^5 cells per mL and grown for 6 days in serum-free Aim V medium (or DMEM-FBS as indicated). Fresh culture medium was provided every other day. The data represent average cell counts ± s.d. based on triplicate or quadruplicate samples. Each cell type was examined at least three times with consistent results.

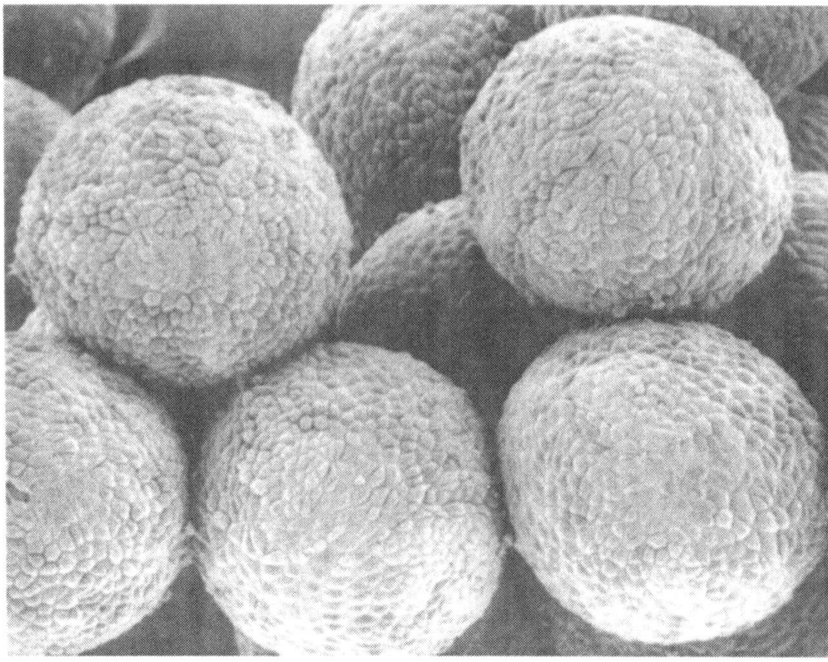

Figure 3. Scanning electron micrograph showing appearance of Vero cells at high-density on CP-TMA microcarriers (Magnification X600).

large CP-TMA microcarriers (average diameter; 150–300 μm) were compared to the standard, intermediate-sized CP-TMA microcarriers (average diameter; 75–150 μm) for ability to support attachment and growth of MDBK and Vero cells. As shown in Figure 4, when normalized for surface area, all three sizes of microcarriers performed equally well. This is of interest because previously it had been thought that microcarriers in the range of approximately 90–225 μm would be optimal with a fall-off in efficiency on either side of this range (Nielson and Johansson, 1980; Gebb, Clark and Hirtenstein, 1982; Obrenovitch, Maintier and Sene, 1982).

Cell detachment from CP-TMA microcarriers

In many large-scale cell culture applications, such as in the animal health industry, cells are never removed from the substrate. Rather, expressed products (such as the feline leukemia virus) are harvested from the culture fluid while cells, cell debris and substrate are discarded together. For such applications, ease of cell detachment from the substate is not an important consideration. However, production of other types of vaccines (for example, Marek's disease vaccine) requires that viable cells be released from the substrate in a viable state. Likewise, 'scale-up' protocols require removal of viable cells from the substrate. With this in mind, experiments were conducted in which MDBK cells, MDCK cells and human diploid fibroblasts were grown for one day on CP-TMA microcarriers. At the end of the incubation period, the culture medium (Aim V) was removed and the cells were exposed to trypsin/EDTA. Cells grown on DEAE-dextran and collagen-coated polystyrene microcarriers served as controls. Consistent with past findings (Varani et al., 1986), cells were readily released from the collagen-coated polystyrene microcarriers but were much more resistant on the DEAE-dextran microcarriers (Figure 5). As can be seen from this figure, cell removal from CP-TMA microcarriers occurred as readily as from the CP-TMA microcarriers as from collagen-coated polystyrene microcarriers.

Production of infectious bovine rhinotracheitis virus by MDBK cells grown on CP-TMA microcarriers

In the final series of experiments, MDBK cells were grown for one day on the CP-TMA microcarriers in serum-free Aim V medium. Cells grown on this substrate as well as on DEAE-dextran and collagen-polystyrene microcarriers were then infected with IBR virus. On the subsequent three days, duplicate cultures were harvested and assayed for infectious IBR virus. As shown in Figure 6, equivalent virus production occurred in cells grown on all three substrates.

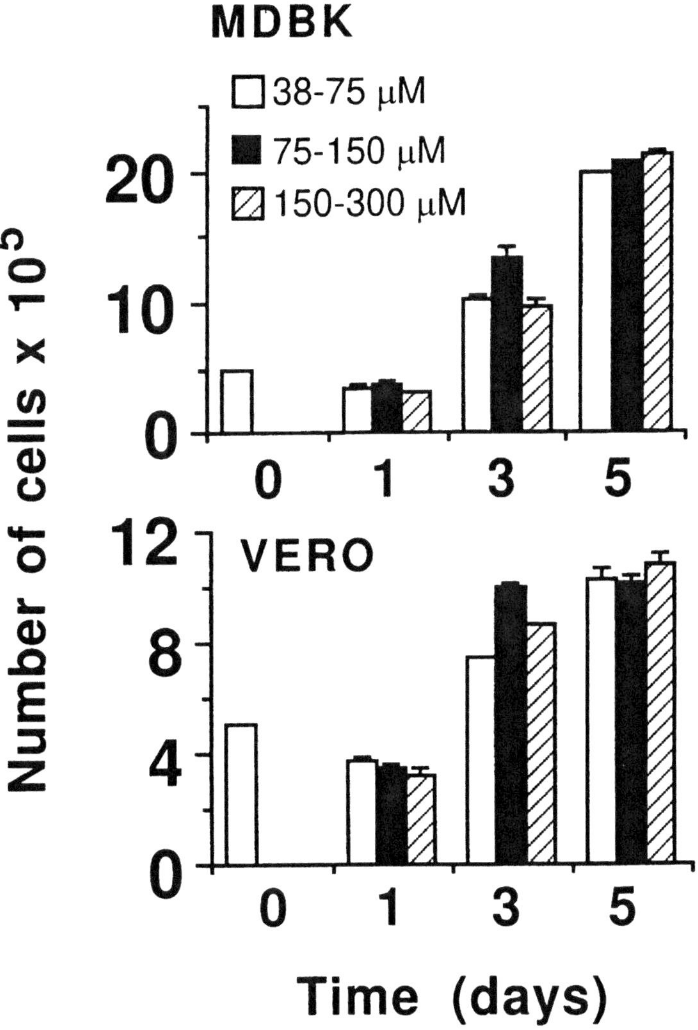

Figure 4. Growth of MDBK and Vero cells on CP-TMA microcarriers of three different sizes ranges. Cells were added to microcarriers at 5×10^5 cells per mL and grown for 5 days in serum-free Aim V. Fresh culture medium was provided every other day. The data represent average cell counts ± s.d. based on triplicate samples. The experiment was repeated three times with similar results.

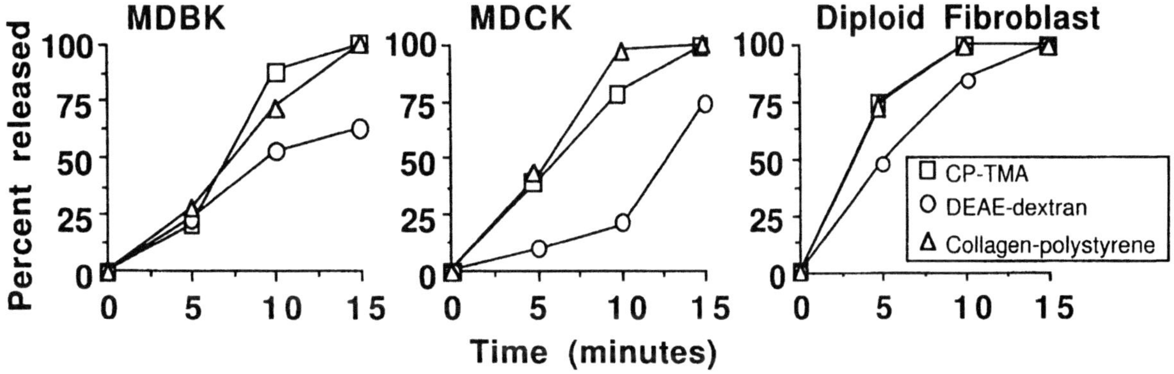

Figure 5. Release of cells from CP-TMA microcarriers in response to trypsin/EDTA. Comparison with release of cells from DEAE-dextran and collagen-coated polystyrene microcarriers. Cells grown for two days on each of the three substrates in Aim V medium were exposed to a solution of trypsin and EDTA. Released cells were separated from microcarriers and counted at each time point. The experiment was repeated three times with similar results.

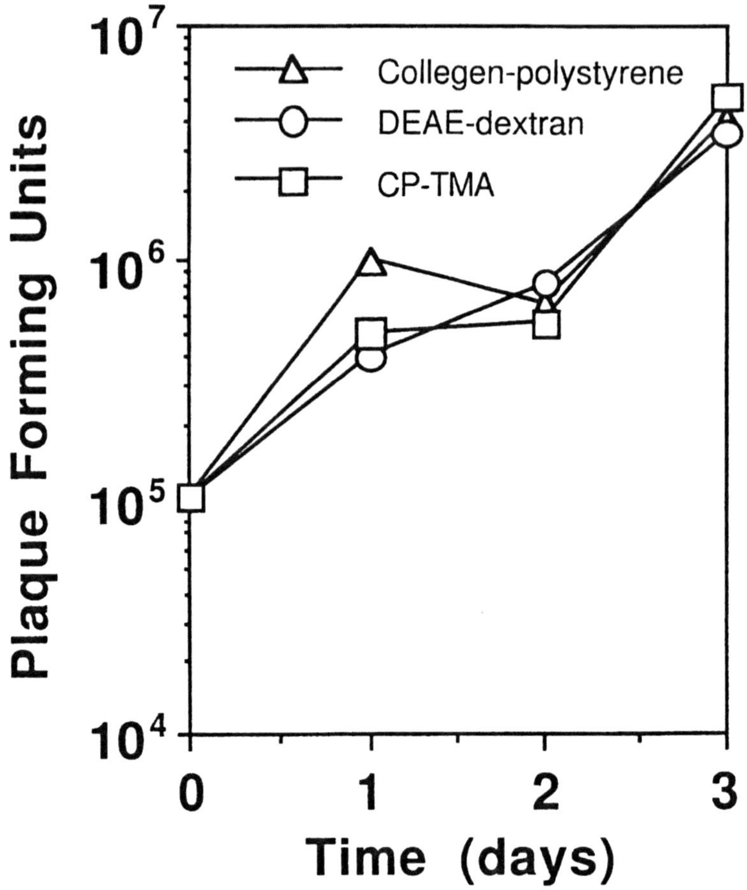

Figure 6. Production of infectious bovine rhinotracheitis (IBR) virus by MDBK cells grown on CP-TMA microcarriers. MDBK cells were grown for one day on CP-TMA microcarriers in serum-free Aim V medium. Collagen-coated polystyrene microcarriers and DEAE-dextran microcarriers served as controls. One day later, 1×10^5 plaque forming units (PFU) of IBR virus was added per mL to the flasks (approximately 0.5 multiplicity of infection). Samples were then obtained over the subsequent three day period and assayed for infectious virus. Values shown represent average number of plaque forming units, based on triplicate samples, where standard deviations were within 15% of the mean values. The experiment was repeated 3 times with similar results.

Discussion

At present, large-scale cultivation of anchorage-dependent cells is done primarily for the production of vaccine strain viruses used in human and animal medicine. However, additional applications, including production of genetically-engineered biologicals, production of virus vectors for gene therapy and production of specific cell types for 'cellular therapies' – may utilize similar technology in the future. In the past, roller bottles have been the most extensively-used technology for large-scale cell production. More recently, other technologies including hollow fiber culture systems and microcarrier-bioreactor systems have replaced roller bottles in some applications. Hollow fiber reactors are useful for growing anchorage-independent cells or loosely adherent cells, but are not optimal for strongly-adherent cells. Microcarrier/bioreactor systems offer the best alternative to roller bottles for the large-scale cultivation of anchorage-dependent cells.

Microcarrier development got its start when Van Wezel (1967) demonstrated that DEAE-dextran beads could be used as a substrate for the growth of anchorage-dependent cells in a suspension culture mode. Since that time, a number of different materials including glass, polystyrene plastic, acrylamide, solid collagen, cellulose and liquid fluorocarbons have been successfully used as microcarriers (Giard, Thilly and Wang, 1977; Nielson and Johansson, 1980; Gebb, Clark and Hirtenstein, 1982; Obrenovitch, Maintier and Sene, 1982; Varani et al., 1983; Keese and Giaever, 1983). In addition, microcarriers have been produced which contain adhesion-supporting extracellular matrix components (or synthetic adhesion-promoting sequences) attached to the surface through covalent or non-covalent linkages (Varani et al., 1993, 1995). Although a large number of different material have been successfully used as microcarriers, only two types of products are widely-used in the industry today. These are charged-surface microcarriers (for example, DEAE-dextran) and matrix-coated microcarriers (for example, collagen – polystyrene or collagen – dextran).

Cells attach to charged-surface microcarriers primarily through electrostatic interactions, while interactions with extracellular matrix components on the microcarrier surface is mediated through specific cell surface receptors (Ginsburg, 1987; Swartz, 1993). Attachment to charged-surface microcarriers occurs, in general, more rapidly than to matrix-coated microcarriers. This is especially true with many of the transformed cell types used for vaccine manufacturing in the animal health industry, since these cells are often deficient in cell surface adhesion receptors (Plantefaber and Hynes, 1989). The high shear forces generated in suspension cultures further exaggerate the differences in attachment characteristics of the two types of microcarriers. Although cells initially attach to charged surfaces, cell spreading and cell growth can be retarded if the electrostatic charge is too great. Furthermore, the electrostatic interactions can prevent release of viable cells from the surface. In contrast, while initial attachment to the substrate through specific receptor-ligand interactions is less efficient, cells rapidly spread and begin proliferating after attachment (Varani et al., 1995).

The CP-TMA microcarriers described here are unique in that they express desirable characteristics of both the charged-surface and matrix-coated microcarriers. A variety of cell types rapidly attach to this material in suspension culture (under serum-free conditions). Once attached, the cells rapidly spread and rapidly begin proliferating. Cells grow to high density, but at the end of the growth phase they can be easily released from the substrate by exposure to trypsin/EDTA. Given these characteristics of the CP-TMA microcarriers, this substrate may provide an ideal replacement for currently available substrates for use in large-scale, serum-free culture systems.

Acknowledgement

This study was supported in part by grant CA 74595 from the USPHS. The authors would like to thank Ms. Sue Ann Stamper for assistance with scanning electron microscopy.

References

Gebb C, Clark JM and Hirtenstein MD (1982) Alternate surface for microcarrier culture of animal cells. Dev Biol Standards 50: 93–102.

Giard DJ, Thilly WG and Wang DIC (1977) Virus production with a newly-developed microcarrier system. Appl Environ Microbiol 34: 668–672.

Ginsburg I (1987) Cationic polyelectrolytes: A new look at their possible role as opsinins, as stimulators of the respiratory burst in leukocytes, in bacteriolysis and as modulators of immune complex disease. Inflammation 11: 489–495.

Keese O and Giaever I (1983) Cell growth on liquid microcarriers. Science 219: 1448–1449.

Nielson V and Johnson A (1980) Biosilon: Optimal culture conditions and various research scale culture techniques. Dev Biol Stand 46: 131–136.

Obrenovitch A, Maintier C and Sene C (1982) Microcarrier culture of fibroblastic cells on modified trisacryl beads. Biol Cell 46: 249–256.

Plantefaber LC and Hynes RO (1989) Changes in integrin receptors on oncogenically transformed cells. Cell 56: 281–290.

Ruoslahti E and Pierschbacher MD (1986) Arg-Gly-Asp: A versitile cell recognition signal. Cell 44517–518.

Swartz MA (1993) Signaling by integrins; implications for tumorigenesis. Cancer Res 53: 1503–1506.

Tamm I, Kikuchi T, Wang E and Pfeffer LM (1984) Growth rate of control and b-interferon-treated human fibroblast populations over the course of their *in vitro* life-spans. Cancer Res 44: 2291–2296.

Van Wezel AL (1967) Growth of cell strains and primary cells on microcarriers. Nature 216: 65–66.

Varani J, Dame M, Beals TF and Wass JA (1983) Growth of three established cell lines on glass microcarriers. Biotech Bioengineer 25: 1359–1372.

Varani J, Bendelow MJ, Chun JH and Hillegas WJ (1986) Cell growth on microcarriers: Comparison of proliferation on and recovery from various substrates. J Biol Stand 14: 331–336.

Varani J, Fligiel SEG, Inman DR, Helmreich DL, Bendelow MJ and Hillegas WJ (1989) Substrate-dependent differences in production of extracellular matrix molecules by squamous carcinoma cells and diploid fibroblasts. Biotech Bioengineer 33: 1235–1241.

Varani J, Inman DR, Fligiel EGS and Hillegas WJ (1993) Use of recombinant and synthetic peptides as attachment factors for cells on microcarriers. Cytotechnology 13: 89–98.

Varani J, Fligiel SEG, Inman DR, Beals TF and Hillegas WJ (1995) Modulation of adhesive properties of DEAE dextran with laminin. J Biomaterials Res 23: 993–997.

Cytotechnology **28:** 111–125, 1998.
© 1998 *Kluwer Academic Publishers. Printed in the Netherlands.*

Regulated multicistronic expression technology for mammalian metabolic engineering

Martin Fussenegger, Samuel Moser & James E. Bailey*
Institute of Biotechnology, Swiss Federal Institute of Technology, ETH Zurich, CH-8093 Zurich, Switzerland

Received 25 August 1998; accepted 25 August 1998

Key words: autoregulation, cell-cycle engineering, eukaryotic operon, *IRES*, multigene engineering, picornavirus, pTRIDENT, regulated expression

Abstract

Contemporary basic research is rapidly revealing increasingly complex molecular regulatory networks which are often interconnected via key signal integrators. These connections among regulatory and catalytic networks often frustrate bioengineers as promising metabolic engineering strategies are bypassed by compensatory metabolic responses or cause unexpected, undesired outcomes such as apoptosis, product protein degradation or inappropriate post-translational modification. Therefore, for metabolic engineering to achieve greater success in mammalian cell culture processes and to become important for future applications such as gene therapy and tissue engineering, this technology must be enhanced to allow simultaneous, in cases conditional, reshaping of metabolic pathways to access difficult-to-attain cell states. Recent advances in this new territory of multigene metabolic engineering are intimately linked to the development of multicistronic expression technology which allows the simultaneous, and in some cases, regulated expression of several genes in mammalian cells. Here we review recent achievements in multicistronic expression technology in view of multigene metabolic engineering.

Introduction

There are two general levels of genetic engineering in which a suitable production cell line is generated; (i) stable introduction of the genetic information for the product protein and (ii) an optional metabolic engineering step to improve cellular activities by the manipulation of enzymatic, transport, and regulatory functions of the cell (Bailey, 1991). Metabolic engineering of animal cells has already been proven useful for improving diverse key characteristics of cultured cells including cell viability (apoptosis engineering: Cotter and Al-Rubeai, 1995; Mastrangelo and Betenbaugh, 1998), product quality (glycosylation engineering: Bailey et al., 1998; Jenkins et al., 1996), product yield (controlled proliferation technology: Fussenegger et al., 1997a; Fussenegger et al., 1998a; Fussenegger et al., 1998b; Papoutsakis, 1998) and growth in protein-free medium (cell-cycle engineering: Renner et al., 1995; Lee et al., 1996; Rivard et al., 1996; Greulich and Erikson, 1998). Most of these successes have been realised by the addition of a single gene to the host cell's genome. However, just as single-gene interpretations of human disease have limited scope (Lander and Schork, 1994), one-gene metabolic engineering cultured cells cannot access anything approaching the full potential set of useful engineered phenotypes (Papoutsakis, 1998).

Owing to the genetic complexity of higher eukaryotic cells and the absence of sophisticated genetic tools (compared to those for several microbial hosts), introduction of heterologous genetic information into mammalian hosts is usually achieved by cotransfection of a selection marker and the gene of interest with subsequent selection for clones containing the marker, and as empirical experience has shown, often also include the cotransfected gene (Kaufman and Sharp, 1982). Many undesired phenomena accompany this haphazard genetic engineering of mammalian cells

* Author for all correspondence.

because of the undefined, mechanistically obscure selection of random integration sites in different stable clones, giving rise to variability in product expression levels, genetic stability, and second order effects on growth, viability, and productivity resulting from disruption of host genes (or regulatory loci) at the integration site. Recently, chromosomal locations of some industrially relevant mammalian cells lines have been found which show high transcription and stability for integration of transgenes (Karreman et al., 1996). Screening for such sites is a time-consuming process that involves establishment of a genetic platform for subsequent targeted integration. However, unlike the situation in mouse ES stem cells (Hicks et al., 1997) gene targeting is difficult to achieve in most industrially relevant cell lines because they seem to lack necessary basic recombination machinery, and therefore they require installation of complex heterologous site-specific recombination systems (Fukushige and Sauer, 1992; Karreman et al., 1996).

Regardless of the method of integration and the chosen combination of product and metabolic engineering genes, it is desirable to manipulate the cell in a minimal number of steps. This goal is addressed by technology for simultaneous cloning and subsequent expression of multiple genes in a desired host. Besides providing a platform for future metabolic engineering breakthroughs, multicistronic expression technology should speed basic functional genomic research and new applications in tissue engineering and gene therapy.

Here we review recent developments in multicistronic expression technology and their use to enable one-step multigene metabolic engineering, positive feedback regulation circuits and auto-selective expression systems.

Internal translation initiation, the key to multicistronic expression technology

Bacteria have evolved expression units called operons which unite functionally related genes under the control of a single promoter, thus enabling coordinated, simultaneous and rapid expression of metabolically coordinated genes in response to specific environmental signals or physiological constraints (the classic example is the lactose operon; Dickson et al., 1975). Individual genes in a bacterial operon are preceded by characteristic sequences, so-called ribosomal binding sites (*RBS*), for translation-initiation at appropriate points within a single mRNA molecule. In contrast to bacterial multigene transcripts, most eukaryotic mRNAs are monocistronic, and optimal translation of the encoded gene relies on a post-transcriptional 5′ modification (capping) for ribosome binding and subsequent AUG-scanning (Shatkin, 1985; Kozak, 1989). However, other cap-independent modes of translation-initiation such as leaky scanning, termination-reinitiation, and internal initiation are used in rare cases (Jackson et al., 1995; Table 1; Figure 1).

As part of their pathogenic life cycle, picornaviruses have evolved specific genetic elements (internal ribosomal entry sites; *IRES*) in their 5′ non-translated leader regions (ntr) which adopt a particular secondary structure able to attract eukaryotic ribosomes and to allow internal translation-initiation (Belsham and Sonenberg, 1996; Table 1). The pivotal role of *IRES* in picornaviral pathogenesis is based on the expression of a viral protease which cleaves the host cap-binding translation-initiation factor eIF4G and allows redirection of host translation machinery for exclusive translation-initiation of *IRES*-containing viral mRNAs (Etchison et al., 1982; Pelletier and Sonenberg, 1988; Jackson et al., 1990; Belsham and Sonenberg, 1996; Rueckert, 1996).

IRES-like elements are present in other viral systems and were recently discovered in eukaryotic cells which give certain mRNA molecules cap-independent translation ability in response to viral infection or stress conditions, as was shown for immunoglobulin heavy chain binding protein (Bip) and the cap-binding protein eIF4G (Macejak and Sarnow, 1991; Gan and Rhoads, 1996) (Table 1). Cap-independent translation can also enforce an alternative translation start site, resulting in translation of different proteins from the same mRNA, such as that mediated by the human fibroblast growth factor 2 (FGF2) (Vagner et al., 1995). Recently, *IRES* elements were also identified in the translation regulation of developmentally regulated genes such as the homeotic gene *Antennapedia* or *Ultrabithorax* of Drosophila (Oh et al., 1992; Ye et al., 1997), the genes for human insulin-like growth factor (IGF-II) (Teerink et al., 1995), and the platelet-derived growth factor B (developmental *IRES* or D-*IRES*; Bernstein et al., 1997). The potential for cap-independent translation-initiation has also been found in yeast and Xenopus oocytes (Iizuka et al., 1994; Kneiper and Rhoads, 1997). Furthermore, the finding of an internal ribosomal entry site in the 5′ untranslated region of c-myc suggests that *IRES*-

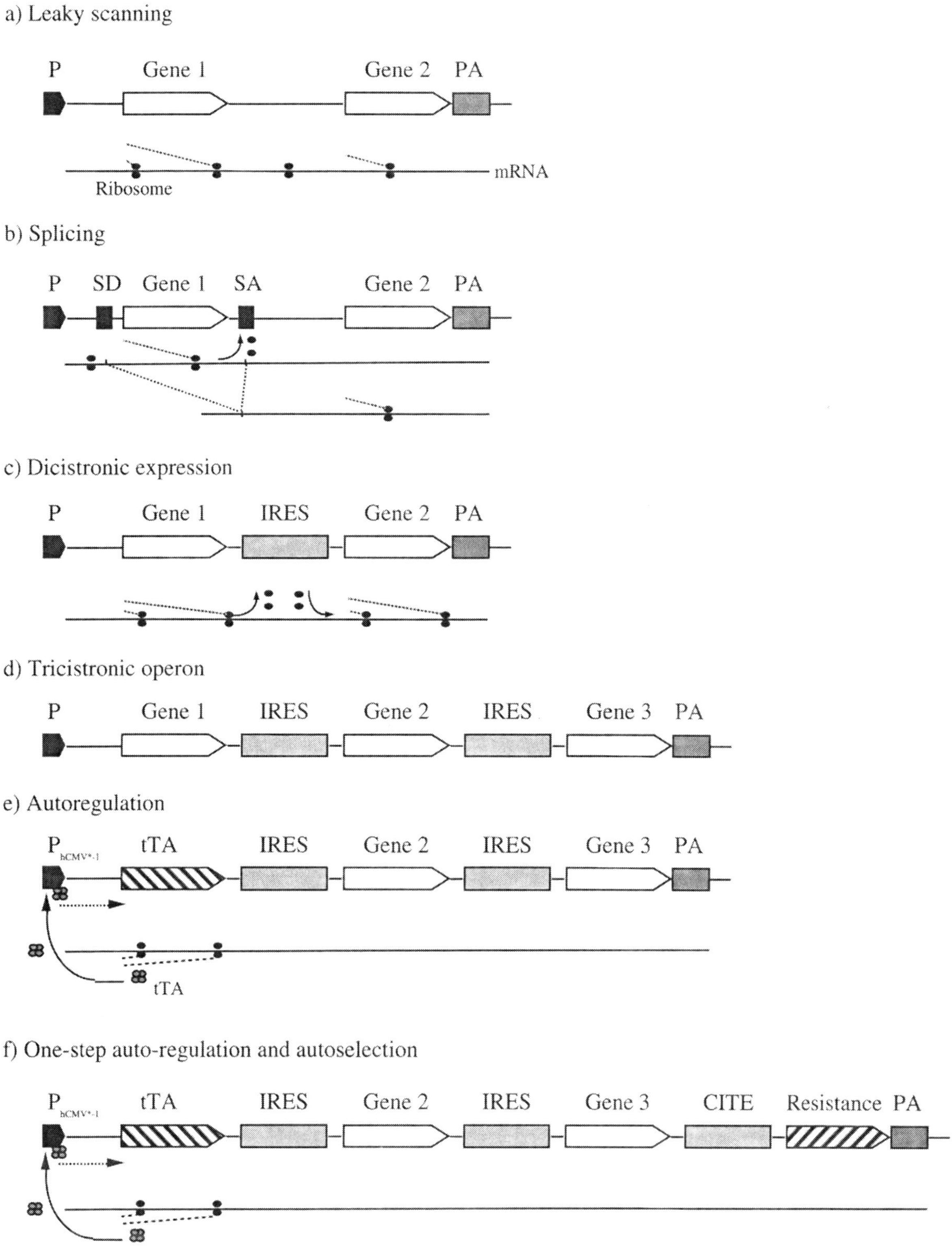

Figure 1. Strategies for simultaneous and in some cases regulated expression of more than one gene in mammalian cells. Key genetic elements for expression in mammalian cells such as the promoter (P), internal ribosomal entry sites of polioviralorigin (*IRES*) or derived from the encephalomyocarditis virus (CITE), the splice donor (SD) and acceptor (SA) and the polyadenylation site (PA) as well as for regulated gene expression including the tetracycline-responsive transactivator (tTA) and the tet-responsive promoter ($P_{hCMV*-1}$) are indicated. In some cases translation is shown below the genetic configuration (mRNA, ribosome, proteins).

Table 1. Origins of different internal ribosomal entry sites (IRES)

Class	Origin	Natural mechanism	Used for dicistronic expression vectors	Reference
Viral	Encephalomyocarditis virus	Viral replication strategy	yes	Jang et al., 1988
	Poliovirus	Viral replication strategy	yes	Pelletier and Sonenberg, 1988
	Foot-and-mouth disease virus	Viral replication strategy	yes	Belsham and Brangwyn, 1990
	Coronavirus infectious Bronchitis virus	Translation of protein 3c	no	Liu and Inglis, 1992
	Hepatitis A virus	Viral replication strategy	yes	Glass et al., 1993
	Hepatitis C virus	Viral replication strategy	yes	Wang et al., 1993
	Leishmania RNA virus 1	Viral replication strategy	no	Maga et al., 1995
	Human rhinovirus 2	Viral replication strategy	yes	Rojas-Eisenring et al., 1995
	Harvey murine sarcoma virus	Translation of VL30 protein	no	Berrlioz et al., 1995
	Mengo virus	Viral replication strategy	yes	Hofmann et al., 1996
	Crucifer infecting Tobamovirus	Translation of CP protein	no	Ivanov et al., 1997
	Reticuloendotheliosis virus type A	Translation of Gag protein	no	Lopez-Lastra et al., 1997
	Classical swine fever virus	Viral replication strategy	yes	Rijnbrand et al., 1997
	Bovine viral diarrhoea virus	Viral replication strategy	yes	Schumacher and Wirth, 1998
Mammalian	Human	Translation of the immunoglobulin heavy-chain binding protein	yes	Macejak and Sarnow, 1990
	Human	Translation of the insulin growth factor	no	Teerink et al., 1995
	Human	Translation of the fibroblast growth factor 2	no	Vagner et al., 1995
	Human	Translation of elongation factor 4G	no	Gan and Rhoads, 1996
	Human	Developmental regulation of the plate-let-derived growth factor B translation	no	Bernstein et al., 1997
	Human	Translation of c-myc proto-oncogene	no	Stoneley et al., 1998
Amphibian	Xenopus	Cap-independent translation in oocytes	no	Kneiper and Rhoads, 1997
Insect	Drosophila	Developmental regulation of the Ultra-bitorax and Antennapedia translation	no	Ye et al., 1997
Fungi	Saccaromyces cerevisiae	Translation of HAP4 and TFIID protein	no	Iizuka et al., 1994

mediated translational control may be vital for higher organisms as aberrant translational regulation of c-myc is likely to play a role in tumorigenesis (Stoneley et al., 1998).

Despite the potential of *IRES* as key elements of operon-like multicistronic expression units in mammalian genomes, such genetic configurations seem to have rarely evolved in a natural context, perhaps because most complex and fine-tuned regulatory circuits in mammalian cells are best configured with independent regulation of individual genes.

Dicistronic expression, a first step towards multicistronic expression systems

Since the transcription and translation of separate co-transfected genes is not strictly correlated, the reliability of product expression based on selection of the cotransfected marker gene can be very low. Further-

Table 2. IRES-based expression technology enables various genetic applications

IRES	Genetic determinants	Application	Reference
PV	luc–seap	Two monocistronic vectors that can be fused to a dicistronic expression system	Dirks et al., 1993
PV	MCS–seap	Cloning vector containing secreted alkaline phosphatase (seap)	Kirchhoff et al., 1995
EMCV	Hepatitis B surface antigen – B7-1 molecule	Adenoviral and retroviral vectors for dicistronic expression	He et al., 1996
EMCV	MCS–hygromycin	Cloning vector	Gurtu et al., 1996
EMCV	MCS-βgal; MCS–neo	Cloning vector	Kobayashi et al., 1996
EMCV	MCS–gfp	Cloning vector	Mosser et al., 1997
EMCV	env–v-fos-CD3	Replication-competent retrovirus for high-level expression of an exogenous gene	Murakami, 1997
HCV	luc–blasticidin S-resistance	Use of the smallest IRES element of HCV to maximize insert size of retroviral vector	Urabe et al., 1997
Mengo virus	lacZ–tTA	Retroviral vector tetracycline autoregulatory cassette	Hofmann et al., 1996
EMCV	tTA–neo	Autoregulated system	Hoshimaru et al., 1996
EMCV	MCS–tTA	Autoregulated system, cloning vector	Zhang et al., 1997
PV	gfp–tTA	pSAM202, autoregulated expression vector	Fussenegger et al., 1997b
PV	MCSI–MCSII–MCSIII	pTRIDENT1, PhCMV∗-1-containing tricistronic expression vector	Fussenegger et al., 1998a
PV/EMCV	MCSI–MCSII–MCSIII	pTRIDENT3, PhCMV∗-1-containing tricistronic expression vector	Fussenegger et al., 1998a
PV	MCSI–MCSII–MCSIII	pTRIDENT2, PSV40-containing tricistronic expression vector	Fussenegger et al., 1998a
PV/EMCV	MCSI–MCSII–MCSIII	pTRIDENT4, PSV40-containing tricistronic expression vector	Fussenegger et al., 1998a
PV	MCSI–MCSII–MCSIII	pTRIDENT7, ecdysone-resp. promoter-containing tricistronic expression vector	Fussenegger et al., 1998a
PV/EMCV	MCSI–MCSII–MCSIII	pTRIDENT8, ecdysone-resp. promoter-containing tricistronic expression vector	Fussenegger et al., 1998a
PV	tTA–MCS–MCS	pTRIDENT-tTA, autoregulated tricistronic expression vector	Fussenegger et al., 1997b
PV/EMCV	tTA–MCS–MCS–neo	pQuattro-tTA, autoregulated quattrocistronic expression vector	Fussenegger et al., 1997b
EMCV	hyg–neo	Retroviral vector to test dicistronic expression	Koo et al., 1992
EMCV	neo–cat	Retroviral vector to test dicistronic expression	Morgan et al., 1992
EMCV, PV	luc, cat, HIV-1 gp120	Retroviral vector with different IRES for test purposes	Alexander et al., 1994
EMCV	MCS–dhfr	Testvector for dhfr-based amplification procedures	Kaufman et al., 1991
EMCV, PV	Poliovirus proteins	Poliovirus with different IRES elements	Schmid et al., 1994
BIP	GP2,HGP2–NA	Retroviral vector for dicistronic expression of an effector and a marker gene	Garcia-Sastre et al. 1994
FMDV	neo, EP, G-CSF, ADA–βgal	Retroviral vector for dicistronic expression of an effector and a marker gene	Ramesh et al., 1996
MLV-VL30	alkaline phosphatase–neo	Retroviral vector for dicistronic expression	Torrent et al., 1996
EMCV	neo–lacZ	Testvector for dicistronic expression in ES cells and mice	Kim et al., 1992
EMCV	MCS–CD4	Surface marker expression	Smarda and Lipsic, 1994
EMCV	FIX–neo	Factor IX production	Chen et al., 1997
EMCV	Light chain–heavy chain	Recombinant antibody production	Kolb and Siddell, 1997
EMCV	G/CSF–tymidine kinase	Granulocyte colony-stimulating factor overexpression in mice	Veelken et al., 1996

Table 2. (continued)

IRES	Genetic determinants	Application	Reference
EMCV	Ebna-1–neo	Human shuttle vector; enforced selection	Ramage et al., 1997
EMCV	NTP–HSVtk	Surface marker in T lymphocytes	Gallardo et al., 1997
EMCV	CFTR–βGeo (βGal:neo)	Detection of CFTR expression	Vasseaux and Huxley, 1997
EMCV	Stem cell-specific transcription factor Oct-4–βgal	Investigation of mammalian emryogenesis	Mountford et al., 1994
EMCV	ADA–βGeo	Dicistronic vetors for knock out constructs due to homologous recombination	Vaulont et al., 1995
EMCV	p53 antisense–resistance	Antisense technology	Yamauchi et al., 1996
EMCV	p16–IAP (phosphatase)	Cytostatic technology	Poulos et al., 1996
PV	seap–p27	pMF113, cytostatic expression vector	Fussenegger et al., 1997a
PV	seap–p53	pMF114, cytostatic expression vector	Fussenegger et al., 1997a
PV	seap–p21	pMF112, cytostatic expression vector	Fussenegger et al., 1997a
PV/EMCV	seap–p27–bcl-xL	pDD6, cytostatic expression vector	Fussenegger et al., 1998c
PV	seap–p21–c/ebpα	pSS5, cytostatic expression vector	Fussenegger et al., 1998c
PV	tTA–p21–gfp	pSAM204, autoregulated expression vector	Fussenegger et al., 1997b
EMCV	Ion channel–gfp	Ion channel characterization	Trouet et al., 1997
EMCV	neo–lacZ	Gene trap vector	Kang et al., 1997
EMCV	MCS-βgal; MCS–lacZ	Genome sequencing/mapping; gene targeting	Yang et al., 1997
EMCV	HSVtk–cytokine	Cancer gene therapy	Castleden et al., 1995
EMCV	dystrophin–βgal	Gene therapy for muscle dystrophy	Fassati et al., 1996
EMCV	MDR–p67	Gene therapy for cronic granulomatous disease	Nunoi et al., 1995
EMCV	Tk1–Na	Cancer gene therapy	Hwang et al., 1996
EMCV	HSVtk–IL2	Gene therapy	Okada et al., 1996
PV	neo–VZVtk	Breast cancer therapy	Grignet and Calberg 1995
EMCV	MCS–gfp	Investigation of micrometastases	Chishima et al., 1997
EMCV	Rev–Lyt2	Inhibition of human immunodeficiency virus replication	Bonyhadi et al., 1997

more, even under high selection pressure, the genetic stability of the expressed product can not be assured in long term cultivations. The combination of product and marker genes on the same vector does not completely alleviate these complications. For these reasons, dicistronic genetic configurations were developed. The first dicistronic constructs used *IRES* elements of picornaviral origin or from encephalomyocarditis virus (EMCV) for cap-independent translation of the second cistron while the first cistron relied on classical cap-dependent translation-initiation (Pelletier and Sonnenberg, 1988; Kaufman et al., 1991). Although the two *IRES* elements differ completely at the sequence level, their secondary structure is very similar and typical for such internal translation initiators. A large number of dicistronic product-marker configurations have since been developed for many different applications. Table 2 gives an overview of recent dicistronic expression vectors. Although genetic combinations used for dicistronic expression vary among different applications, EMCV and picornaviral *IRES* remained the most popular cap-independent translation-initiators for dicistronic configurations because these elements function in a wide variety of cell lines including the industrially relevant CHO and BHK cell lines (Borman et al., 1997). However, recent reports of varying translation-initiation capabilities of *IRES* in different host cell environments and discovery of new *IRES* elements is stimulating new development to apply these *IRES* elements for multicistronic expression (Bernstein et al., 1997; Schumacher and Wirth, 1997).

Dicistronic genetic configurations which contain the marker gene in the second cistron enable autoselective expression in addition to simultaneous and coordinated gene expression. Resistance to the marker gene or expression of the reporter gene is only possible if all 5′-encoded genetic elements are intact. This intrinsic self-selective program was found to be very reliable, with nearly all of the resistant cells also expressing the desired product gene (Gurtu et al., 1996; Rees et al., 1996). Furthermore, product-marker configurations can also be used for efficient generation and screening of high producing cell clones: *IRES*-based translation-initiation of the second cistron is usually less efficient compared to cap-dependent translation, and can be decreased further by loss-in-function mutations of the *IRES* or the marker gene itself. The overall lower translation efficiency or activity is then compensated under high selective pressure by integration of the dicistronic expression unit into chromosomal sites with high transcriptional activity (Kaufman et al., 1991; Gurtu et al., 1996; Rees et al., 1996). Certainly, simultaneous expression of two genetic traits can also be achieved by gene fusions (Krömer et al., 1997; pTracer plasmids of Clontech) or recently developed splicing expression technology (Lucas et al., 1996; Figure 1), but gene fusion strategies are limited in functional applications or may lead to fusion products with altered physiologic specificities, and splicing-based two-gene expression leads to unequal expression levels of both proteins. Only *IRES*-based dicistronic expression guarantees simultaneous and coordinated expression of both transgenes at comparable levels for multi-subunit proteins (for example antibodies) which enables genetic configurations for a wide variety of contemporary research and development applications which are also listed in Table 2 (Dirks et al., 1993; Dirks et al., 1994; Fussenegger et al., 1997a). Furthermore, *IRES*-mediated expression systems can be extended beyond the dicistronic level to tri- or even quattrocistronic artificial eukaryotic operons (Fussenegger et al., 1997b; Fussenegger et al., 1998c).

pTRIDENT, tricistronic artificial mammalian operons

Despite the numerous expression vectors available containing dicistronic expression units (Table 2), most of these expression systems express a marker or reporter gene in a fixed configuration, leaving only one cistron free for heterologous gene expression. However, for one-step transfection of a product protein, metabolic engineering, and a selection marker in a single expression unit, multicistronic artificial mammalian operons with 3 or even 4 cistrons are desirable. We recently reported the construction of a novel vector family, pTRIDENT, for tricistronic gene expression in mammalian cells (Fussenegger et al., 1998c; Figure 2). A single promoter allows high level expression and, in some vectors, adjustable transcription of all three genes. Whereas the first cistron is translated in a classical cap-dependent manner, translation-initiation of the subsequent two cistrons rely on *IRES* elements of picornaviral (*IRES*; pTRIDENT1) and EMCV origin (denoted here *CITE*, cap-independent translation enhancer; third cistron; pTRIDENT3; Fussenegger et al., 1998c).

Tricistronic pTRIDENT1- and pTRIDENT3-derived test vectors encoding the model product gene SEAP (secreted alkaline phosphatase; first cistron), a metabolic engineering determinant (the cyclin-dependent kinase inhibitor p21 (CDI) second cistron), and the reporter gene GFP (green fluorescent protein; third cistron) were transfected into a CHO cell derivative which allows tetracycline-responsive gene expression (Fussenegger et al., 1998c). Both tricistronic configurations were stable in CHO cells and showed strict simultaneous, coordinated as well as regulated expression of all three cistrons. The expression levels of individual cistrons were assessed by comparison to respective values of isogenic monocistronic expression vectors. Although expression levels of genes encoded on different cistrons are largely dependent on the overall stability of the polycistronic mRNA and therefore a direct function of the genetic configuration of encoded genes, our test vectors showed similar expression levels to those provided by the monocistronic vector on the first two cistrons and approximately 35% (*CITE*) to 50% (*IRES II*) lower expression levels on the third cistron. For enhanced translation-initiation of the third cistron, *CITE* was specially mutated (*CITE*∗) to avoid erroneous translation-initiation at upstream ATG start codons (Jackson et al., 1990; Kaufman et al., 1991; Davies and Kaufman, 1992; Rees et al., 1996; Fussenegger et al., 1998c). Initially, the use of pTRIDENT3 derivatives (*IRES-CITE*) was preferred over double *IRES*-containing counterparts because pTRIDENT3-based vectors show a slightly higher translation efficiency of the third cistron, and they contain no duplicated sequence elements (*IRESI-IRESII*; pTRIDENT1) which bear the risk of recombination-mediated deletion of the second cistron. However, genetic rearrangements or deletions in double *IRES*-containing pTRIDENT1 derivatives were never observed during cloning steps in *recA*$^-$ *E.coli* nor in mammalian cells (Fussenegger et al., 1998a and 1998c).

pTRIDENT vector backbones encode a bacterial ampicillin resistance and origin of replication (*ori*)

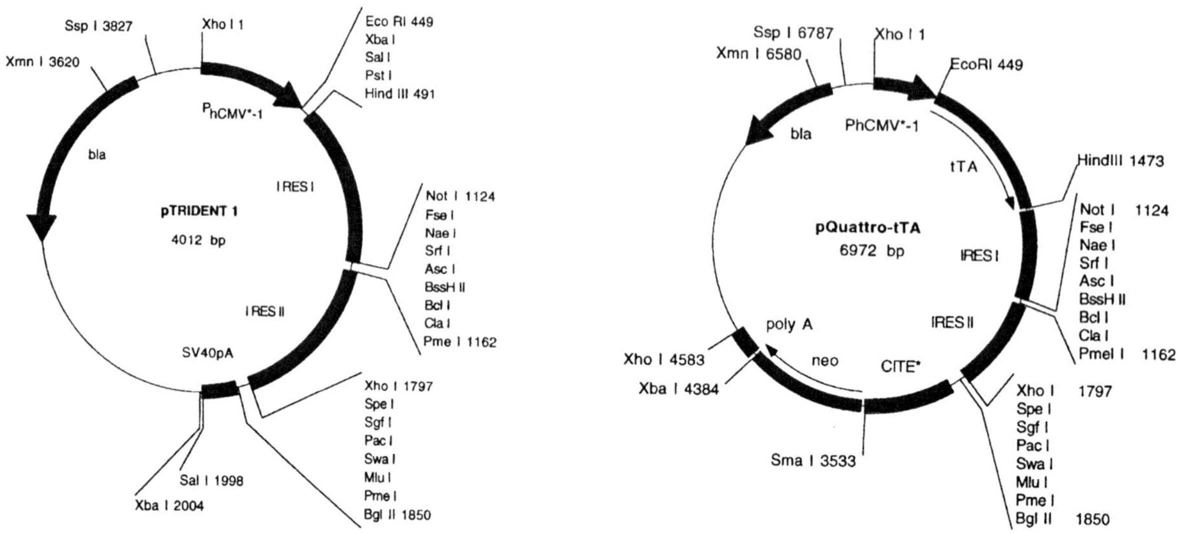

Figure 2. Examples for multicistronic expression in mammalian cells: pTRIDENT1 and pQuattro-tTA. Both vectors allow tetracycline-regulated expression of all transgenes. Since pQuattro-tTA encodes all genetic elements for regulated gene expression in its multicistronic expression unit, pQuattro-tTA allows autoregulation and can be used to achieve one-step regulated expression the genes of interest in any cell line where the internal ribosomal entry sites of polioviral origin function (*IRES, CITE*).

for high copy number amplification of these plasmids (Figure 2). High copy number amplifications in bacterial hosts is a prerequisite for large-scale transient transfection protocols which are becoming increasingly popular for industrial R&D applications (Fussenegger et al., 1997a). The tricistronic expression unit contains three multiple cloning sites (MCS) with up to 18 unique restriction sites, many for 8 bp-targeting, rare-cutting enzymes to allow sequential, complication-free cloning of all three transgenes into pTRIDENT. The general modular set-up of the key genetic elements in the pTRIDENT series, including, promoter, *IRES* elements, polyadenylation site, and vector backbone with their well selected flanking (or sometimes internal) restriction sites or MCS allows straightforward elimination or exchange of cistrons between existing conventional monocistronic or pTRIDENT expression vectors. Also, the modular set-up enables rapid adaptation of the pTRIDENT vector concept for special applications and stimulates future developments in expression vector design. Based on the compatibility of pTRIDENT to existing vector families, for example the one presented by Dirks et al. (1993 and 1994), recent developments of the growing pTRIDENT family resulted in tricistronic vector derivatives with various constitutive (P_{SV40}, P_{MPSV}), tetracycline- and ecdysone- regulated promoters ($P_{hCMV*-1}$; P_{EC}) and in construction of auto-regulated, self-selective, one-step transfection systems described below.

pTRIDENT-based multigene metabolic engineering greatly improves cytostatic production technology

Pioneering reports by Suzuki and Ollis (1990) and Al-Rubeai et al. (1992) showing increased specific productivity of growth-inhibited hybridoma cells stimulated research on chemical culture additives to arrest cell growth and initiated efforts to control cell growth by controlled overexpression or inhibition of selected genes. Three successful one-gene metabolic engineering strategies have been developed to reversibly control mammalian cell growth: (i) estrogen-regulated overexpression of the interferon-responsive factor (IRF-1), a transcription factor which is upregulated by interferons as response to viral cell invasion, in BHK cells (Koester et al., 1995); (ii) dexamethasone-inducible suppression of the key transcription factor c-jun by antisense technology in Friend murine erythroleukemia cells (F-MEL) (Kim et al., 1998); and (iii) tetracycline-regulated overexpression of negative key regulators of the cell-cycle including the tumor suppressor p53 and the CDIs p21 and p27 in CHO cells (Fussenegger et al., 1997a; Fussenegger et al., 1998a and 1998b). Overexpression of IRF-1 resulted in cell-cycle-independent growth arrest, but heterologous gene expression was not enhanced unless the exogenous genes were placed under control of IRF1-responsive promoters. Furthermore, IRF-1-overexpressing BHK cells rapidly die, probably by an apoptosis-independent pathway (Koester et al., 1995; Müller et al., 1998). On the contrary, c-jun suppression leads to sustained G0-phase arrest of F-MEL cells for over two weeks and protects these cells against apoptosis (Kim et al., 1998). Unfortunately, this promising antisense technology remains to be assessed in an industrially relevant cell line and in connection with cloned protein production. However, G0-arrested cells have previously shown to produce exogenous protein at a lower rate (Kim et al., 1998).

In another strategy, transient tetracycline-responsive overexpression of p53, p21 or p27 in a dicistronic configuration (SEAP-p53; SEAP-p21; SEAP-p27) led to G1-specific cell-cycle arrest, and in each case was accompanied by an up to 4-fold increase in SEAP production compared to proliferation-competent control cells (Fussenegger et al., 1997a). These results compare favourably with those from G1-arrested, temperature-sensitive CHO cells generated by random mutagenesis, which also showed a 3-4-fold higher heterologous protein production upon growth arrest but retained low cell viability at elevated permissive temperatures (Jenkins and Hovey, 1993).

However, in a stable genetic configuration in CHO cells, only SEAP-p27 overexpression lead to a significant increase in productivity, with specific SEAP productivity increasing by 15-fold compared to control cells (Fussenegger et al., 1998b). Intracellular p21 levels were probably insufficiently high to cause significant growth inhibition, and p53-based cell-cycle arrest led to rapid decrease in cell viability accompanied by cell morphologies indicative of apoptosis, even when achieved by overexpression of the apoptosis-deficient mutant p53175P (Rowan et al., 1996), a phenomenon which could not be observed with p27-induced G1-arrest (Fussenegger et al., 1998b).

The failure to produce stable growth-controllable CHO cells by p21-mediated overexpression exemplifies current limitations of one-gene metabolic engineering strategies. Although global regulatory proteins certainly exist, such key metabolic effectors are rare, difficult to find and their overexpression may imbalance fine-tuned interconnected cellular circuits, as seems the case with overexpression of p53.

Using tricistronic expression technology we extended the SEAP-p21-encoding dicistronic configuration by an additional cistron harbouring the differentiation factor, CCAAT/enhancer binding protein α (C/EBPα) (pSS5; Figure 3). C/EBPα has been shown to stabilise p21 at the protein level and also to induce endogenous p21 alleles (Timchenko et al., 1996). Using this tricistronic set-up for metabolic engineering, the induction of conditional growth arrest of CHO cells was successful, and the sustained cell-cycle arrest achieved was accompanied by an up to 15-fold higher specific SEAP productivity compared to proliferation-competent control cells, similar to that achieved by p27-based one-gene metabolic engineering (Fussenegger et al., 1998a and 1998c).

In a further preventive measure against possible apoptosis, which was strongly suggested by morphologies of p53 overexpressing cells, we linked SEAP-p27 expression with the expression of the survival gene *bcl-x*$_L$ in a tricistronic configuration (pDD6; Figure 3). *bcl-x*$_L$ belongs to the family of *bcl*-2 anti-apoptosis genes which have been successfully used to suppress apoptosis in production cell lines (Cotter and Al-Rubeai, 1995; Mastrangelo and Betenbaugh, 1998). Although overexpression of SEAP-p27-bcl-x$_L$ induced sustained growth arrest in CHO cells like its dicistronic counterpart, the specific SEAP productivity of arrested cells was increased by an additional factor of three, which corresponds to 30-times higher specific SEAP productivity than respective proliferation-competent control cell lines (Fussenegger et al., 1998a). This unexpected effect of *bcl-x*$_L$ expression cannot be explained based on the current knowledge of cell-cycle and apoptosis regulatory pathways, and further investigations are needed to reveal the mechanism of this new, apparently apoptosis-unrelated effect of *bcl-x*$_L$.

Thus, using controlled proliferation technology as an example, multigene metabolic engineering has proven to be useful for achieving difficult-to-attain cell culture states, and the combinatorial expression of an intuitively unrelated gene revealed previously unknown functions and potential molecular links of complex cellular pathways.

Mammalian expression systems for one-step cloning of autoregulated product gene expression

There is much current interest in the development of regulatable expression systems in basic functional genomic research, since externally regulated transcription enables the effects of a particular gene product

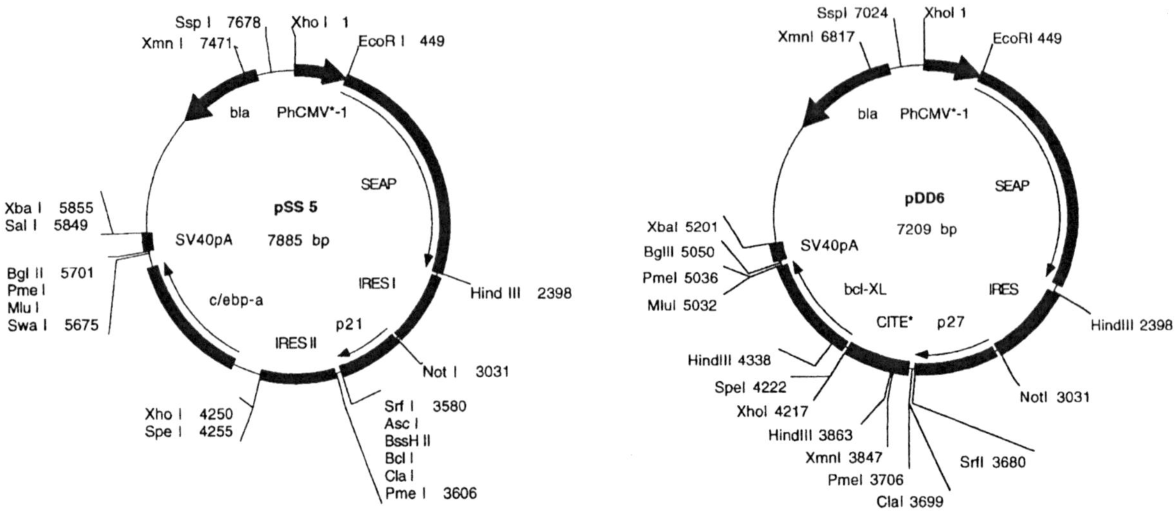

Figure 3. Tricistronic expression vectors enabling multigene metabolic engineering. Both vectors express the model product gene, the secreted alkaline phosphatase (SEAP) and one of the cell-cycle inhibitors p21 (pSS5) and p27 (pDD6). While the expression of p27 is sufficient to cause cell-cycle arrest and result in enhanced specific SEAP productivity which is additionally increased by coexpression of the anti-apoptosis gene *bcl-x*$_L$, cell-cycle arrest using p21 is only effective when the differentiation factor *c/ebpα* is coexpressed and stabilizes p21.

to be assessed in an identical genetic background. Regulated gene expression is also gaining increasing importance for biotechnological applications since it allows conditional metabolic engineering and achievement of specific cell culture states in a timely manner (Fussenegger et al., 1997a; Fussenegger et al., 1998a). For example, regulated metabolic engineering in a multicistronic configuration allows differentiation of a cell culture process into two stages: a non-productive growth phase in which the cells are rapidly expanded to the desired cell density, and a subsequent non-proliferating production phase where the cells can devote all of their metabolic capabilities to the production of protein instead of biomass.

Several *in vivo* regulated eukaryotic promoters have been described (Schweinfest et al., 1988; Israel and Kaufman, 1989; Ko et al., 1989; Hu and Davidson; 1990; Mattioni et al., 1994) and used for regulated gene expression. However, as most of these regulated promoters are derived from regulatory circuits which mediate metabolic responses, the corresponding regulating external stimuli may lead to undesired pleiotropic effects. More successful transcriptional regulation circuits rely on basic regulatory machineries of heterologous origin which are genetically adapted for use in mammalian cells. Besides the lac switch (Fieck et al., 1992) and the ecdysone-responsive system (No et al., 1996), the tetracycline-regulatable expression system (tet system; Gossen and Bujard, 1992) is by far the most popular. The tet system consists of two separate genetic entities, the tet-responsive transactivator (tTA) and the tTA responsive promoter, $P_{hCMV*-1}$ each of which represents a chimeric genetic configuration composed, respectively, of a protein fusion between the bacterial tet repressor and the VP16 domain of the herpes simplex virus (tTA) and a genetic fusion which places a tet operator adjacent to a minimal cmV promoter ($P_{hCMV*-1}$). While the bacterial parts are responsible for promoter recognition and integrate responsiveness to the external stimulus tetracycline, the viral parts initiate transcription (Gossen and Bujard, 1992).

Despite its success story, the tet system has two major limitations. First, prior to introduction of the regulated transgene, each host cell line must be engineered to express tTA in a fashion which affords efficient tetracycline-mediated control of $P_{hCMV*-1}$-initiated transcription. Screening for this phenotype is tedious and time-consuming. Cotransfection of the tTA-expression plasmid and the vector encoding the gene of interest or the transfection of a single vector with a combination of both genes, are not recommended since, in either case, the close proximity of the two genes in the host chromosome may cause the enhancer of tTA-driving promoter expression to interfere with $P_{hCMV*-1}$ thus leading to hardly regulatable configurations (Gossen and Bujard, 1992). Second, a transcriptional 'squelching' effect by the

VP16 transactivator domain may be lethal for the host cell, even at moderate expression levels (Gill and Ptashne, 1988). Consequently, since the activity of $P_{hCMV*-1}$ is proportional to the intracellular tTA levels, moderate tTA expression may lead to apparently low expression levels of regulated transgenes (Furth et al., 1994). Several improvements have been made to alleviate these complications including (i) fusion of the tTA to the ligand-binding domain of the estrogen receptor to control the transfer of tTA into the nucleus (Iida et al., 1996), (ii) fusion of tTA to a nuclear localisation signal enabling tight regulation and high-level induction (Yoshida and Hamada, 1997), (iii) construction of a regulatory cascade by controlling tTA expression by another higher-order control system, for example, by the lac switch (Aubrecht et al., 1996), (iv) construction of $P_{hCMV*-1}$ derivatives harbouring minimal promoters of various sources which show altered regulatory features, promoter strength and tTA responsiveness (Hoffmann et al., 1997), and (v) placing tTA expression itself under control of $P_{hCMV*-1}$ to prevent accumulation of toxic tTA levels prior to induction (Shocket et al., 1995). However, all of these improvements still require two rounds of transfection for their implementation.

We recently reported one-step, auto-regulated and auto-selective multicistronic mammalian expression systems which included the tTA in a multicistronic, pTRIDENT-based or quattrocistronic configuration (pQuattro-tTA; Fussenegger et al., 1997b; Figure 2). Since the tTA gene is encoded on the multicistronic expression unit itself, little or no tTA is expressed under repressive conditions. This genetic configuration alleviates intracellular accumulation of toxic tTA levels. However, when the auto-regulated system is induced, the few tTA molecules originating from the leakiness of $P_{hCMV*-1}$ activate this tet-responsive promoter (Figure 1). Since tTA is itself encoded on the artificial operon, every round of transcription generates also an additional tTA message resulting in a positive feedback regulation system with high tTA levels and consequently high level expression of all cocistronically expressed transgenes. Since all genetic elements essential for regulated gene expression are united in a single vector, these autoregulated pTRIDENT derivatives mediate one-step regulated gene expression in various cell lines including CHO, BHK and HeLa cells (Fussenegger et al., 1997b). Previously, HeLa cells have been reported to be very sensitive to squelching, and prolonged screening procedures are usually necessary to select HeLa clones with moderate tTA expression to avoid this problem (Gossen and Bujard, 1992). To our knowledge, no convincing tet-regulation gene expression has previously been established in BHK cells apart from a recent report by Sekigushi and Hunter (1998) which shows very high background under repressed conditions and only 10-fold induction. On the contrary, our positive feedback regulation system showed both tight regulation as well as high levels of tet-responsive gene expression in all these cell lines with no signs of squelching. The lack of squelching is rather surprising, considering that the positive feedback circuit is expected to produce high intracellular levels of tTA. However, recent experiments with a monocistronic positive feedback configuration in transgenic animals also showed no detrimental effects (Shocket et al., 1995).

Positive feedback configurations with tTA in the last cistron consist of both essential and interdependent elements for regulated expression, with $P_{hCMV*-1}$ and tTA at the perimeters of the multicistronic expression unit. This set-up harbours an intrinsic, auto-selective program which guarantees full length transcripts and maintains the functional integrity of all genetic elements encoded on this autoregulatory operon. Recently, a similar autoregulated dicistronic expression system was reported (Shocket et al., 1995; Hofmann et al., 1996; Zhang et al., 1997), but only pTRIDENT-based or pQuattro-tTA systems allow one-step, autoregulated and auto-selective multigene metabolic engineering in industrially-relevant cell lines (Fussenegger et al., 1997b).

Conclusions

Since one-gene metabolic engineering will necessarily reach its limits when coping with today's increasingly complex challenges, the recent development of artificial eukaryotic operons enables effective multigene metabolic engineering of mammalian cells. This greatly expands possibilities to reprogram interconnected cellular networks in desired ways to improve key characteristics of mammalian cells. Besides use in next-generation multigene metabolic engineering, multicistronic expression units are expected to have great impact on very specific applications including (i) straightforward combinatorial evaluation of gene functions and metabolic networks, (ii) one-step transfection, selection and maintenance of difficult-to-express (multi-subunit) proteins, (iii) selection of high producer cell lines, and (iv) genetic immunisation and gene therapy in combination using sense, antisense or ribozyme technology.

Acknowledgement

This work was supported by the Bundesamt für Bildung und Wissenschaft (BBW) within the Framework IV Biotechnology Program of the European Commission and by the Swiss Priority Program in Biotechnology (SPP BioTech).

References

Alexander L, Lu HH, Gromeier M and Wimmer E (1994) Dicistronic polioviruses as expression vectors for foreign genes. AIDS Res Hum Retroviruses 10 Suppl 2: 57–60.

Al-Rubeai M, Emery AN, Chalder S and Jan DC (1992) Specific monoclonal antibody productivity and the cell-cycle comparisons of batch, continuous and perfusion cultures. Cytotechnology 9: 85–97.

Aubrecht J, Manivasakam P and Schiestl RH (1996) Controlled gene expression in mammalian cells via a regulatory cascade involving the tetracycline transactivator and lac repressor. Gene 172: 227–231.

Bailey JE (1991) Towards a science of metabolic engineering. Science 252: 1668–1675.

Bailey JE, Prati E, Jean-Mairet J, Sburlati A and Umaña P (1998) Engineering gylcosylation in animal cells. Animal cell technology, Kluwer academic publishers, The Netherlands (in press).

Belsham GJ and Brangwyn JK (1990) A region of the 5′ noncoding region of foot-and-mouth disease virus RNA directs efficient internal initiation of protein synthesis within cells: involvement with the role of L protease in translational control. 64: 5389–5395.

Belsham GJ and Sonenberg N (1996) RNA-Protein Interactions in Regulation of Pocornavirus RNA translation. Microbiological Reviews 60: 449–511.

Berlioz C, Torrent C and Darlix JL (1995) An internal ribosomal entry signal in the rat VL30 region of the Harvey murine sarcoma virus leader and its use in dicistronic retroviral vectors. 69: 6400–6407.

Bernstein J, Sella O, Le S-Y and Elroy-Stein O (1997) PDGF2/C SIS mRNA leader contains a differentiation linked internal ribosomal entry site (D IRES). J Biol Chem 272: 9356–9362.

Bonyhadi ML, Moss K, Voytovich A, Auten J, Kalfoglou C, Plavec I, Forestell S, Su L, Boehnlein E and Kaneshima H (1997) RevM10-expressing T-cells derived in vivo from transduced human hemapoietic stem-progenitor cells inhibit human immunodeficiency virus replication. J Virol 71: 4707–4716.

Borman AM, Le Mercier P, Girard M and Kean KM (1997) Comparison of picornaviral *IRES*-driven internal initiation in cultured cells of different origins. Nucl Acids Res 25: 925–932.

Castleden S, Chong H, Suszynski A, Hart I and Vile RG (1995) Co-expression of the HSV tk and cytokine genes for the gene therapy of cancer. Cancer Gene Ther 2: 246.

Chen L, Perlick H and Morgan RA (1997) Comparison of retroviral and adeno-associated viral vector designed to express human clotting Factor-IX. Human Gene Ther 8: 125–135.

Chishima T, Miyagi Y, Wang X, Yamaoka H, Shimada H, Moossa AR and Hoffman RM (1997) Cancer invasion and micrometastasis visualized in life tissue by green fluorescent protein expression. Cancer Res 57: 2042–2047.

Cotter TG and Al-Rubeai M (1995) Cell death (apoptosis) in cell culture systems. Trends in Biotechnol 13: 150–155.

Davies MV and Kaufman RJ (1992) The sequence context of the intitiation codon in the encephalomyocarditis virus leader modulates efficiency of internal translation initiation. J Virol 66: 1924–1932.

Dickson R, Abelson J, Barnes W and Reznikoff W (1975) Genetic regulation: the lac control region. Science 187: 27–35.

Dirks W, Wirth M and Hauser H (1993) Dicistronic transcription units for gene expression in mammalian cells. Gene 128: 247–249.

Dirks W, Schaper F, Kirchhoff S, Morelle C and Hauser H (1994) A multifunctional vector family for gene expression in mammlian cells. Gene 149: 387–388.

Etchison D, Milburn S, Edery I, Sonenberg N and Hershey JWB (1982) Inhibition of HeLa cell protein synthesis following poliovirus infection correlates with the proteolysis of a 220,000 dalton polypeptide associated with eukaryotic initiation factor 3 and a cap-binding complex. J Biol Chem 257: 14806–14810.

Fassati A, Serpente P, Wells DJ, Walsh FS and Dickson G (1996) Retroviral-mediated gene transfer into murine and human skeletal muscle for the correction of dystrophin deficiency. Biochem Soc Trans 24: 275.

Fieck A, Wyborski DL and Short JM (1992) Modification of the *E. coli* Lac repressor for expression in eukaryotic cells: effects of nuclear signal sequences on protein activity and nuclear accumulation. Nucl Acids Res 20: 1785–1791.

Fukushige S and Sauer B (1992) Genomic targeting with a positive-selection lox integration vector allows highly reproducible gene expression in mammalian cells. Proc Natl Acad Sci USA 89: 7905–7909.

Furth PA, Onge LS, Boger H, Gruss P, Gossen M, Kistner A, Bujard H and Hennighausen L (1994) Temporal control of gene expression in transgenic mice by a tetracycline-responsive promoter. Proc Natl Acad Sci USA 91: 9302–9306.

Fussenegger M, Mazur X, Bailey JE (1997a) A novel cytostatic process enhances the productivity of Chinese hamster ovary cells. Biotechnol Bioeng 55: 927–939.

Fussenegger M, Moser S, Mazur X, Bailey JE (1997b) Autoregulated multicistronic expression vectors provide one-step cloning of regulated product gene expression in mammalian cells. Biotech Prog 13: 733–740.

Fussenegger M, Schlatter S, Dätwyler D, Mazur X and Bailey JE (1998a) Controlled proliferation by multigene metabolic engineering enhances the productivity of chinese hamster ovary cells. Nature Biotechnology 16: 468–472.

Fussenegger M, Mazur X, Renner WA and Bailey JE (1998b) Higher productivity of growth-arrested Chinese hamster ovary (CHO) cells expressing tumor suppressor gene p27. Biotechnol Prog (in press).

Fussenegger M, Mazur X and Bailey JE (1998c) pTRIDENT, a novel vector family for tricistronic gene expression in mammalian cells. Biotech Bioeng 57: 1–10.

Gallardo HF, Tan C and Sadelain M (1997) The internal ribosomal entry site of the encephalomyocarditis virus enables reliable co-expression of two transgenes in human primary T lymphocytes. Gene Ther 4: 1115–1119.

Gan W and Rhoads RE (1996) Internal initiation of translation directed by the 5′-untranslated region of the mRNA for eIF4G, a factor involved in the picornavirus-induced switch from cap-dependent to internal initiation. J Biol Chem 271: 623–626.

Garcia-Sastre A, Muster T, Barclay WS, Percy N and Palese P (1994) Use of a mammalian internal ribosomal entry site element for expression of a foreign protein by a transfectant influenza virus. J Virol 68: 6254–6261.

Gill G and Ptashne M (1988) Negative effect of the transcriptional activator GAL4. Nature 334: 721–724.

Glass MJ, Jia XY and Summers DF (1993) Identification of the hepatitis A virus internal ribosome entry site: in vivo and in vitro analysis of bicistronic RNAs containing the HAV 5′ noncoding region. Virology 193: 842–852.

Gossen M and Bujard H (1992) Tight control of gene expression in mammalian cells by tetracycline-responsive promoters. Proc Natl Acad Sci USA 89: 5547–5551.

Greulich H and Erikson RL (1998) An analysis of Mek1 signaling in cell proliferation and transformation. J Biol Chem 273: 13280–13288.

Grignet C and Calber-Bacq CM (1995) A dicistronic vector carrying varicella-zoster virus thymidine-kinase confers BVDU sensitivity to breast cancer cells in vitro and in vivo. Cancer Gene Ther 2: 246–247

Gurtu V, Yan G and Zhang G (1996) IRES bicistronic expression vector for efficient creation of stable mammalian cell lines. Biochem Biophys Res Commun 229: 295–298.

He XS, Rivikina M and Robinson WS (1996) Construction of adenoviral and retroviral vectors coexpressing the genes encoding the hepatitis B surface antigen and B7-1 protein. Gene 175: 121–125.

Hicks GG, Shi EG, Li XM, Li CH, Pawlak M and Ruley HE (1997) Functional genomics in mice by tagged sequence mutagenesis. Nat Genet 16: 338–344.

Hoffmann A, Villalba M, Journot L and Spengler D (1997) A novel tetracycline-dependent expression vector with low basal expression and potent regulatory properties in various mammalian cell lines. Nucl Acids Res 25: 1078–1079.

Hofmann A, Nolan GP and Blau HM (1996) Rapid retroviral delivery of tetracycline-inducible genes in a single autoregulatory cassette. Proc Natl Acad Sci USA 93: 5185–5190.

Hoshimaru M, Ray J, Sah DW and Gage FH (1996) Differentiation of the immortalized adult neuronal progenitor cell line HC2S2 into neurons by regulatable suppression of the v-myc oncogene. Proc Natl Acad Sci USA 93: 1518–1523.

Hwang JJ, Scuric Z and Anderson WF (1996) Novel retroviral vector transferring a suicide gene and a selectable marker gene with enhanced gene expression by using a tetracycline-responsive expression system. J Virol 70: 8138–8141.

Hu MCT and Davidson N (1990) A combination of derepression of the lac operator-repressor system with positive induction by glucocorticoid and metal ions provides a high-level-inducible gene expression system bases on the human metallothioninie IIA promoter. Mol Cell Biol 10: 6141–6151.

Iida A, Chen ST, Friedmann T and Yee JK (1996) Inducible gene expression by retrovirus-mediated transfer of a modified tetracycline-regulated system. J Virol 70: 6054–6059.

Ivanov PA, Karpova OV, Skulachev MV, Tomashevskaya OL, Rodinova NP, Dorokhov YuL and Atabekov JG (1997) A tobamovirus genome that contains an internal ribosome entry site functional in vitro. Virology 232: 32–43.

Iizuka N, Najita L, Franzusoff A and Sarnow P (1994) Cap-dependent and cap-independent translation by internal initiation of mRNAs in cell extracts prepared from Saccharomyces cerevisiae. Mol Cell Biol 14: 7322–7330.

Israel DI and Kaufman RJ (1989) Highly inducible expression from vectors containing multiple GRES in CHO cells overexpressing the glucocorticoid receptor. Nucl Acids Res 12: 4589–4604.

Jackson RJ, Howell MT and Kaminski A (1990) the novel mechanism of initiation of picornavirus RNA translation. TIBS 15: 477–483.

Jackson RJ, Hunt SL, Reynolds JE and Kaminski A (1995) Cap-dependent and cap-independent translation: operational distinctions and mechanistic interpretations. Curr Top Microbiol Immunol 203: 1–30.

Jang SK, Krausslich HG, Nicklin MJ, Duke GM, Palmenberg AC and Wimmer E (1988) A segment of the 5′ nontranslated region of encephalomyocarditis virus RNA directs interneal entry of ribosomes during in vitro translation. Journal of Virology 62: 2636–2643.

Jenkins N and Hovey A (1993) Temperature control of growth and productivity in mutant Chinese hamster ovary cell synthesising a recombinant protein. Biotechnol. and Bioeng 42: 1029–1036.

Jenkins N, Parekh RB and James DC (1996) Getting the glycosylation right: implications for the biotechnology industry. Nature Biotechnology 14: 975–981.

Kang HM, Kang NG, Kim DG and Shin HS (1997) Dicistronic tagging of genes active in embryonic stem cells of mice. Mol Cells 7: 502–508.

Karreman S, Hauser H and Karreman C (1996) On the use of double FLB recognitions targets (FRTs) in the LTR of retroviruses for the construction of high producer cell lines. Nucleic Acids Res 24: 1616–1624.

Kaufman RJ and Sharp PA (1982) Amplification and expression of sequences cotransfected with a modular dihydrofolate reductase complementary DNA gene. J Mol Biol 159: 601–621.

Kaufman RJ, Davies MV, Wasley LC and Michnick D (1991) Improved vectors for the stable expression of foreign genes in mammalian cells by use of the untranslated leader sequence from EMC virus. Nucleic Acids Res 19: 4485–4490.

Kim DG, Kang HM, Jang SK and Shin HS (1992) Construction of a bifunctional mRNA in the mouse by using internal ribosomal entry site of the encephalomyocarditis virus. Mol Cell Biol 12: 3636–3643

Kim YH, Iida T, Fujita T, Terada S, Kitayama A, Ueda, H, Prochownik EV and Suzuki E (1998) Establishment of an apoptosis-resistant and growth-controllable cell line by transfecting with inducible antisense c-jun gene. Biotechnol Bioeng 58: 65–72.

Kirchhoff S, Koester M, Wirth M, Schaper F, Gossen M, Bujard H and Hauser H (1995) Identification of mammalian cell clones exhibiting highly regulated expression from inducible promoters. Trends Genet 11: 219–220.

Kneiper BD and Rhoads RE (1997) Cap-independent translation initiation in xenopus oocytes. Nucl Acids Res 25: 395–402.

Ko MS, Sugiyama SH and Takano T (1989) An auto-inducible vector conferring high glucocorticoid inducibility upon stable transformant cells. Gene 84: 383-389.

Kobayashi M, Yamauchi J, Tanaka A and Shimamura S (1996) Improved dicistronic mRNA expression vectors for efficient selection of transfectants highly expressing foreign genes. Biotechniques 21: 398–402.

Kolb AF and Siddell SG (1997) Expression of a recombinant monoclonal antibody from a bicistronic mRNA. Hybridoma 16: 421–426.

Koo HM, Brown AM, Kaufmann RJ, Prorock CM, Ron Y and Dougherty JP (1992) A spleen necrosis virus-based retroviral vector which expresses two genes from a dicistronic mRNA. Virology 186: 669–675.

Köster M, Kirchhoff S, Schaper F and Hauser H (1995) Proliferation control of mammalian cells by the tumor suppressor IRF-1. Cytotechnology 18: 67–75.

Krömer WJ, Carafoli E, Bailey JE (1997) Purification of the cardiac sarcoplasmic reticulum membrane protein phospholamban from recombinant *E. coli.* Eur J Biochem 248: 814–819.

Kozak M (1989) The scanning model for translation: an update. J Cell Biol 108: 229–241.

Lander ES and Schork NJ (1994) Genetic dissection of complex traits. Science 265: 2037–2048.

Lee KH, Sburlati A, Renner WA and Bailey JE (1996) Deregulated expression of cloned transcription factor E2F-1 in chinese hamster ovary cells shifts protein patterns and activates growth in protein-free medium. Biotechnol Bioeng 50: 273–279.

Liu DX and Inglis SC (1992) Internal entry of ribosomes on a tricistronic mRNA encoded by infectious bronchitis virus. Journal of Virology 66: 6143–6154.

Lopez-Lastra M, Gabus C and Darlix JL (1997) Characterization of an internal ribosomal entry segment within the 5′ leader of avian reticuloendotheliosis virus type A RNA and development of novel MLV-REV-based retroviral vectors. Human Gene Therapie 8: 1855–1865.

Lucas BK, Giere LM, DeMarco RA, Schien A, Chisholm V and Crowley CW (1996) High-level production of recombinant proteins in CHO cells using a dicistronic DHFR intron expression vector. Nucl Acids Res 24: 1774–1779.

Macejak DG and Sarnow P (1990) Translational regulation of the immunoglobulin heavy-chain binding protein mRNA. Enzyme 44: 310–319.

Macejak DG and Sarnow P (1991) Internal initiation of translation mediated by the 5′ leader of a cellular mRNA. Nature 353: 90–95.

Maga JA, Widmer G and LeBowitz JH (1995) Leishmania RNA virus 1-mediated cap independent translarion. Molecular and Cellular Biology 15: 4884–4889.

Mastrangelo AJ and Betenbaugh MJ (1998) Overcoming apoptosis: new methods for improving protein-expression systems. TIBTECH 16: 88–95.

Mattioni J, Louvion J and Picard D (1994) Regulation of protein activities by fusion to steroid binding domains. Methods Cell Biol 43: 335–352.

Morgan RA, Couture L, Elroy-Stein O, Ragheb J, Moss B and Anderson WF (1992) Retroviral vectors containing putative internal ribosome entry sites: development of a polycistronic gene transfer system and applications to human gene therapy. Nucleic Acids Res 20: 1293–1299.

Mosser DD, Caron AW, Bourget L, Jolicoeur P and Massie B (1997) Use of a dicistronic expression cassette encoding the green fluorescent protein for the screening and selection of cells expressing inducible gene products. Biotechniques 22: 150–154.

Mountford P, Zevnik B, Duwel A, Nichols J, Li M, Dani C, Robertson M, Chambers I and Smith A (1994) Dicistronic targeting constructs: reporters modifiers of mammalian gene expression. Proc Natl Acad Sci USA 91: 4303–4307.

Murakami M, Watanabe H, Niikura Y, Kameda T Saitoh K, Yamamoto M, Yokouchi Y, Kuroiwa A, Mizumoto K and Iba H (1997) High-level expression of exogenous genes by replication-competent retrovirus vectors with an internal ribosomal entry site. Gene 202: 23–29.

Müller P, Kirchhoff S and Hauser H (1998) Sustained expression in proliferation controlled BHK-21 cells. In: New Developments and New Applications in Animal Cell Technology, Merten OW, Perrin P and Griffiths JB (eds.) Kluwer Academic Publishers (in press).

No D, Yao T-P and Evans RM (1996) Ecdysone-inducible gene expression in mammalian cells and transgenic mice. Proc Natl Acad Sci USA 93: 3346–3351.

Nunoi H, Sugimoto J, Tsuruio T and Kanegasaki S (1995) Construction of a bicistronic retrovirus vector, pHa-MDR-IRES-p67, for gene therapy of p67-phox deficiency in chronic granulomatous disease. Gene Ther 2: 674.

Oh SK, Scott MP and Sarnow P (1992) Homeotic gene antennapedia messenger-RNA contains 5′-noncoding sequences that confer translational initiation by internal ribosome binding sites. Genes Dev 6: 1643–1653.

Okada H, Miyamura K, Itoh T, Hagiwara M, Wakagayashi T, Mizuno M, Colosi P, Kurtzmann G and Yoshida J (1996) Gene therapy against an experimental glyoma using adeno-associated virus vectors. Gene Ther 3: 957–964.

Papoutsakis ET (1998) Express together and conquer. Nature Biotechnology 16: 416–417.

Pelletier J and Sonenberg N (1988) Internal initiation of translation of eukaryotic mRNA directed by a sequence derived from polyovirus RNA. Nature 334: 320–325.

Poulos NE, Farmer AA, Chan KW and Stanbridge EJ (1996) Design of an novel bicistronic expression vector with demonstration of a p16INK4-induced G(1)-S block (1). Cancer Res 56: 1719–1723.

Ramage AD, Clark AJ, Smith AG, Mountford PS and Burt DW (1997) Improved EBV-based shuttle vector system: dicistronic mRNA couples the synthesis of the Epstein-Barr nuclear antigen-1 protein to neomycin resistance. Gene 197: 83–89.

Ramesh N, Kim ST, Wei MQ, Khalighi M and Osborne WR (1996) High-titer bicistronic vectors employing foot-and-mouth disease virus internal ribosome entry site. Nucleic Acids Res 24: 2697–2700.

Rees S, Coote J, Stables J, Goodson S, Harris S and Lee MG (1996) Bicistronic vector for the creation of stable mammalian cell lines that predisposes all antibiotic-resistant cells to express recombinant protein. BioTechniques 20: 48–56.

Renner WA, Lee KH, Hatzimanikatis V, Bailey JE and Eppenberger HM (1995) Recombinant cyclin E expression activates proliferation and obviates surface attachment of chinese hamster ovary (CHO) cells in protein-free medium. Biotechnol Bioeng 47: 467–482.

Rijnbrand R, Van der Straaten T, Van Rijn PA, Spaan WJ and Bredenbeek PJ (1997) Internal entry of ribosomes is directed by the 5′ noncoding region of classical swine fever virus and is dependent on the presence of an RNA pseudoknot upstream of the initiation codon. J Virol 71: 451–457.

Rivard N, L'Allemain G, Bartek J and Pouyssegur J (1996) Abrogation of p27Kip1 by cDNA antisense suppresses quiescence (G_0) state in fibroblasts. J Biol Chem 271: 18337–18341.

Rojas-Eisenring IA, Cajero-Juarez M and Angel RM (1995) Cell proteins bind to a linear polypyrimidine-rich sequence within the 5′ untranslated region of rhinovirus 14 RNA. J Virol 69: 6819–6824.

Rowan S, Ludwig RL, Haupt Y, Bates S, Lu X, Oren M and Vousden KH (1996) Specific loss of apoptotic but not cell-cycle arrest function in a human tumor derived p53 mutant. EMBO J 15: 827–838.

Rueckert RR (1996) Picornaviridae the viruses and their replication. In Fields BN, Knipe DM, Howley PM et al., (eds.) Virology 3rd ed. Lippincott-Raven Publishers, Philadelphia, pp. 609–654.

Schmid M and Wimmer E (1994) IRES-controlled protein synthesis and genome replication of poliovirus. Arch Virol Suppl 9: 279–289.

Schweinfest CW, Jorcyk CL, Fujiwara S and Papas TS (1988) A heat shock-inducible eukaryotic expression vector. Gene 71: 207–210.

Schumacher L and Wirth M (1998) Polycistronic expression vectors for biotechnology and gene therapy: the characterization of the internal ribosomal entry site (IRES) of bovine viral diarrhea virus RNA in cultured cells. Animal cell technology, Kluwer academic publishers, The Netherlands (in press).

Sekiguchi T and Hunter T (1998) Induction of growth arrest and cell death by overexpression of the cyclin-cdk inhibitor p21 in hamster BHK21 cells. Oncogene 16: 369–380.

Shatkin AJ (1985) mRNA cap binding proteins: essential factors for initiating translation. Cell 40: 223–224

Shocket P, Difilippantonio M, Hellman N and Schatz DG (1995) A modified tetracycline-regulated system provides autoregulatory, inducible gene expression in cultured cells and transgenic mice. Proc Natl Acad Sci USA 92: 6522–6526.

Smarda J and Lipsick JS (1994) Dicistronic selection for nuclear proteins in living animal cells. Gene 137: 145–149.

Stoneley M, Paulin FE, Le Quesne JP, Chappell SA and Willis AE (1998) C-myc 5′ untranslated region contains an internal ribosome entry segment. Oncogene 16: 423–428.

Suzuki E and Ollis DF (1990) Enhanced antibody production at slowed growth rates: experimental demonstration and a simple structured model. Biotechnol Prog 6: 231–236.

Teerink H, Voorman HO and Thomas AA (1995) The human insulin-like growth factor II leader 1 contains an internal ribosomal entry site. Biochim Biophys Acta 1264: 403–408.

Timchenko NA, Wilde M, Nakanishi M, Smith JR and Darlington GJ (1996) CCAAT/enhancer-binding protein α (C/EBPα) inhibits cell proliferation through the p21 (WAF-1/CIP-1/SDI-1) protein. Genes and Dev 10: 804–815.

Torrent C, Berlios C and Darlix JL (1996) Stable MLV-VL30 dicistronic retroviral vectors with a VL30 or MoMLV sequence promoting both packaging of genomic RNA and expression of the 3′ cistron. Human Gene Therapie 7: 603–612.

Trouet D, Nilius B, Voets T, Droogmans G and Eggermont J (1997) Use of a bicistronic GFP-expression vector to characterize ion channels after transfection in mammalian cells. Pfluggers Arch 434: 632–638.

Urabe M, Hasumi Y, Ogasawara Y, Matsushita T, Kamoshita N, Nomoto A, Colosi P, Kurtzman GJ, Tobita K and Ozawa K (1997) A novel dicistronic AAV vector using a short IRES segment derived from hepatitis C virus genome. Gene 200: 157–162.

Vagner S, Gensac M-C, Maret A, Bayard F, Amalric F, Prats H and Prats AC (1995) Alternative translation of human fibroblast growth factor 2 mRNA occurs by internal entry of ribosomes. Mol Cell Biol 15: 35–44.

Vasseaux G and Huxley C (1997) A dicistronic construct allows easy detection of human CFTR expression from YAC DNA in human cells. Nucl Acids Res 25: 4167–4168.

Vaulont S, Daines S and Evans M (1995) Disruption of the adenosine deaminase (ADA) gene using a dicistronic promoterless construct: production of an ADA-deficient homozygote ES cell line. Transgenic Res 4: 247–255.

Veelken H, Re D, Kulmburg P, Rosenthal FM, Mackensen A, Mertelsmann R and Lindemann A (1996) Systematic evaluation of chimeric marker genes on dicistronic transcription units for regulated expression of transgenes in vitro and in vivo. Hum Gene Ther 7: 1827–1836.

Wang C, Sarnow P and Siddiqui A (1993) Translation of human hepatitis C virus RNA in cultured cells is mediated by an internal ribosome-binding mechanism. Journal of Virology 67: 3338–3344.

Yamauchi Y, Kobayashi M and Tanaka A (1996) Application of a dicistronic mRNA expression vector to antisense RNA expression in mammalian cells. Antisense NuclAcid Drug Dev 6: 305–309.

Yang XW, Model P and Heintz N (1997) Homologous recombination based modification in E. coli and germline transmission in transgenic mice of a bacterial artificial chromosome. Nature Biotchnol 15: 859–865.

Ye X, Fong P, Iizuka N, Choate D and Cavener DR (1997) Ultrabitorax and Antennapedia 5′ untranslated regions promote developmentally regulated internal translation initiation. Molecular and Cellular Biology 17: 1714–1721.

Yoshida Y and Hamada H (1997) Adenovirus mediated inducible gene expression through tetracycline controllable transactivator with nuclear localization signal. Biochem Biophys Res Comm 230: 426–430.

Zhang Y, Katakura Y, Ohashi H and Shirahata S (1997) An autocatalytic expression system for regulated production of recombinant protein in mammalian cells. Anal Biochem 252: 286–292.

Cytotechnology **28:** 127–138, 1998.
© 1998 *Kluwer Academic Publishers. Printed in the Netherlands.*

Design, characterization and application of a minibioreactor for the culture of human hematopoietic cells under controlled conditions

Antonio De León[1], Héctor Mayani[2] & Octavio T. Ramírez[1]*

[1] *Departamento de Bioingeniería, Instituto de Biotecnología, Universidad Nacional Autónoma de México, Apdo, Postal 510-3, Cuernavaca, Morelos 62250, México*
E-mail: tonatiuh@ibt.unam.mx
[2] *Hospital de Oncología, Centro Médico Nacional Siglo XXI, México, D.F., México*

Received 27 August 1998; accepted 27 August 1998

Key words: expansion, hematopoiesis, monitoring, oxygen uptake rate, redox potential, umbilical cord blood

Abstract

The *in vitro* culture of human hematopoietic cells has recently received considerable attention due to its clinical importance. Most studies of the culture and expansion of hematopoietic cells have been performed in static cultures but only very few reports exist on the use of bioreactors where strict control of environmental variables is maintained. In this work, the design, characterization and application of a fully instrumented minibioreactor for the culture of human hematopoietic cells from umbilical cord blood is presented. The system consists of a stirred-tank reactor where cells are maintained in suspension in an homogeneous environment and without the need of a stromal feeding layer. The minibioreactor was coupled to a data acquisition and control system which continuously monitored pH, dissolved oxygen and redox potential. When operated at 75 rpm with a hanging magnetic bar (impeller-to-tank diameter ratio of 0.57), the dead and mixing times were 120 and 80 s, respectively, and the maximum response rate and volumetric oxygen transfer coefficient were 0.8 mM O_2 hr^{-1}, and 1.8 hr^{-1}, respectively. Such characteristics allowed a tight control of pH(until day 11) and dissolved oxygen at predetermined set-points, and up to a 7-fold expansion of hematopoietic progenitors was possible in cultures maintained at 20% dissolved oxygen with respect to air saturation. Growth phase and cell concentration could be inferred on-line through determinations of oxygen uptake rate and culture redox potential. Oxygen uptake rate increased during exponential growth phase to a maximum of 40 μM hr^{-1}. Such an increase closely followed the increase in concentration of hematopoietic progenitors. In contrast, culture redox potential decreased during exponential growth phase and then increased during death phase. The designed system permits not only the maintenance of controlled environmental conditions and on-line identification of fundamental culture parameters, but also the application of control strategies for improving expansion of hematopoietic cells.

Abbreviations: CFU – colony forming units; CRP – culture redox potential; t_d – dead time of the system; DO – dissolved oxygen; HHC – human hematopoietic cells; MNC – mononucleated cells; MRR – maximum response rate; t_m – mixing time; OUR – oxygen uptake rate; k_La – volumetric oxygen transfer coefficient

Introduction

In recent years, *in vitro* culture of human hematopoietic cells (HHC) has experienced a growing interest which has been fostered by various applications where the expansion of stem and progenitor cells or the generation of mature cells, such as erythrocytes and platelets, is required. A special attention has received the blood from umbilical cord due to its important advantages over other HHC sources (Mayani and Lansdorp, 1998), such as bone marrow and peripheral blood. Most of the information in the field of expansion of HHC has been generated using static

* Author for all correspondence.

cultures, usually multiwell plates, dishes, and flasks of culture volumes ranging between 0.1 to 10 mL. Particularly, the so called 'long term culture' has been widely used as an experimental tool for the study of hematopoiesis *in vitro* (Mayani et al., 1990, 1993, and 1998). However, the utility of all such systems for quantitative kinetic and metabolic studies is limited due to the reduced volume available and the impossibility of maintaining controlled and homogeneous conditions of environmental variables, including pH and dissolved oxygen (DO).

Expansion of HHC in various bioreactor configurations has been recently demonstrated. For instance, Sardonini and Wu (1993) reported the expansion of HHC in small-scale, non-instrumented air-lift, microcarrier, and agitated suspension bioreactors. In particular, their results showed that only the suspension bioreactor gave a better expansion than static cultures. Zandstra et al. (1994) also found that the use of microcarriers did not improve the performance of suspension cultures. Encapsulation in alginate beads (Levee et al., 1994) or immobilization in hollow fibers (Sardonini and Wu, 1993) has also been explored for human bone marrow cultures. However, separation of cells from these systems, as well as from microcarriers, can represent an important problem when harvesting HHC. Perfusion has been a preferred mode of operation, as depleted nutrients are continuously provided to the cells while metabolized medium is removed (Koller et al., 1993; Sandstrom et al., 1996). Although perfusion systems have been shown to perform adequately in the presence or absence of a stromal layer, several limitations and disadvantages persist. For instance, costly medium is inefficiently used and the systems are inherently heterogeneous, which can result in concentration gradients and difficulty in monitoring, controlling, cell sampling, and cell harvesting.

In contrast to static cultures and immobilized bioreactor systems, instrumented stirred-tank bioreactors are especially suited for maintaining a constant and homogeneous environment through the application of simple control algorithms. Nevertheless, very scarce information exists on the use of computer-controlled and instrumented stirred-tank bioreactors for the culture of HHC, probably due to concerns of cell fragility and complications when a stromal layer is required. To our knowledge, the only report of the culture of HHC in a stirred-tank bioreactor system is a recent study by Collins et al. (1998). A similar control system was used by Pierson et al. (1996) for the *ex vivo* expansion of human natural killer cells. In addition to the inherent concerns of using stirred-tank systems, for routine experimental purposes, the high cost of culture medium (as high as \$2 mL^{-1}, if costs of cytokines are included) needed for HHC expansion makes prohibitive the use of commercially available stirred-tank bioreactors since relatively large volumes are required. In this work, the design characterization, and application of an instrumented stirred-tank minibioreactor for the culture of HHC maintained in an homogeneous suspension without a stromal layer, is presented. The system advantages include simple automatic measurement and control of the main culture variables through a computerized algorithm, and direct cell sampling and cell harvesting. The reactor volume was minimized in order to serve as an experimental tool where costly culture medium is optimized but enough volume remains to perform detailed kinetic characterizations. The utility of the developed system is exemplified by the expansion of HHC from umbilical cord blood cells. It is shown that on-line monitoring of oxygen uptake rate and culture redox potential can give fundamental information of the stage of the culture and cell concentration.

Materials and methods

Design considerations

The number of nucleated cells recovered from an umbilical cord is very limited, typically in the range of 40 to 200 $\times 10^6$ cells. Accordingly, the main design consideration was to keep the bioreactor volume as small as possible in order to have an initial mononucleated cell concentration of at least 0.5×10^6 cells mL^{-1}. Previous reports have shown the importance of initiating cultures with a critical minimum inoculum concentration (Levee et al., 1994; Collins et al., 1996). A minimum bioreactor volume would also optimize expensive media usage. On the other hand, the bioreactor should be large enough in order to accommodate commercial-size autoclaveable pH, DO, and redox potential sensors, and to allow repeated medium sampling for detailed analytical characterization. Based on such constraints, a bioreactor was constructed in a local workshop with a nominal volume of 500 mL and a working volume in the range of 75 to 250 mL. The resulting liquid height to vessel diameter ratio was between 0.45 to 1.2. The vessel (7 cm diameter, 13.5 cm total height) was constructed

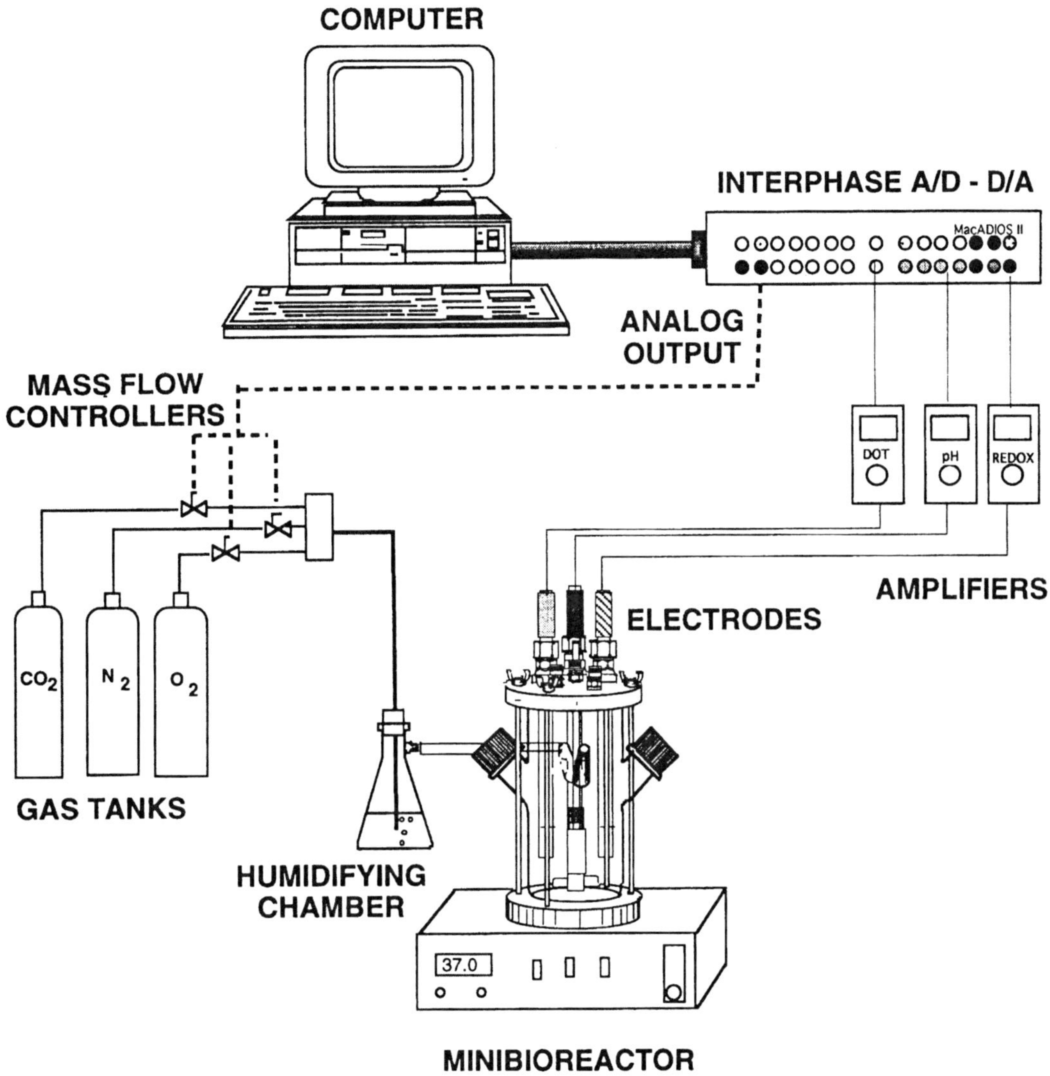

Figure 1. Schematic diagram of the minibioreactor and data acquisition and control system used for expansion of human hematopoietic cells.

in borosilicate glass and the head plate in stainless steel. Ports (including those for electrodes) were made from Swagelok-type fittings.

Bioreactor characterization

The bioreactor mixing time (t_m) was determined with a pH sensor (Ingold; 20 s response time), following the response curve after acid and base pulses. The t_m was defined as the time needed to reach 95% of final pH reading after acid or base additions. The volumetric oxygen transfer coefficient (k_La), maximum response rate (MRR), and dead time of the system (t_d) were determined from the response curve of the system as measured from DO readings upon a step change in the oxygen partial pressure in the reactor headspace, as described elsewhere (Court, 1988). The physical characterization of the bioreactor was performed in cell-free medium with agitation rates between 0 and 175 rpm, using a magnetically driven suspended bar (impeller-to-tank diameter ratio of 0.57).

Cells and culture medium

Umbilical cord blood, obtained from the Regional General Hospital No. 1 (Instituto Mexicano del Seguro Social) was processed in a Ficoll gradient as described by Koller et al. (1993), to separate mononucleated cells (MNC). MNC were seeded in 25 cm^2 T-flasks and placed in an incubator (Precision) for

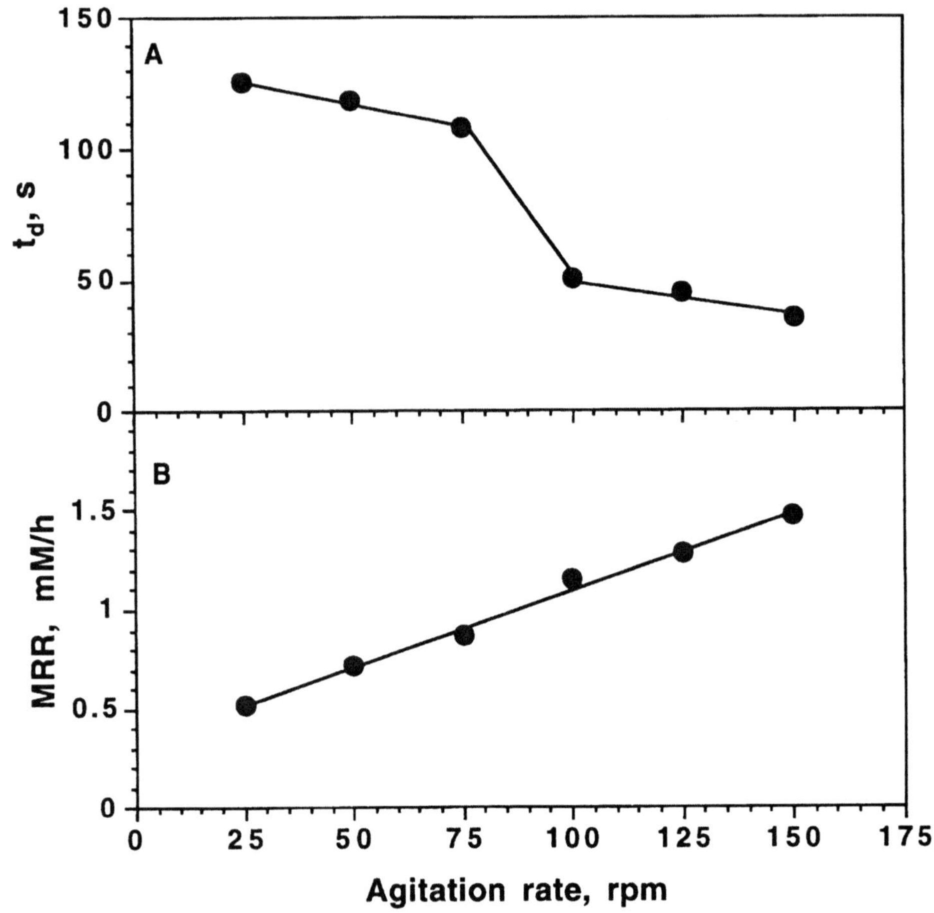

Figure 2. Effect of agitation rate on dynamic parameters of the system. A) dead time and B) maximum response rate. Continuous lines drawn only to show trend.

3 days at 37 °C and 5% CO_2, prior to their inoculation into the bioreactor or the control culture. Such a procedure was implemented as a shortened quarantine to screen for possible microbial or fungal contamination in UCB samples. The medium used was IMDM (Sigma) supplemented with 10% fetal bovine serum (Gibco), 1 ng mL^{-1} interleukin-3 (IL-3, Promega), 5 ng mL^{-1} stem cell factor (SCF, Peprotech), 1 ng mL^{-1} granulocyte-macrophage colony stimulating factor (GM-CSF, Probiomed), 3 U mL^{-1} erythropoietin (EPO, Cilag), 0.1 mg mL^{-1} streptomycin, 100 U mL^{-1} penicillin, and 0.25 μg mL^{-1} of amphotericin B (Sigma). Control cultures consisted of static 1-mL wells of a 24-well plate, inoculated from the same umbilical cord blood sample used in the bioreactor and performed simultaneously to the bioreactor cultures.

Bioreactor cultures

Batch cultures were performed using a working volume of 120 mL, agitated at 75 rpm, and maintained at 37 °C. DO was measured with a polarographic sensor (Ingold) and is reported as percentage with respect to air saturation. DO was controlled at a predetermined value of 20% with respect to air saturation by a PID algorithm which varied the oxygen composition in inlet gas through mass flow controllers for N_2, O_2 and CO_2 (Brooks Inc.) while maintaining the total gas flow rate constant at 250 mL min^{-1}. pH was measured with an Ingold electrode and was controlled at 7.2 by manipulation of CO_2 partial pressure in the inlet gas flow by a proportional control algorithm. Culture redox potential was measured with a platinum sensor (pHoenix). DO, pH, CRP, and individual gas flow rates were acquired on-line with a Macintosh II-SI computer via a MacADIOS 411 (GW Instruments) A/D D/A inter-

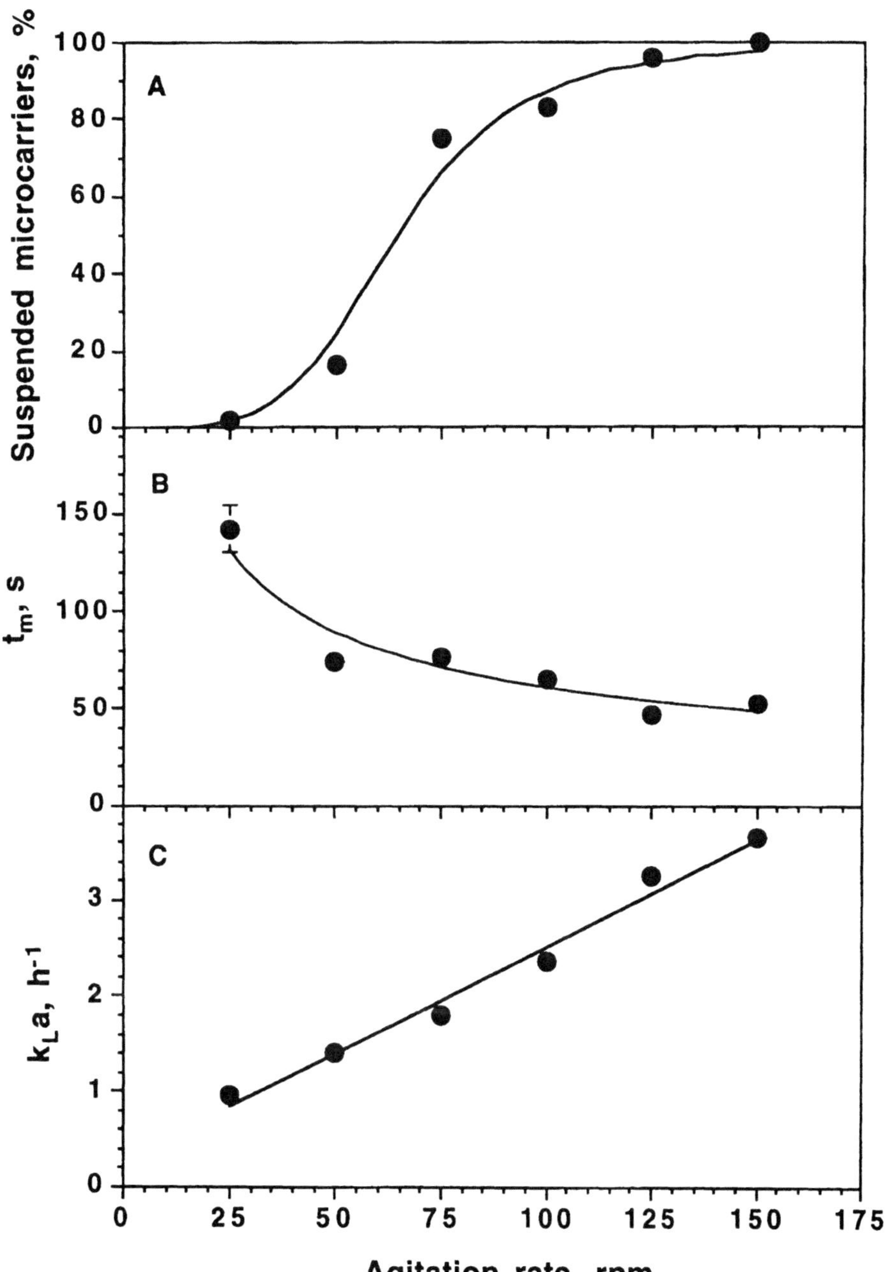

Figure 3. Effect of agitation rate on: A) percentage of suspended microcarriers (1 g L^{-1} Cytodex-1); B) mixing time; and C) volumetric oxygen transfer coefficient. The gas flow was fixed to 250 mL min^{-1}. Continuous lines drawn only to show trend. Data points correspond to average of two determinations. Error bars, except in one data point, are smaller than data symbols.

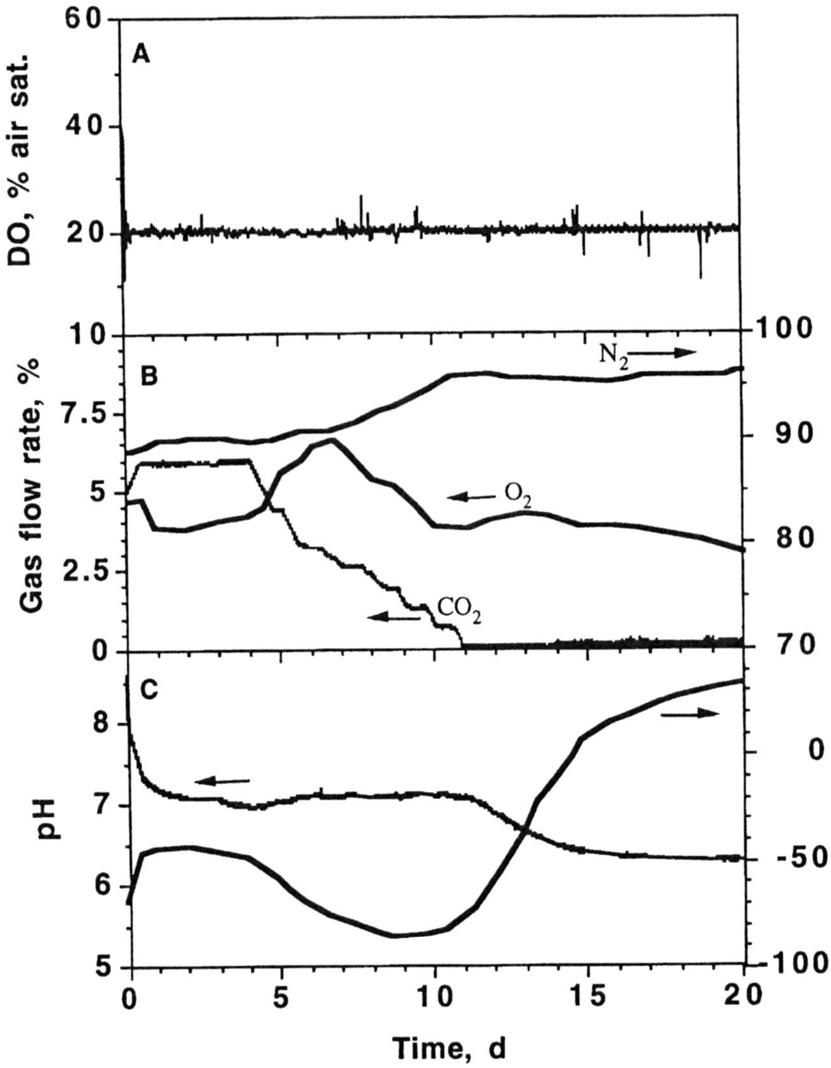

Figure 4. Typical behavior of a batch culture at 20% DO of human hematopoietic cells from umbilical cord blood. On-line acquired variables: A) DO; B) Inlet gas composition (% with respect to total gas flow rate of 250 mL min^{-1}); C) CRP and pH profiles.

phase. The control algorithms for DO and pH were written in Quick Basic® (Microsoft) and executed through the same A/D D/A interphase. Oxygen uptake rate (OUR) was determined on-line from a mass balance in the liquid phase according to Equation (1):

$$\mathrm{OUR} = \frac{k_L a}{2}\left[\left(C_i^* + C_{i+1}^*\right) - \left(C_{L_i} + C_{L_{i+1}}\right)\right] - \frac{C_{L_{i+1}} - C_{L_i}}{t_{i+1} - t_i} \quad (1)$$

where C_L and C^* are the oxygen concentration in the liquid phase and in equilibrium with the gas phase, respectively, t is time and i is the time discrete element. C_L is given directly by the DO electrode, while C^* can be calculated from the O_2 partial pressure in the bioreactor headspace. Detailed description of assumptions leading to Equation (1) can be found elsewhere (Palomares and Ramírez, 1996; Higareda et al., 1997).

Analytical methods

Clonogenic assay for progenitor cells

Number of hematopoietic colonies was determined from methylcellulose-based semisolid cultures (Metho Culture; StemCell Technologies) containing 50 ng mL^{-1} SCF, 10 ng mL^{-1} IL-3, 10 ng mL^{-1} GM-CSF, and 3 U mL^{-1} EPO. Plates were inoculated with

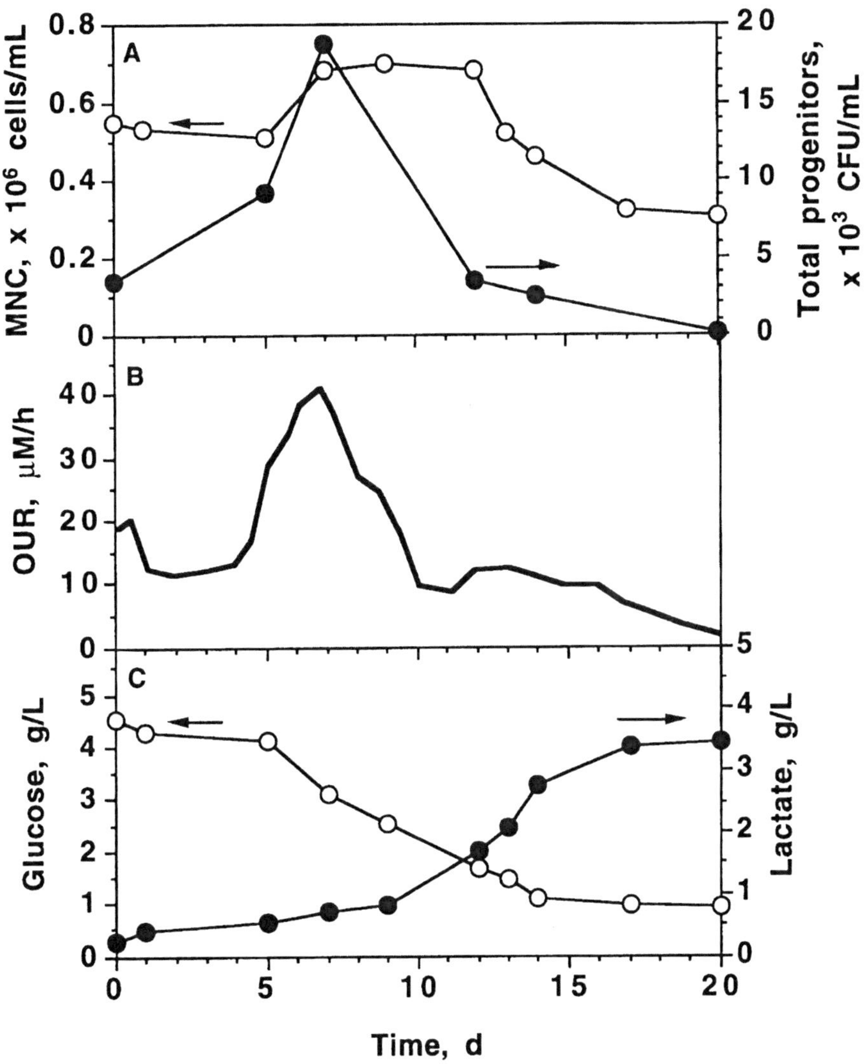

Figure 5. Typical behavior of a batch culture at 20% DO of human hematopoietic cells from umbilical cord blood. A) viable mononuclear cell concentration (○) and total progenitor concentration (●); B) oxygen uptake rate; C) glucose (○) and lactate (●) concentration.

10 000 to 40 000 cell mL^{-1} and incubated for 14 days at 37 °C and 5% CO_2. Hematopoietic colonies were classified as described previously (Mayani et al., 1990). A detailed description of the assay can be found elsewhere (Mayani et al., 1998).

Total cell concentration was determined in a Coulter Counter (Multisizer, Coulter Electronics) and viability by Trypan Blue exclusion using an hemacytometer. Glucose and lactate were determined from culture supernatants using a YSI 2700 (Yellow Spring Instruments) analyzer.

Results and discussion

In Figure 1, a schematic of the minibioreactor/data acquisition and control system designed is shown. The glass vessel included two side ports for inoculation and sampling, and two side arms for medium addition and removal for optional continuous or perfusion operation. The small volume used and the relatively large gas flow rate to volume ratio, needed for accurate OUR determinations (see Palomares and Ramírez, 1996; and Higareda et al., 1997), resulted in large amounts of medium being evaporated. This was prevented with a humidifying chamber placed at the inlet gas stream. For applications requiring the generation

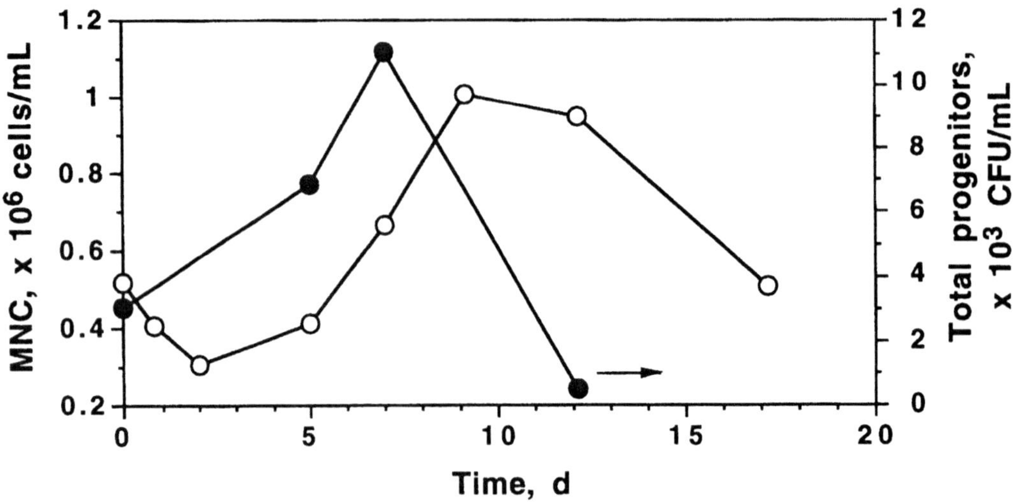

Figure 6. Static control culture performed in 1-mL wells of a 24-well-plate. Culture performed simultaneously and with same inoculum and medium as that in Figures 4 and 5. Viable mononuclear cell concentration (○) and total progenitor concentration (●). Total MNC concentration not determined.

of mature cells, such as transfusions of erythrocytes or platelets (McAdams et al., 1996b) or research involving mast cells (Saito et al., 1996), the stirred-tank configuration offers the advantage of a straightforward scale-up. In cases where the goal is transplants of hematopoietic cells in adults, then an expansion of at least 2×10^9 mononucleated cells is required (calculated from Gluckman et al. 1989 and Kohli-Kumar et al., 1993). As shown below, based on the maximum oxygen transfer rate of the minibioreactor and the specific oxygen uptake rate of HHC, a single 250-mL bioreactor culture would be enough to generate the number of mononucleated cells required for an allogenic transplant. Accordingly, for transplant applications, translation of the process to a clinical stage will not necessarily imply a volumetric increase of the culture, but rather warranting acceptability in safety and validation issues, including reproducibility. Such aspects are particularly satisfied in an homogeneous stirred-tank configuration.

Bioreactor characterization

The degree of homogeneity and the maximum oxygen transfer capacity are fundamental parameters of bioreactor performance. Therefore, a dynamic characterization of the minibioreactor was performed. The effect of agitation rate on the dead time of the system (t_d) and the maximum response rate (MRR), is shown in Figure 2. MRR and t_d indicate the rate at which the system responds, and the lag time before the system responds, respectively, to a perturbation. It can be seen that above 75 rpm, t_d decreased from around 110 s, to about 40 s, and then remained relatively constant at such a value (Figure 2a). In contrast, MRR linearly increased as agitation rate increased (Figure 2b). Determination of MRR and t_d is important since such parameters define the PID control loop constants when using classical tunning methods such as the Ziegler and Nichols technique (Court, 1988). For an agitation rate of 75 rpm, the PID tuning resulted in 2.2, 216 s, and 54 s, for the gain, integral, and derivative constants, respectively. Interestingly, the gain is similar to values reported for *E. coli* cultures, whereas the integral and derivative constants were about 1 order of magnitude higher than typical bacterial cultures (Court, 1988).

To investigate if the impeller configuration chosen was enough to maintain the cells in suspension, 1 g L^{-1} of Cytodex-1 (Pharmacia) was placed in the minibioreactor and the percentage of suspended microcarriers was measured as a function of agitation rate. As seen in Figure 3a, 25% of the microcarriers remained settled at 75 rpm, while above 125 rpm all microcarriers were suspended. The effect of agitation rate on the mixing time (t_m), and volumetric oxygen transfer coefficient (k_La) is shown in Figures 3b and c, respectively. It can be seen that t_m decreased from 140 to 50 s as agitation rate increased from 25 to 125 rpm. Similarly to MRR, the k_La linearly increased as agitation rate increased, reaching values as high as 3.5 hr^{-1} at 150 rpm. These k_La values compare very favorably with those reported for animal cell culture agitated-bioreactors in the 1- to 2-L scale range (Hu et al., 1986; Lavery and Nienow, 1987; Palomares and Ramírez, 1996). For a culture maintained at 20% DO using an O_2-enriched air source, the k_La of 1.8 hr^{-1} obtained at 75 rpm is enough to maintain a concentration of at least 8×10^6 cell mL^{-1} of cells respiring at a specific rate of 0.6×10^{-10} mmol $cell^{-1}$ h^{-1}. This specific oxygen uptake rate corresponds to the maximum value obtained in this work for total nucleated cells (see below). From the results shown in Figures 2 and 3, an agitation rate of 75 rpm was selected to perform batch cultures of HHC. Such agitation rate corresponded to an acceptable threshold value of homogeneity and dynamic characteristics of the system, but was low enough to minimize any possible damage to cells caused by excessive agitation, although this was not tested. Nevertheless, by comparing the lag phase and progenitor cells expansion data of static and bioreactor cultures, shown in the following sections, a possible hydrodynamic damage in the stirred-tank system can be discarded.

On-line monitoring and control of HHC cultures

The variables acquired on-line for a typical batch culture of HHC are shown in Figure 4. As seen in Figure 4a, the PID tunning constants derived from data in Figure 2 proved to be adequate for maintaining a tight control in DO, where only very minor perturbations were observed, usually related to external factors such as sampling. DO has been shown to be a fundamental variable which can determine differentiation of progenitor cells, proliferation of colony-forming cells, and production of mature cells and progenitors of different lineages, among other effects (Bradley, et al., 1978; Koller et al., 1992; McAdams et al., 1996a). However, very few reports, as the work shown here, have performed HHC cultures at a constant DO using a feedback control algorithm. Collins et al. (1998) and Pierson et al. (1996) have previously expanded HHC and human natural killer cells, respectively, in stirred-tanked bioreactors maintained at constant DO. It should be noted that most of the studies of the effect of DO are based on maintaining a gaseous overlay of various O_2 concentrations and without any actual measurement of DO (Bradley, et al., 1978; Koller et al., 1992). However, this is not sufficient to maintain the DO neither at a constant nor at a predetermined value. Consequently, most literature reports can only assume that a 'low' or 'high' DO exists in the HHC cultures, but the actual DO value and its variations remain unknown. Accordingly, the system presented here constitutes an important experimental tool for investigating the effect of constant DO on HHC cultures.

The inlet gas flow rates of O_2, N_2 and CO_2 are shown in Figure 4b, while the culture redox potential (CRP) and pH profiles are shown in Figure 4c. It can be seen that during the lag phase (first 3 days) the gas composition and CRP remained relatively constant. At the fourth day, upon initiation of the exponential growth phase (see progenitor cells in Figure 5a), the control loop automatically increased the O_2 flow rate in order to maintain the DO at the desired set-point and in response to the increased oxygen demand. After reaching a maximum at 7 days of culture, the O_2 flow rate decreased to 4.0% towards the end of the culture. Between days 4 and 11, accumulation of organic acids, particularly lactate, could be inferred from the decrease in CO_2 flow rate as the controller acted to maintain pH at the desired set-point (see also Figure 5c). After CO_2 depletion, control in pH was lost and decreased by almost 1.0 units. Base addition at this moment should prevent such a pH drop. This important decrease emphasizes the need for pH control, such as the one used here, as most HHC cultures performed in static or non-instrumented bioreactors are maintained in a constant CO_2 atmosphere and lack any type of pH control. In particular, it has been demonstrated that an optimum pH range exists for differentiation of HHC to specific lineages (McAdams et al., 1996a). Accordingly, the system presented here constitutes also an important tool for assessing the effect of pH in HHC cultures, and represents one of the few automatic feedback pH control system reported to date for such cells. It should be noted that HHC and human natural killer cells have previously been maintained at controlled pH in stirred-tank bioreactors by Collins et al. (1998) and Pierson et al. (1996), respectively. Finally, N_2 flow rate changed throughout the culture to maintain a constant total gas flow rate.

The various growth phases could also be identified from changes in the CRP profile. Initiation of the exponential growth phase could be inferred from a decrease in the CRP. For the culture illustrated in Figure 4, CRP decreased from –50 to –81 mV during the exponential growth phase and then increased during the death phase to a maximum of +34 mV. Similar characteristic variations in CRP and its correlation with viable cell concentration have been reported previously for hybridoma cultures at constant pH and DO (Hwang and Sinskey, 1991; Eyer and Heinzle,

1996; Higareda et al., 1997). In particular, Higareda et al. (1997) demonstrated the utility of simultaneous measurements of CRP and oxygen uptake rate (OUR) for discriminating real metabolic events from operational failures. To our knowledge, this is the first time that CRP measurements are determined, and its utility demonstrated, for a HHC culture. It should be noted that when DO and pH are maintained constant, changes in CRP solely reflect the metabolic activity of the cells. Hwang and Sinskey (1991) reported that the characteristic CRP variations could be attributed to thiol groups being excreted by animal cells. Thus, the results shown in Figure 4c are consistent with the idea that thiol production is a generalized phenomena in mammalian cell cultures. Furthermore, as established by Nernst Equation, CRP is inversely related to pH. Thus, the increase in CRP after the 11th day of culture is partially related to the decrease in pH.

In Figure 5, growth kinetics, OUR, glucose consumption, and lactate production of the same HHC culture as the one shown in Figure 4, are presented. The static control culture is shown in Figure 6. Both cultures showed a long lag phase (ca. 8 days if the preincubation in T-flask is considered). Such long lag phases have also been observed by others (Sardonini and Wu, 1993). The apparent absence of a lag phase for the progenitor cells is due to the limited number of assays performed during the first 5 days of culture. Only a small increase of mononucleated cells (MNC), was observed in the bioreactor (from 0.55 to 0.7×10^6 cell mL^{-1}) and control culture (from 0.55 to 1.01×10^6 cell mL^{-1}). Viability remained above 98% since inoculation until the end of exponential growth phase, whereas it remained between 94 and 80% during stationary phase. During death phase, viability decreased continuously until less than 20% at the end of the culture. Although cell expansion was very low for the cultures illustrated in Figures 4, 5, and 6, an expansion of MNC as high as 14-fold was possible in the minibioreactor system using a different cytokine combination (data not shown). Hematopoietic progenitors were determined for only a fraction of the samples. As seen in Figure 5a, progenitors concentration increased from 3 to 18×10^3 CFU mL^{-1} during the first 7 days of culture, and then rapidly decreased to undetectable levels towards the end of the culture. A similar behavior has been reported by Collins et al. (1997) for a variety of non-instrumented cultures of peripheral blood and umbilical cord blood. It should be noted that even though only a ca. 1.3-fold expansion in total cells was observed, a 6-fold expansion of progenitor cells was attained (Figure 5a). Furthermore, expansion of progenitor cells in the static control culture was only 3.6 fold (Figure 6), i.e., lower than in the stirred-tank system. Thus, caution should be taken when evaluating the performance of a cultivation system only on the basis of total cell expansion. For instance, a total cell expansion as high as 27-fold has been reported for human bone marrow in a suspension bioreactor, but only less than 3-fold expansion of progenitor cells could be attained (Sardonini and Wu, 1993).

OUR, as determined on-line by Equation (1) from DO and inlet gas composition data, is shown in Figure 5b. As it can be seen, OUR remained relatively constant during the first 4 days, corresponding to the lag phase of the culture. During exponential growth phase, OUR increased exponentially from 12 μM hr^{-1} to a maximum of 42 μM hr^{-1}, closely following the concentration of total hematopoietic progenitors. As hematopoietic progenitors concentration decreased, OUR also decreased to a final negligible value. A similar behavior of OUR with respect to the various growth phases of the culture has been previously reported by our group for hybridoma (Higareda et al., 1997) and insect cell cultures (Palomares and Ramírez, 1996). The importance of results shown in Figure 5 are several-fold. Firstly, OUR monitoring can be used as a very convenient tool for on-line assessment of the various growth phases of the culture. In addition, OUR has been shown to be an effective tool for application of nutrient supply strategies for optimizing hybridoma and insect cell cultures (Palomares and Ramírez, 1996; Higareda et al., 1997). Accordingly, similar strategies could also be applied in HHC cultures by using OUR as the feed-back control variable. Finally, due to their close and direct correlation, OUR can be used as a simple technique for assessing hematopoietic progenitors concentration. Collins et al. (1998) have recently shown that maximum specific oxygen uptake rate closely corresponded with the maximum percentage of progenitor or colony-forming cells. It should be noted that determination of hematopoietic progenitors is performed by a time-consuming and laborious clonogenic assay (14 days). Thus, on-line and real-time estimation of hematopoietic progenitors concentration is a very valuable tool for control and optimization of HHC cultures.

Only very scarce information of OUR determinations in HHC cultures is available. For instance, Sardonini and Wu (1993) were not able to detect OUR in hollow fiber perfusion bioreactors due to the small DO differences between the inlet and outlet streams. Peng and Palsson (1996), measured OUR in a specially designed six-well plate insert. Although accurate values were measured, the dynamic method used only allowed discrete OUR determinations for a limited number of samples during the whole culture period. Furthermore, due to the small volume of the system, each determination originated from a different well which had to be sacrificed. Similarly, Kimura et al. (1987) also determined OUR by the dynamic method from samples of a HHC line taken from a 100-mL spinner flask. Such determinations suffer from the same limitations associated to OUR measurements by the dynamic method (Higareda et al., 1997). Peng and Palsson (1996) have also determined OUR of HHC in a radial-flow parallel-plate perfusion system, but no continuous OUR data was reported. The specific OUR values, corresponding to the OUR data shown in Figure 5 (0.4 to 0.6×10^{-10} mmol cell^{-1} hr^{-1}), compare very well with data reported by Collins et al. (1998) and Peng and Palsson (1996) for other HHC cultures but are 2 to 3 times lower than the values given by Kimura et al. (1987) for a permanent HHC cell line which may have altered metabolic responses. Likewise, specific OUR for HHC obtained here are 2 to 5 times lower than for other mammalian and insect cell cultures (Palomares and Ramírez, 1996; Higareda et al., 1997).

Finally, the onset of the stationary phase with respect to MNC and sharp decrease in total progenitors occurred at glucose and lactate concentrations of 3 and 0.7 g L^{-1}, respectively (Figure 5c), and coincided with the sudden decrease in OUR (ca. 6.8 days). Thus, such a decrease in OUR cannot be attributed to glucose limitation. Furthermore, Kimura et al. (1987) have shown that specific OUR is not affected by lactate concentrations as high as 13 g L^{-1}. Accordingly, a possible inhibitory effect on OUR by the lactate concentrations obtained in this work can be ruled-out. The OUR behavior can therefore be indicative of another substrate or toxic by-product limiting HHC cultures at around the 7 day of culture, but this remains to be proved. Collins et al. (1997) have reported that specific lactate production rate, and thus the specific glucose consumption rate, was three orders of magnitude higher for hematopoietic progenitors than for cells that do not generate colonies. This suggests that OUR must be mainly influenced by the hematopoietic progenitors.

Conclusions

Blood from umbilical cord represents an exceptional source of HHC for various clinical applications if a suitable *in vitro* expansion and differentiation system is available. The minibioreactor/data acquisition and control system presented here is an important experimental tool for studying cultures of HHC from umbilical cord blood, and can constitute a very attractive system for scale-up of such cultures to a clinical stage. The main advantages of the system include an homogeneous environment and the possibility of a strict control and on-line monitoring of the various environmental variables, including pH and DO, while optimizing medium usage. In particular, characteristic variations of oxygen uptake rate and culture redox potential were determined for HHC batch cultures maintained at a constant and predetermined pH and DO values. In addition, the utility of on-line determinations of OUR and CRP was demonstrated. On-going research in our laboratory has shown that expansion of CFU-GM, total CFU, and MNC, of 10-, 9.2-, and 14-fold can be achieved in the minibioreactor system presented in this paper, which compare very favorably with other systems reported in the literature.

Acknowledgments

Technical support by M. Martínez, A. Gómez, and N. Osorio is gratefully acknowledged. Helpful discussions with E. López-Chalini and C. Berdugo. We thank the staff from the Regional General Hospital No. 1 (IMSS) for providing umbilical cord blood samples. This work was supported by Consejo Nacional de Ciencia y Tecnología (CONACYT) grant No. 102 PM, and by DGAPA IN119398 of the National University of Mexico. A.D. León acknowledges CONACyT 84201, DGAPA-UNAM and PADEP-UNAM 030527 and 202338 for financial support during his graduate studies. We are grateful to Probiomed S.A. de C.V. for donation of cytokines.

References

Bradley TR, Hodgson GS and Rosendaal M (1978) The effect of oxygen tension on haemopoietic and fibroblast cell proliferation *in vitro*. J Cell Physiol 97: 517–522.

Collins PC, Papoutsakis ET and Miller WM (1996) Adaptation of hematopoietic cell culture to stirred systems. Paper 57a. American Institute of Chemical Engineers Annual Meeting, Chicago IL, U.S.A.

Collins PC, Nielsen LK, Wong CK, Papoutsakis ET and Miller WM (1997) Real-time method for determining the colony-forming cell content of human hematopoietic cell cultures. Biotechnol Bioeng 55: 693–700.

Collins PC, Nielsen LK, Patel SD, Papoutsakis ET and Miller WM (1998) Characterization of hematopoietic cell expansion, oxygen uptake, and glycolysis in a controlled, stirred-tank bioreactor system. Biotechnol Prog 14: 466–472.

Court JR (1988) Computers in fermentation control: laboratory applications. In: Bushell ME (ed.) Computers in Fermentation Technology. Progress in Industrial Microbiology, Vol. 25, Elsevier, Amsterdam, The Netherlands, pp. 1–45.

Eyer K and Heinzle E (1996) On-line estimation of viable cells in a hybridoma culture at various DO levels using ATP balancing and redox potential measurement. Biotechnol Bioeng 49: 277–283.

Gluckman E, Broxmeyer HE, Auerbarch AD, Friedman HS, Douglas GW, Devergie A, Esperou H, Thierry D, Socie G, Lehn P, Cooper S, English D, Kurtzberg J, Bard J and Boyse EA (1989) Hematopoietic reconstitution in a patient with Fanconi's anemia by means of umbilical-cord blood from a HLA-identical sibling. The New England J Med 321: 1174–1178.

Higareda AE, Possani LD and Ramírez OT (1997) The use of culture redox potential and oxygen uptake rate for assessing glucose and glutamine depletion in hybridoma cultures. Biotechnol Bioeng 56: 555–563.

Hu WS, Meier J and Wang DIC (1986) Use of surface aerator to improve oxygen transfer in cell culture. Biotechnol Bioeng 28: 122–125.

Hwang, C and Sinskey AJ (1991) The role of oxidation-reduction potential in monitoring growth of cultured mammalian cells. In: Spier RE, Griffiths JG and Meignier B (eds.) Production of biologicals from animal cells in culture, Butterworth-Heinemann, Oxford, U.K., pp. 548–568.

Kimura T, Lijima S and Kobayashi T (1987) Effects of lactate and ammonium on the oxygen uptake rate of human cells. J Ferm Technol 65: 341–344.

Kohli-Kumar M, Shahidi NT, Broxmeyer HE, Masterson M, Delaat C, Sambrano J, Morris C, Auerbach AD and Harris RE (1993) Haemopoietic stem/progenitor cell transplant in Fanconi anemia using HLA-matched sibling umbilical cord blood cells. British J Haematol 84: 419–422.

Koller MR, Bebder JG, Miller WM and Papoutsakis ET (1992) Reduced oxygen tension increases hematopoiesis in long-term culture of human stem and progenitor cells from cord blood and bone marrow. Exp Hematol 20: 264–270.

Koller MR, Emerson SG and Palsson BO (1993) Large-scale expansion of human stem and progenitor cells from bone marrow mononuclear cells in continuous perfusion culture. Blood 82: 378–384.

Lavery M and Nienow AW (1987) Oxygen transfer in animal cell culture medium. Biotechnol Bioeng 30: 368–373.

Levee MG, Lee GM, Paek SH and Palsson BO (1994) Microencapsulated human bone marrow cultures: a potential culture system for the clonal outgrowth of hematopoietic progenitor cells. Biotechnol Bioeng 43: 734–739.

Mayani H, Guilbert LJ and Janowska-Wieczorek A (1990) Modulation of erythropoiesis and myelopoiesis by exogenous erythropoietin in human long-term marrow cultures. Exp Hematol 18: 174–179.

Mayani H, Dragowska W and Lansdorp PM (1993) Characterization of functionally distinct subpopulations of $CD34^+$ cord blood cells in serum-free long-term cultures supplemented with hematopoietic cytokines. Blood 82: 2664–2672.

Mayani H, Guitiérrez-Rodríguez M, Espinoza L, López-Chalini E, Huerta-Zepeda A, Flores E, Sánchez-Valle E, Luna-Bautista F, Valencia I and Ramírez OT (1998) Kinetics of hematopoiesis in Dexter-type long-term cultures established from human umbilical cord blood cells. Stem Cells 16: 127–135.

Mayani H and Lansdorp PM (1998) Biology of human umbilical cord blood-derived hematopoietic stem/progenitor cells. Stem Cells (in press).

McAdams TA, Miller WM and Papoutsakis ET (1996a) Hematopoietic cell culture therapies (Part I): cell culture considerations. TIBTECH 14: 341–349.

McAdams TA, Winter JN, Miller WM and Papoutsakis ET (1996b) Hematopoietic cell culture therapies (Part II): clinical aspects and applications. TIBTECH 14: 388–396.

Palomares LA and Ramírez OT (1996) The effect of dissolved oxygen tension and the utility of oxygen uptake rate in insect cell culture. Cytotechnol 22: 225–237.

Pierson BA, Europa AF, Hu WS and Miller JS (1996) Production of human natural killer cells for adoptive immunotherapy using a computer-controlled stirred-tank bioreactor. J Hematother 5: 475–483.

Peng CA and Palsson BO (1996) Determination of specific oxygen uptake rates in human hematopoietic cultures and implications for bioreactor design. Annals Biomed Eng 24: 373–381.

Sandstrom CE, Bender JG, Miller WM and Papoutsakis, ET (1996) Development of novel perfusion chamber to retain nonadherent cells and its use for comparison of human 'mobilized' peripheral blood mononuclear cell cultures with and without irradiated bone marrow stroma. Biotechnol Bioeng 50: 493–504.

Sardonini CA and Wu YJ (1993) Expansion and differentiation of human hematopoietic cells from static cultures through small-scale bioreactors. Biotechnol Prog 9: 131–137.

Saito H, Ebisawa M, Tachimoto H, Scichijo M, Fukagawa K, Matsumoto K, Likura Y, Awaji T, Tsujimoto G, Yanagida M, Uzumaki H, Takahashi G, Tsuji K and Nakahata T (1996) Selective growth of human mast cells induced by *steel* factor, Il-6, and prostaglandin E_2 from cord blood mononuclear cells. J Immunol 157: 343–350.

Zandstra PW, Eaves CJ and Piret JM (1994) Expansion of hematopoietic progenitor cell population in stirred suspension bioreactors of normal human bone marrow cells. Biotechnol 12: 909–914.

Cytotechnology **28:** 139–145, 1998.
© 1998 *Kluwer Academic Publishers. Printed in the Netherlands.*

Historical reflections on cell culture engineering

Anthony S. Lubiniecki
Biopharmaceutical Development, SmithKline Beecham Pharmaceuticals, 709 Swedeland Road, King of Prussia, PA 19406-0939, U.S.A.

Received 26 August 1998; accepted 26 August 1998

Key words: biopharmaceuticals, cell culture engineering, commercial manufacturing

Abstract

Cell culture engineering has enabled the commercial marketing of about a dozen human therapeutic products derived from rDNA technology and numerous monoclonal antibody products as well. A variety of technologies have proven useful in bringing products to the marketplace. Comparisons of the technologies available 15 years ago are contrasted with those available today. A number of improvements in unit operations have greatly improved the robustness of the processes during the past 15 years. Further evolution of the technology is expected in several directions driven by commercial and regulatory pressures. Some problems remain for the next generation of cell culture engineers to solve.

Abbreviations: BSE – bovine spongiform encephalopathy; CHO – Chinese hamster ovary; CCL – continuous cell line; CJD – Jakob Creutzfeld disease; HIV – human immunodeficiency virus; Mab – monoclonal antibody; rDNA – recombinant DNA; tPA – tissue plasminogen activator; TSE – transmissible spongiform encephalopathy.

Introduction

This article will attempt to provide a historical perspective of cell culture engineering from the viewpoint of an industrial manager trained as a cell biologist who gets up in the morning to put novel biotechnology products on the market. The term 'novel biotechnology' is used to describe those technologies which came about as a result of scientific developments in the 1970s such as hybridoma preparation and what we know today as genetic engineering. This concept of novel biotechnology also serves to distinguish these newer tools from conventional biotechnology such as fermentation, which was successfully used for preparation of medicinals, vaccines, foods and beverages, and for sanitary engineering (Miller, 1995).

The author's conventional biotechnology experiences started in the 1960s with natural interferons and virus vaccines, and his exposure to novel biotechnology started in the 1970s with hyridomas, continued with recombinant DNA (rDNA) technology in cell culture in the 1980s, and rDNA monoclonal antibodies (Mabs) in the 1990s. In that time interval, many things have changed, mostly for the better, and this can be seen in the content of this series of meetings on cell culture sponsored by the Engineering Foundation since 1984. It would now be appropriate to describe some of what has changed, what enabled the changes, and what progress has been made as an industry in terms of putting novel products on the market.

Then

Compare snapshots of where the industrial use of cell culture technology stood 15 years ago (when the first Cell Culture Engineering meeting was being planned) and today. Then, no marketed products existed from continuous cell lines (CCLs). The first rDNA (rDNA) -derived product to be approved for marketing had just happened in 1982 (human insulin) using *Escherichia coli* as an expression system. There was substantial doubt that cell culture would survive as a manufacturing technology, and there was a feeling that cell culture as a source of viral vaccines (and its many technological problems) would soon be superceded by *E. coli*-expressed rDNA products. The author joined Genentech, Inc at that time, when it was then very much a company in which *E. coli* was the predominant technology. An example of this occurred shortly after the author joined the company; when he was introduced to the head of process engineering, who asked, 'Tell me, what exactly is it that you cell culture people do with cow blood?' The nature of this question informed the author that he had a substantial education gap which would have to be closed in order to succeed!

The enthusiasm of the early days of microbial expression of rDNA proteins also reflected a general lack of appreciation for the need for large proteins to be correctly folded in order to be active, and for the ability of animal cells to fold and process highly complex proteins in ways that microbes could not do well. A famous biochemical engineer stated at a prestigious biochemical engineering meeting of the times that he had a quarter kilo of tissue plasminogen activator (tPA) in a jar which he held up for public view, which had been expressed in *E. coli*, and all of it was enzymatically inactive. In 1984, Genentech, Inc. entered human clinical trials with rDNA-derived tPA expressed in Chinese hamster ovary (CHO) cells, which was enzymatically and biologically active because it was properly folded and processed. Despite this technical success which enabled small scale human trials, it was far from clear in 1983 that cost–effective large scale animal cell technology was possible to implement for preparation of rDNA proteins. An established biochemical engineering consultant offered the opinion that six major technological miracles were needed to make it work at large scale.

Even if the technological miracles happened, it was also far from clear whether the regulatory community would approve the product for market. Major concerns existed over perceived risks of tumorigenicity for recipients of products derived from CCLs (Lubiniecki, 1987; Petricciani, 1985; Hopps, 1985). In fact, one senior regulator of the time confided to the author (years afterwards) that when he first heard the author describe the results of characterization of the CHO cells expressing tPA (basically, CHO cells are tumorigenic in nude mice and express retrovirus-like particles), he feared that the desire to investigate these products in humans would rekindle the spirited debates of 1954–1975 on the use of CCLs to prepare biologicals for human use. The then-recent discovery of human immunodeficiency virus (HIV) and its transmission to recipients of blood and plasma products and their derivatives (Curran et al., 1985) fueled concern in the regulatory community over the possibility of iatrogenic transmission of pathogenic viruses, especially retroviruses. There was also theoretically-based concern in a few ex-US regulatory agencies over the possibility of transmissible spongiform encephalopathies (TSEs) as a potential contaminant of medical products based on earlier studies of ovine scrapie and human kuru. These concerns were heightened because the unknown causative agent seemed unusually resistant to chemical or physical inactivation agents. This particular concern became more than theoretical a few years later in 1989 when bovine TSE or bovine spongiform encephalopathy (BSE) was discovered. Another problem was that no 'Points to Consider' document or ex-U.S.A. equivalent existed at that time for rDNA products, for monoclonal antibodies, or for products derived from CCLs. In short, there was no regulatory roadmap for the development of tPA as a biological product for human use. The only tools available were those of science and common sense (which fortunately were up to the task).

Now

If one were now to fast forward into the present, in 1998, most of these situations which have just been described look very different. Currently, there are 15 licensed/approved rDNA products plus 10 Mabs for *in vivo* diagnostic or therapeutic use in man which are derived from CCLs (Tables 1 and 2). In addition, there are four Mabs licensed for use in preparation of natural or rDNA-derived biologicals (Table 2). In addition, there are over 100 purified protein products created by novel biotechnology methods which are currently in clinical trials. Other things have also changed.

Some of the early Cell Culture Engineering meetings devoted substantial time to discussion of which technology was the best suited to prepare the products of novel biotechnology. In general, there were a variety of systems to choose from; sometimes it seemed as though there was at least one system for each presenter at the meeting. If one revisits these debates about bioreactor design and strategy, it is now possible to conclude that many of the beliefs about the adequacy of some of the proposed systems were perfectly correct based on the fact that multiple systems are actually used to prepare commercial supplies for regulated distribution (Table 3). Clearly, it is not only possible to use many of these systems, it is also possi-

Table 1. Licensed/approved rDNA biologicals expressed by mammalian cells

Product	Protein	Cells	Year[a]	Licensed
Activase/Actilyse	tPA	CHO	1987	Broadly
Epogen/Procrit/Eprex	Epo	CHO	1989	Broadly
Epogin/Recormon	Epo	CHO	1990	Japan, Europe
Saizen	hGH	C127	1989	Broadly
GenHevac B Pasteur	HBsAg	CHO	1989	France
HB Gamma	HBsAg	CHO	1990	Japan
Granocyte	G-CSF	CHO	1991	Japan, Europe
Recombinate	F VIII (80+90 kDa)	CHO	1992	Broadly
Kogenate	F VIII (80+90 kDa)	BHK-21	1993	Broadly
Pulmozyme	DNase I	CHO	1993	Sweden, U.S.A., Switzerland
Cerezyme	Glucocerebrosidase	CHO	1994	U.S.A., Austria, New Zealand
Gonal-F	FSH	CHO	1995	Sweden, Finland
Puregon	FSH	?	1996	Denmark
Novo Seven	F VIIa	BHK	1996	Switzerland, Europe
Avonex	IFN β	CHO	1996	U.S.A.
Bene Fix	FIX	CHO	1997	U.S.A.

[a] First licensure/approval.

Table 2. Licensed/approved MAb biologicals

Category	Product	Immunogen	Indication	Year	Licensed
Therapeutic	OKT3	CD3	GVHR	1986	Broadly
	Centoxin	Endotoxin	Sepsis	1990	Europe[a]
	ReoPro	Platelet IIb/IIIa	MI	1994	U.S.A.
	Panorex	?	Colorectal Cancer	1995	Germany
	Zenapax	IL2R	GVHR	1997	U.S.A.
	Rituxan	CD20	Non-Hodgkin B cell lymphoma	1997	U.S.A.
In vivo diagnostic	Oncoscint O/V	CEA	Cancer	1990	Europe, U.S.A.
	Myoscint	Myosin	Cardiac muscle necrosis	1989	Europe, U.S.A.
	Prosta Scint	?	Prostate cancer	1996	U.S.A.
	Verluma	?	Small cell lung cancer	1996	U.S.A.
	CEA-Scan	CEA	Colorectal cancer metastases	1996	U.S.A.
Preparative	Roferon A	IFNα2A	Purification from cell lysate	1986	Broadly
	Monoclate	Factor VIII	Purification from plasma	1987	U.S.A.
	MonoNine	Factor VIII	Purification from plasma	1992	U.S.A.
	Kogenate	Factor VIII	Purification from conditioned medium	1993	U.S.A.
In vitro diagnostic	⊛100	Various	Various	1980	Broadly

[a] Withdrawn from marketing in 1993.

ble to conclude that they are adequately cost-effective otherwise the products would probably not remain in commercial distribution. However, this does not mean that the bioreactor systems are all equal, or that any one of them is capable of making all the products – there are no data to support such conclusions. It should also be remembered that some of the technologies discussed in the past are no longer used, and some of the companies founded upon them have disappeared. A few of these now defunct companies were Endotronix, Helix Biocore, Verax, Invitron, Damon Biotech, and BioResponse. In general, it seems to be true that the large volume products tend to be made from some form of suspension culture, whether perfused or batch. It also seems clear that simpler technologies perform more consistently and robustly at large scale than more complicated technologies, again giving an advantage to batch suspension culture as typically the simplest of all the major technologies in current use at large scale.

Several technologies have been critical to the development of robust cell culture technology, upon which the current rDNA and Mab product portfolio depends. Key examples include the development of 0.1 micron cartridge filtration in the 1970s to effectively remove mycoplasma contamination from animal serum. Prior to this, large scale cell culture meant either periodic contamination or use of irradiation (often ineffective) or alkylation agents (effective but not without risk) to reduce mycoplasma infectivity. Contamination by small bacteria such as pseudomonads can also be effectively controlled by use of 0.1 micron filters. Another major advance in the 1980s was the development of serum-free medium by a number of companies to support large scale growth of hybridomas and CHO cells. Developments in sanitary design technology such as diaphragm valves and steam block devices, steam-in-place, and clean-in-place equipment also improved the robustness of large scale cell culture. Additionally, widespread use of affinity chromatography and ion exchange resins capable of base sanitization and the availability of excellent analytical tools for proteins have added further robustness and reliability to these processes and products.

On the regulatory front, substantial progress has also been made since 1983. At this time, there is no serious concern by most regulators over potential tumorigicity risks from purified products with low levels of redsidual cellular DNA derived from CCLs. World Health Organization, which pioneered scientific risk assessment for residual cellular DNA (Petricciani and Regan, 1986), recently advocated a standard of 10 ng DNA per dose (WHO Expert Committee on Biological Standardization, 1998). This is 1000-fold higher than what was acceptable in 1983. Fermentation and purification processes for some products would be designed in a substantially different manner in 1998 than they were in 1983, and some would now have substantially higher yields from process designs based on the current recommendations. Many regulatory bodies now publish useful guidance documents for novel biotechnology products, and have undertaken efforts with industry trade associations to harmonize the regulations in the United States, Europe, and Japan for these products as part of the International Conference on Harmonization (Lubiniecki, 1997). Six such harmonized guidance documents have been prepared to date for novel biotechnology products, of which five are official in those geographic areas.

HIV concerns in 1983 turned out to be well-founded. Over 300 000 US residents have died from HIV infection and sequellae; over 6000 acquired HIV infection from contaminated blood and plasma-derived products. Recently, regulatory officers and health ministry officials in several countries have been imprisoned for a perceived lack of timely governmental action to protect the blood and plasma supply and hence the public health. Product liability settlements for HIV contamination of plasma-derived factor VIII alone has totaled over $1 billion. Naturally, such actions tend to focus regulatory attention as well as industry concern on preventing recurrence of this type of problem. This concern has also carried over to the use of human plasma proteins for preparation of other products, including use in cell culture, purification, and formulation of novel biotechnology products. For example, products for approval in the European Community may use only blood or plasma products which have come from materials tested as if the materials were to be used for blood donation *by the currently recommended test*. Advances in test technology lead to periodic changes in the nature of the recommended test, which may affect products in preparation, storage, or distribution which have been prepared with blood and plasma derivatives tested under the previous version of the recommended test. Rejection of product lots or product recalls from commercial distribution have resulted from such events without substantial warning.

Recently, concerns over transmission of BSE have been heightened by discovery of a new variant of Jakob Creutzfeld disease (CJD) in relatively young residents of the United Kingdom and France, presumed to be associated with the recent epizooitic of BSE in that geographic area (Will et al., 1996). Biochemical evidence is also beginning to appear which supports the similarity of the agents causing both the human and bovine diseases (Bruce et al., 1997; Hill et al., 1997). The European Union has created legislation to ban the use of specified raw materials (essentially, neurological tissue) from medicinal products for human or animal use. No distinction is currently made over whether the specified raw materials are product excipients or used earlier in bulk manufacturing. This concern not only applies to animal–derived proteins used for cell culture, but may also be applied to gelatin and tallow derivatives used for most orally administered pharmaceuticals. At this time, there is some question as to when and how this legislation will be implemented.

In addition to concerns over BSE and new variant CJD, new information suggesting the possibility of iatrogenic transmission of CJD by blood and perhaps plasma products donated by CJD patients prior to diagnosis has led to regulatory actions. The US Food and Drug Administration has withdrawn from distribution over 1600 lots of plasma derivatives containing materials derived from CJD patients or their close relatives during the past several years, and *any other products made with those 1600 lots of plasma derivatives*. Since plasma derivatives are prepared typically from large pools of up to 40,000 donors, even a rare disease like CJD with an incidence rate of one per 10^6 per year can potentially affect many product lots. Thus, regulatory concerns over CJD and BSE have changed the debate over the use of human and animal–derived raw materials (e.g., serum, transferrin, albumin) from a theoretical discussion in 1983 to the intensely practical problem of many product lots being recalled or destroyed in 1998.

In summary, the environment in which novel biotechnology products are developed, manufactured, and marketed has changed significantly in the past 15 years. Technologies have become proven, and many regulatory issues have been largely put to rest. A few old problems like HIV still provide concerns for human cell lines, and for cell culture performed with human raw materials like transferrin. CJD and BSE have provided a further basis for concern over human and animal–derived raw materials, especially those processed under relatively mild conditions.

Table 3. Technologies used to prepare commercial products

Technology	Products
• Roller Bottles	Epo, hGH, HBsAg
• Microcarriers	Glucocerebrosidase
• Suspension	tPA, FVIII, FIX, alpha IFN, several MAbs
• Perfused suspension	FVIII, several MAbs
• Hollow fiber	*In vivo* diagnostic MAb
• Ascites	OKT3

Table 4. Licensed/approved purified natural proteins

Product	Protein	Cells	Year	License
Welferon	IFNα	Namalva	1985	Europe, Japan
Several	IFNβ	HDF	1980	Japan, Germany
Abbokinase	Urokinase	HDF	1990	Japan

Cell culture products in commercial distribution

As mentioned above, a number of purified protein products from novel biotechnology have been commercialized in the past 15 years. Table 1 shows that there are currently 15 licensed/approved rDNA products expressed in cell culture representing 12 nominal molecular entities distributed under 20 tradenames. Table 2 shows that there are six therapeutic Mabs, 5 *in vitro* diagnostic Mabs, and 4 preparative Mabs which have been licensed/approved. In 1997, the first chimeric and first humanized Mabs were licensed; these were created and developed specifically to overcome the human immune response to murine Mab determinants which limited the therapeutic utility of some early Mab investigational products used in a repeated dose regimen. Also in 1997. the first cellular therapy product (Carticell) was licensed. Table 3 shows that there are several cytokine preparations and one enzyme prepared from CCLs. In addition to these products of novel biotechnology, it should be remembered that there are also 12 viral vaccines still made from conventional biotechnology.

The first Mab product, OKT3, was licensed in 1986, and the first cell culture rDNA product, tPA, was licensed in 1987. Currently, about half of all commercialized rDNA products are expressed in cell culture, and well over half of those in the clinic are expressed in cell culture. Cell culture technology has been employed to express some very large and complicated proteins, including, for example, Factor VIII (the size of the encoded protein is 330 kDa) and Mabs (heterotetramers) and tPA (17 disulfide bonds) which cannot be expressed in active form in microbial systems at commercially feasible levels. Cell culture technology has proven adequately robust and efficient to create these commercial markets, and is no longer thought of as a curious technique associated vaguely with cow blood and headed towards extinction.

The slow but steady growth of cell culture biotechnology and biotechnology products has resulted in significant sales. In 1996, an estimated \$ 3.5 billion worth of rDNA and Mab products were sold, and about \$ 1.0 billion worth of conventional viral vaccines were sold. This represents about 2% of the worldwide pharmaceutical sales (around \$ 275 billion). In 1986, only the conventional viral vaccines were in commercial distribution. Of today's figure of \$ 3.5 billion, \$ 2.2 billion comes from a single product, erythropoeitin. The preparation and properties of many of these cell culture-expressed rDNA products have been reviewed recently elsewhere (Lubiniecki and Lupker, 1994).

It is also true that modern cell culture technology is capable of providing products which are safe in terms of risk of viral contamination. Previous generations of conventional biotechnology products (viral vaccines and plasma derivatives) were not free of such problems (Parkman, 1996). So far, there have been no product failures due to process safety concerns, reflecting a combination of cell bank characterization, raw material certification, process validation, quality control testing, and GMP compliance. While there have been a handful of incidents where viral contamination of in process materials was detected, these viral contaminants were not detectable in final product (Garnick, 1996; Burstyn, 1996). Such episodes have stimulated development of rapid PCR-based screening technology to ascertain the presence of specific viral contaminants prior to harvesting a given production fermentor (Garnick, 1996). Thus, current technology is quite robust and provides safe products.

Future trends

The recent clinical success of chimeric and humanized Mabs in overcoming the human anti-mouse antibody (HAMA) response problem, which limited treatment efficacy, is an example of ingenuity and science triumphing over natural limitations. It paves the way for additional products of this type in the future. There are also a number of truly human Mabs currently in clinical trial. One human Mab, Centoxin, was licensed in Europe in 1991, but was subsequently withdrawn for safety problems. This product was prepared using so-called trioma technology. Other technologies employed for creating human Mabs currently in clinical trial include immunization in mice whose immunoglobulin gene repertory has been knocked out and transgenically reconstituted with human immunoglobulin regions. Additional innovations will undoubtedly arise.

A number of investigational products are currently being prepared in transgenic animals. It seems likely that technical barriers to economical production systems in animals will be overcome. What is less clear at this time is what quality systems will be required to assure product quality, especially from an adventitious agent standpoint. It will take time to evaluate these methods and gain experience, but in principle there is no known reason why this cannot occur. Work is also ongoing with plant expression systems, but current product titers per unit biomass appear somewhat lower than those obtained in transgenic animals or in cell culture.

Genetic therapies based on genetic modification of cultured cells and/or genetic modifications of cells *in situ* by rDNA-modified viruses or plasmids continue to be developed. Successes are being reported with various systems, but so far, plasmid expression seem to be transient. It is not yet clear whether these approaches for therapeutic applications will lead to therapies, or whether the therapies will lead to products. Prophylactic applications such as vaccines are also under study, and here product efficacy may not be limited by transient expression.

The recent regulatory concern over CJD, TSEs, and HIV will probably continue to shift medium component selection away from animal and human materials toward chemically defined synthetic medium components. Required proteins such as insulin can be obtained from microbially-expressed microbial sources. Synthetic chemicals and plant-derived materials can substitute for natural products derived from animals or humans. While not yet regulatory requirements, concern over these risk factors can take on an unpredictable aspect. Those who have already removed the putative risk factor from their processes and products will be in a superior competitive position to those who have not yet done so if/when regulatory controls are imposed in the future.

Another trend among newer products, especially Mabs, is that doses are becoming larger. Most of the early rDNA products, especially the growth factor and hormone products, were administered in sub-milligram amounts. Many of the current new products are administered in tens or hundreds of milligrams, and a few are nearing the marketplace with regimens requiring over a gram of protein. This trend seriously moves the bar for process economics and process quality. The former puts pressure on cost of goods to keep selling price at appropriate levels, while the latter calls for higher purities for contaminants with fixed specifications per dose (e.g., DNA, endotoxin). For example, a regulatory specification of not more than 350 Endotoxin Units (EU) per 70 kg adult per 8 hr (or less) of treatment means a product specification of 350 EU mg^{-1} for a 1 mg hormone dose, but a product specification of 0.35 EU mg^{-1} for a 1 gram dose of Mab. Consistently attaining less than 0.35 EU mg^{-1} probably means having a target of 0.1 EU mg^{-1} or less, which requires substantial attention to sanitary equipment design, raw material specifications, and processing conditions to achieve on a routine basis. These trends will continue to drive the search for ever more efficient ways to grow cells, express protein, and purify protein drugs.

A final thought is that current cell culture technology allows the manufacture of every biopharmaceutical for which efficacy has been shown. In other words, there is no case where the technology has failed to deliver a commercializable product at an affordable price. Despite concerns whether cell culture technology would be cost effective, no product with good clinical data was ever kept off the market by high production costs. As stated above, this has been true in an era where doses for most biopharmaceuticals were at or below a mg per day. However, the current trend is for higher doses, especially for some Mabs, some of which are studied at doses of one gram per day. It is hoped that cell culture engineering methods will continue to improve and to be capable of providing the biopharmaceuticals of the future in a cost effective manner.

References

Bruce ME, Will RG, Ironside JW, McConnell I, Drummond D, Suttie A, McArdle L, Chree A, Hope J, Birkett C, Cousens S, Fraser H and Bostock CJ (1997) Transmission to mice indicate that 'new variant' CJD is caused by the BSE agent. Nature 389: 498–501.

Burstyn DG (1996) Contamination of genetically engineered Chinese Hamster Ovary cells. Dev Biol Standard 88: 199–203.

Curran JW, Morgan WM, Hardy AM, Jaffe HW, Darrow WW and Dowdle WR (1985) The epidemiology of AIDS: current status and future prospects. Science 229: 1352–1357.

Garnick RL (1996) Experience with viral contamination in cell culture. Dev Biol Standard 88: 49–56.

Hill AF, Desbruslais M, Joiner S, Sidle KCL, Gowland I, Collings J, Doey LJ and Lantos P (1997) The same prion strain causes vCJD and BSE. Nature 389: 448–450.

Hopps HE (1985) Cell substrate issues – A historical perspective, In: Hopps HE and Petricciani JC (eds.) Abnormal Cells, New Products, and Risk Tissue Culture Association, Gaithersburg, pp. 13–17.

Lubiniecki AS (1987) Safety considerations for cell culture-derived biologicals, In: Lydersen BK (ed.) Large Scale Cell Culture, Carl Hanser, Munich, pp. 231–247.

Lubiniecki AS (1997) Potential influence of international harmonization of pharmaceutical regulations on biopharmaceutical development. Curr Opin Biotechnology 8: 350–356.

Lubiniecki AS and Lupker JH (1994) Purified protein products of rDNA technology expressed in animal cell culture. Biologicals 22: 161–169.

Miller HI (1995) In: Lubiniecki AS and Vargo SA (eds.) Regulatory Practice for Biopharmaceutical Production, Wiley-Liss, New York, pp. 13–32.

Parkman PD (1996) Safety of biopharmaceuticals: a current perspective. Dev Biol Standard 88: 5–7.

Petricciani JC (1985) Regulatory considerations for products derived from the new biotechnology. Pharmaceutical Manufacturing 9: 31–34.

Petricciani JC and Regan PJ (1986) Risk of neoplastic transformation from cellular DNA: Calculations using the oncogene model. Dev Biol Standard 68: 43–49.

Will RG, Ironside JW, Zeidler M, Cousens SN, Estibeiro K and Alperovitch A (1996) A new variant of CJD in the UK, Lancet 347: 921–925

WHO Expert Committee on Biological Standardization, 47th Report, World Health Organization, Geneva.

Cytotechnology 28: 147–155, 1998.
© 1998 Kluwer Academic Publishers. Printed in the Netherlands.

Optimization of transient gene expression in mammalian cells and potential for scale-up using flow electroporation

Janet H. Parham*, Marie A. Iannone, Laurie K. Overton & Jeff T. Hutchins
Glaxo Wellcome, 5 Moore Drive, Research Triangle Park, NC 27709, U.S.A.
E-mail: parham-jh@glaxo.com

Received 25 August 1998; accepted 25 August 1998

Key words: cell cycle, electroporation, Epstein Barr nuclear antigen, flow electroporation, mammalian transfection, T-antigen

Abstract

The goals of this study were to identify mammalian cell lines which could be efficiently transiently-transfected and scaled-up for protein production. The transfection efficiencies of eight cell lines (NSO, NSO-TAg, CV-1, COS-7, CHO, CHO-TAg, HEK 293, and 293-EBNA) were measured using electroporation for DNA delivery and green fluorescent protein (Evans, 1996) as the reporter gene. In addition, we have evaluated the effects of stable expression of viral proteins, cell cycle manipulation, and butyrate post-treatment in small scale experiments. The cell lines varied widely in their GFP transfection efficiencies. Stable expression of simian virus 40 large T-antigen or Epstein Barr nuclear antigen failed to significantly increase transfection efficiency above that seen in the parental lines. Aphidicolin (a DNA polymerase inhibitor), which blocked cells from S or G2/M, brought about an increase in transfection efficiency in two cell lines. The primary effect of butyrate (a histone deacetylase inhibitor) post-treatment was an increased intensity of the fluorescent signal of green fluorescent protein, as measured by flow cytometry (1.0 to 4.2-fold, depending on the cell line). The combined use of aphidicolin pretreatment followed by butyrate treatment post-electroporation yielded increases in fluorescence intensities ranging from 0.9 to 6.8-fold. Based on their high transfection efficiencies in small scale experiments, rapid growth, and ability to grow in suspension culture, CHO, CHO-TAg, and 293-EBNA were selected to assess the feasibility of using flow electroporation for large-scale transfections. Using secreted placental alkaline phosphatase as a reporter, 293-EBNA cells produced the highest protein levels in both the presence and absence of butyrate. These data indicate that flow electroporation provides an efficient method of DNA delivery into large numbers of cells for mammalian protein production.

Abbreviations: APH – aphidicolin; BT – butyrate; EBNA – Epstein Barr nuclear antigen; GFP – green fluorescent protein; MESF – molecules of equivalent soluble fluorochrome; SPAP – secreted placental alkaline phosphatase; SV40 – Simian virus 40; TAg – T-antigen; TE – transfection efficiency.

Introduction

The advent of genetics strategies, mechanism-based drug design and high through-put screening in the bio-pharmaceutical industry has generated an increased need for the rapid, large-scale production of mammalian recombinant proteins. In this context, modest levels of protein are needed for antibody production, target compound screening, and *in vitro/in vivo* validation studies. Achieving high levels of protein production is dependent on efficient methods of DNA delivery and gene expression, RNA stability, translation efficiency, and proper post-translational modification. DEAE-dextran and calcium phosphate precipitation were two of the early methods developed for DNA delivery (Pari, 1997). A number of commercial cationic lipid products that facilitate the

* Author for all correspondence.

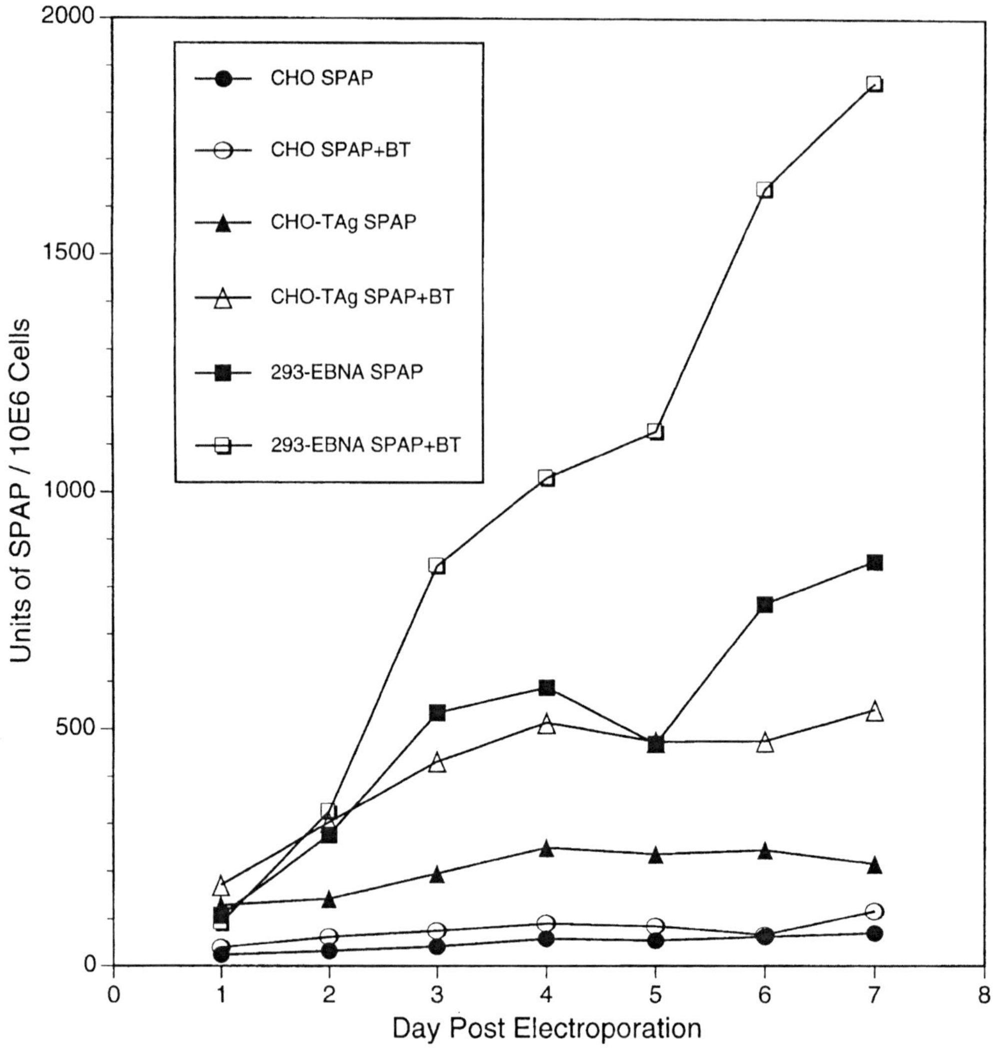

Figure 1. Large-scale SPAP electroporations were performed as described in Section Materials and Methods. Medium was sampled and filtered daily for seven days following electroporation. Samples were refrigerated and assayed at one time. Results are the average of two experiments.

uptake of DNA into mammalian cells have been developed (Hawley-Nelson, 1996; Schenborn, 1998). Other compounds currently being used to deliver DNA into cells include polyethylenimine (Boletta, 1997), polybrene/DMSO (Aubin, 1997), and viral vectors such as Semliki Forest (Berglund, 1998), adenovirus (Konesdi, 1997), and baculovirus (Possee, 1997). Alternatively, electroporation offers a non-chemical method of DNA delivery for mammalian cell transfection (Lurquin, 1997). Several of these methods, while very efficient on a small scale, are not practical for use in large scale transfections.

Mammalian cell lines vary widely in their ability to be transfected with recombinant DNA. While fibroblasts and myeloid lines are known to be difficult to transfect, cell lines such as CHO, COS-7, 293-EBNA, and BHK are in common use in transfection experiments because they take up DNA and express recombinant proteins very efficiently. A number of other factors may influence gene expression following transfection; for example, cell cycle manipulation has been shown to enhance the transfectability of cells (Nicolau, 1982; Strain 1985). In addition, SV40 large T-antigen (TAg) and Epstein Barr nuclear antigen (EBNA), multi-functional proteins which immortalize mammalian cells and increase growth potential, can also increase transcription, thereby enhancing gene expression (Harvey, 1997; Manfredi, 1994; Parker,

Table 1. Cell line comparison of transfection efficiency (TE) and fluorescence intensity (MESF) using GFP as a reporter gene

Cell line	Doubling time (hr)	Max. express.	TE (%)	Fold-increase	Fluor. intensity (MESF × 10^6)	Fold-increase
NSO	20.8	Day 1	8.43		0.22	
NSO-TAg	25.4	Day 1	9.82	1.2	0.22	1.0
CV-1	44.8	Day 2	61.03		12.67	
COS-7	41.1	Day 3	71.23	1.2	8.00	0.6
CHO	19.1	Day 2	62.42		7.08	
CHO-TAg	19.2	Day 2	79.22	1.3	8.29	1.2
293	46.7	Day 3	68.74		12.75	
293-EBNA	14.2	Day 2	82.99	1.2	5.71	0.4

Three pairs of cell lines (+ and – stably expressed T-antigen) and one pair of cell lines (+ and – stably expressed EBNA-1 of Epstein Barr virus) were compared for their transfection efficiency using GFP as a reporter gene. Doubling times were calculated over a 48 hr period. Electroporation data are expressed as the percentage of total cells expressing GFP (TE). Fold-increase represents the increase in transfection efficiency or fluorescence intensity over the parental cell line, presumably due to the presence of T-antigen or EBNA. Transfection efficiency and MESF were measured 17 hr (Day 1), 41 hr (Day 2), and 65 hr (Day 3) following electroporation. The transfection efficiency presented for each cell line was the maximum measured for that cell line over the three day post-transfection period. Results are the average of two experiments.

1996; Rice, 1993). Furthermore, protein production following transfection can be enhanced by histone deacetylase inhibitors such as butyrate and trichostatin (Kruh, 1982; Tang, 1994; Yoshida, 1990; Chen, 1997).

In these experiments we have evaluated the potential of several commercially available cell lines, as well as two cell lines developed internally, for potential use in large-scale electroporation experiments. In addition, we examined the effects of viral proteins, cell cycle manipulation, and the post-electroporation effects of butyrate on transfection efficiency and protein expression. The conditions for optimal protein production, using SPAP as a reporter gene and flow electroporation as the method of gene delivery, are described for the CHO, CHO-TAg and 293-EBNA cell lines. The selection criteria for large-scale protein production by these cell lines were rapid growth (doubling time <24 hr), high transfection efficiency by electroporation (>60%) and ability to grow in serum-free suspension culture.

Materials and methods

Cell lines

Chinese Hamster Ovary (CHO, ATCC) and CHO-TAg cells (J. Northrop, E. Tate; Affymax, Palo Alto, CA) were routinely maintained in DMEM/F12 (Life Technologies; Gaithersburg MD) + 5% Fetal Clone II (HyClone; Logan UT). African Green Monkey kidney cells (CV-1, ATCC), SV 40 transformed African Green Monkey kidney cells (COS-7, ATCC), human embryonic kidney (HEK 293, ATCC) and 293-EBNA (InVitrogen; Carlsbad CA) were maintained in DMEM/F12 + 10% Fetal Clone II. Mouse myeloma (NSO – CellTech; Berkshire UK) and NSO-TAg cells (isolated in our laboratory following transfection with the T antigen construct from J. Northrop) were grown in a defined growth medium (RD002) supplemented with 2X glutamine, lipids, peptone, and B cyclodextrin.

Chemicals

Aphidicolin (Sigma; St. Louis MO), prepared as a 1000X stock in DMSO was added to logarithmically growing cells for 24 hr. Following a 6 hr rest period, cells were harvested for cell cycle analysis or electroporation. Butyrate (Sigma), prepared as a 100X stock in distilled water and used at a final concentration of 2 mM, was added to cultures immediately following electroporation.

Table 2. Effects of aphidicolin (APH) pretreatment on cell cycle progression

Cell line	Treatment	Cell cycle phase		
		G0/G1	S	G2/M
		(%)		
NSO	Control	57.1	27.6	15.3
	APH	26.5	55.5	18.1
NSO-TAg	Control	62.4	26.9	10.7
	APH	29.3	54.8	15.8
CV-1	Control	64.2	18.4	17.4
	APH	21.2	69.4	9.4
COS-7[a]	Control	66.2	23.0	10.7
	APH	44.1	55.4	0.47
CHO	Control	54.6	28.1	17.3
	APH	0.75	81.0	13.2
CHO-TAg	Control	27.3	56.1	16.6
	APH	3.8	82.5	13.7
HEK 293	Control	56.1	19.4	24.5
	APH	55.1	34.4	10.5
293-EBNA	Control	44.5	45.1	10.4
	APH	14.6	82.3	3.1

Log-phase cultures were pre-treated with aphidicolin for 24 hr. Aphidicolin was removed and the cells were allowed to recover for 6 hr before being trypsinized for electroporation. Cells were harvested for cell cycle analysis immediately prior to electroporation. Results are average of two experiments.

[a] Cell cycle analysis demonstrated that the COS-7 cells were aneuploid, with two separate cycling populations. The COS-7 results presented in the table represent the percentages of the cell population with higher DNA content.

Electroporations

Small-scale electroporation experiments were performed using a Bio-Rad Gene Pulser (BioRad; Hercules CA). Twenty four μg of GFP (pGreen Lantern – Life Technologies; Gaithersburg MD) DNA was added to 6×10^6 cells in 0.4 mL of cold RPMI (Life Technologies) with 20% Fetal Clone II and 10 mM HEPES (Life Technologies) in 0.4 cm BioRad cuvettes. Optimal electroporation conditions for each cell line had been previously determined and were as follows: COS-7 and CV-1 (250 V, 960 μF); CHO and CHO-TAg (290 V, 960 μF); 292 and 293-EBNA (300 V, 500 μF); and NSO and NSO-TAg (1500 V, 3 μF, 2 pulses).

Large-scale electroporations were performed using an Electro Flow Porator T9000 (BTX Inc., a Division of Genetronics; San Diego CA). The pump was calibrated to pump 18.5 mL min^{-1} and the flow electroporation chamber holds 1.54 mL of medium. The instrument can be programmed to deliver an infinite number of sequential pulses. The voltage was set at 400 V and the capacitance at 2100 μF for CHO and CHO-TAg (time constant of ~11 mseconds) and at 500 V, 800 μF for 293-EBNA (~4.5 mseconds). 10^8 cells were suspended with 200 μg of plasmid DNA in 1.54 mL of electroporation medium for each pulse. The electroporation medium consisted of cold RPMI with 20% Fetal Clone II and 10 mM HEPES. Following electroporation the cells were placed in suspension culture in Techne stir flasks in 50 mL of either CHO-S-SFM II (Life Technologies) or IS 293 (Irvine Scientific; Irvine CA) at a density of 10^6 cells mL^{-1}.

Flow cytometric analysis

Cultures were harvested for analysis at various times following electroporation and cells were counted in a coulter counter. 1×10^6 cells were resuspended in 0.5% paraformaldehyde for flow cytometric analysis using a FACStar^{PLUS} (Becton Dickinson; San Jose CA) equipped with an INNOVA 90 laser (Coherent; Palo Alto CA) operated at 200 mW and tuned to an excitation wavelength of 488 nm. GFP fluorescence was determined in FL1 using a 510/20 bandpass filter. Cells were evaluated for transfection efficiency (percentage of total cells expressing GFP), as well as the fluorescence intensity of GFP. Quantitative flow cytometric analysis of GFP was conducted using Quantum Fluorescence for MESF Units of FITC (Sigma). When GFP fluorescence intensity was brighter than the standards, a neutral density filter (0.1, 1.0 or 2.0; Oriel Instruments, Straford, CT) was placed before the FL1 bandpass filter such that GFP light signal was on the same scale as the bead standards (same instrument settings). The GFP fluorescence values were converted to MESF values using a calibration plot. The filtered GFP MESF values were corrected mathematically after measuring the actual percent light of the brightest bead standard in the presence or absence of the neutral density filter.

Cells were immunofluorescently-stained for TAg using a modification of Su et al. (1997). A monoclonal antibody to TAg (isolated in house)

Table 3. Effects of aphidicolin (APH) pretreatment on electroporation efficiency and GFP fluorescence intensity

Cell line	Treatment	TE (%)	Fold-increase	Fluor. intensity	Fold-increase
NSO	Control	8.43		0.22	
	APH	16.46	2.0	0.25	1.1
NSO-TAg	Control	9.82		0.22	
	APH	23.73	2.4	0.33	1.5
CV-1	Control	61.03		12.67	
	APH	49.53	0.8	12.84	1.0
COS-7	Control	71.23		8.00	
	APH	59.22	0.8	10.55	1.3
CHO	Control	63.64		7.08	
	APH	73.12	1.2	10.87	1.5
CHO-TAg	Control	79.22		8.29	
	APH	88.59	1.1	19.79	2.4
HEK 293	Control	68.74		12.75	
	APH	61.00	0.9	15.13	1.2
293-EBNA	Control	82.99		5.71	
	APH	65.34	0.8	9.83	1.7

Log-phase cells were treated for 24 hr with 2 μg mL^{-1} aphidicolin and then allowed to recover for six hours prior to electroporation. The GFP percentages shown represent the maximal value for each cell line over the three day expression period. Fold-increase represents the increase in the total cell GFP expression or fluorescence intensity after aphidicolin pretreatment compared to the untreated control cells. Fluorescence intensity was measured as indicated in Materials and Methods and is expressed as MESF $\times$ 10^6.

was used as the primary antibody and a goat-antimouse phycoerythrin-conjugated antibody (Southern Biotechnology Associates, Inc; Birmingham AL) was used for detection and quantitation as described above. Cultures to be used for cell cycle analysis and electroporation were in log phase (50–70% confluent) at the beginning of the experiment. Cells were harvested immediately prior to electroporation, fixed in 70% ethanol, and stained later with propidium iodide (50 μg mL^{-1}) containing DNAse-free RNAse (1.0 mg mL^{-1}) in PBS. The various cell cycle phases were measured and separated according to DNA content using a FACScalibur analytical cytometer (Becton Dickinson; San Jose CA) using Modfit LT software (Verity Software House Inc.; Topsham ME).

SPAP assays

SPAP was assayed (hydrolysis of the p-nitro-phenyl phosphate substrate) using a modified version of the assay described by Berger et al. (1988). Unknown samples were read against an SPAP standard curve containing known amounts of SPAP in a SpectraMax 250 spectrophotometer and data was analyzed using SpectraMax Pro software.

Results and discussion

Cell line comparison of transfection efficiency

Eight mammalian cell lines were evaluated for their growth characteristics and transfection potential using GFP as a reporter gene. NSO, CHO, and CV-1 were

Table 4. Effects of butyrate (BT) post-treatment and combined APH pretreatment and BT post-treatment on transfection efficiency (TE) and fluorescence intensity of GFP

Cell line	Treatment	TE (%)	Fold-increase	Fluor. intensity	Fold-increase
NSO	Control	8.43		0.22	
	BT	29.60	3.5	0.29	1.3
	APH+BT	25.23	3.0	0.33	1.8
NSO-TAg	Control	9.82		0.22	
	BT	36.36	3.7	0.38	1.7
	APH+BT	35.30	3.6	0.63	2.9
CV-1	Control	61.03		12.67	
	BT	53.90	0.9	13.03	1.0
	APH+BT	42.89	0.7	10.37	0.8
COS-7	Control	71.23		8.00	
	BT	57.44	0.8	15.45	1.9
	APH+BT	35.04	0.5	13.77	1.7
CHO	Control	62.74		7.08	
	BT	64.79	1.0	12.58	1.8
	APH+BT	66.55	1.1	33.18	4.7
CHO-TAg	Control	79.22		8.29	
	BT	77.66	1.0	34.08	4.1
	APH+BT	87.17	1.1	56.77	6.8
HEK 293	Control	68.74		12.75	
	BT	63.50	0.9	31.58	2.5
	APH+BT	53.53	0.8	35.86	2.8
293-EBNA	Control	82.99		5.71	
	BT	82.97	1.0	24.09	4.2
	APH+BT	60.15	0.7	10.97	1.9

Cells were post-treated with butyrate or pretreated with aphidicolin and post-treated with butyrate as described in Materials and Methods. Transfection efficiencies (TE) shown were the maximum obtained for each cell line over the three day expression period. Fold-increase represents the increase in transfection efficiency or GFP fluorescence intensity with butyrate treatment or combined aphidicolin/butyrate treatment compared to untreated control cells. Results are the average of two experiments.

the immortalized, parental lines for NSO-TAg, CHO-TAg, and COS-7 respectively, all of which stably expressed SV40 large TAg. 293 was the immortalized, parental line for 293-EBNA which had been stably transfected with EBNA-1. GFP expression was measured by flow cytometry for three days following electroporation to determine the time of maximal expression for each cell line. The maximal transfection efficiency for each cell line was found to vary widely (Table 1). The highest transfection efficiencies were obtained with 293-EBNA and CHO-TAg (82.99 and 79.22%, respectively), followed by COS-7 (71.23%), 293 (68.74%), CHO (62.42%), and CV-1 (61.03%). NSO and NSO-TAg had very low transfection efficiencies (8.43 and 9.82%, respectively).

SV40 TAg and EBNA are origin-binding proteins that have been used to transform and immortalize cells (Manfredi and Prives, 1994; Grossman and Laimins, 1996). These processes usually involve an enhancement of growth potential, loss of dependence on growth factors, expanded life span in culture, and alterations in cell cycle – all of which may affect gene expression. CHO and NSO had doubling times of 20 hr or less and expression of TAg had no effect on doubling times (Table 1). The expression of TAg in COS cells caused only a slight growth enhancement (i.e. a small decrease in doubling time). However, expression of EBNA in 293 cells reduced the doubling time from ~47 to 14.2 hr (Table 1). The abnormally low doubling time of 293-EBNA was averaged over a 48 hr period in which there was a very rapid burst of proliferation in the first 24 hr, followed by approximately one doubling in the second 24 hr period (data not shown).

In addition, the stable expression of TAg has been reported to increase the transient expression of genes from transfected vectors (Northrop et al., 1993; Alwine, 1985; Harvey et al., 1997) and has been implicated as being a universal activator of many different transcriptional promoters (Alwine, 1985; Rice, 1993). TAg expression also amplifies the copy number of the expression plasmids boosting the protein expression level of the transfected plasmid (Harvey, 1997). In contrast, EBNA has not been shown to be associated with universal transactivation, having more in common with papillomavirus proteins than with TAg (Parker, 1996; Grossman, 1996). EBNA has been reported to enhance the replication of plasmids which include EBNA–1 and oriP (Yates, 1985). In these experiments, the stable expression of TAg in three cell lines failed to significantly enhance transfection efficiency or the fluorescence intensity of GFP (Table 1). Similarly, the expression of EBNA had no significant effect on 293 cells (Table 1). However, only one EBNA-containing cell line was tested. It should also be noted that the GFP plasmid used for the transient transfections did not contain the Epstein Barr virus origin of replication, but rather an SV40 ori and the cmv promoter region.

The fact that all four parental cell lines were already transformed and immortalized may explain why TAg and EBNA failed to affect transfection efficiencies. The expression of these viral proteins in nontransformed cells may have generated more striking differences. However, for our purposes, nontransformed cell lines were not deemed practical for use in large scale transfections.

Quantitation of T-antigen expression

To determine whether there was a correlation between the level of stable TAg expression and transfection efficiency, TAg expression was quantitated by flow-cytometric analysis concurrently with the initial transfection experiments. COS-7 showed the highest level of TAg expression (582,136 MESF), followed by CHO-TAg (181,863 MESF), and NSO-TAg (8,061 MESF). We hypothesized that cell lines expressing higher levels of T antigen would have higher transfection efficiencies than the parental lines; however (as was shown in the previous section), T antigen had no effect on transfection efficiencies.

Effects of viral proteins and aphidicolin on cell cycle and GFP expression

In murine melanocytes (Zepter, 1995), stable expression of TAg increased the percentage of cells in G2/M when compared to the parental line, and has also been reported to induce S-phase in 3T3 cells (Ogris, 1993). As shown in Table 2, TAg had no effect on the cell cycle progression of NSO cells which may be a function of the low level of expression of TAg in NSO cells since varying amounts of TAg are required to induce different cellular functions (Ogris, 1993; Price, 1994). COS-7, CHO-TAg and 293-EBNA cells showed a 2-fold increase in the percentage of cells in S phase compared to the parental lines. This increase in the percentage of S-phase cells was not accompanied by a similar increase in transfection efficiency (Table 1). It should be noted that the numbers for COS cells reflect the percentages in aneuploid cells that made up the majority of the population.

Studies have shown that cells in G2/M demonstrated enhanced transfectability (Nicolau, 1982; Strain, 1985). By blocking cell cycle progression with aphidicolin (a DNA polymerase inhibitor), we sought to determine whether increasing the percentage of cells in S or G2/M would translate into increased transfection efficiency or fluorescence intensity in our cell lines. Table 2 represents a 24 hr pretreatment with aphidicolin followed by a 6 hr rest, demonstrating the movement of the majority of the cells out of G0/G1 in the 8 cell lines tested (Table 2). Aphidicolin pretreatment enhanced transfection efficiency in only two of the eight cell lines tested – by 2.0- and 2.4-fold in NSO and NSO-TAg cells, respectively (Table 4), correlating with the 2-fold increase in S-phase cells (Table 2). Although COS, CHO and CHO-TAg cells had 2.4, 2.9-fold and 1.5-fold respective increases in the percentage of S phase cells (Table 2), their transfection efficiencies were statistically unchanged (Table 3). The fluorescence intensity, as a measure of GFP expression levels, did increase in the NSO-TAg, COS-7, CHO, CHO-TAg, HEK-293, and 293-EBNA cell lines (Table 3).

Effects of butyrate and combined aphidicolin and butyrate treatment on GFP expression

Sodium butyrate has been reported to produce many morphological and biochemical changes in cells, including increased stable and transient expression of recombinant proteins (Gorman, 1983), which may be linked to butyrate's inhibition of histone deacetylase (Kruh, 1982). In this context, we tested the effects of post-electroporation treatment with butyrate alone and in combination with aphidicolin on GFP expression levels in our eight cell lines. When compared to untreated controls, butyrate treatment alone increased transfection efficiency in two of eight cell lines (NSO – 3.5-fold and NSO-TAg – 3.7-fold), as well as GFP fluorescence intensity, indicating increased production of the GFP protein (Table 4). Five other cell lines showed increases in the fluorescent intensity of GFP (MESF) ranging from 1.3-fold in NSO to 4.1 and 4.2-fold in CHO-TAg and 293-EBNA (Table 4). Combined aphidicolin pre-treatment and butyrate post-treatment further increased the GFP fluorescence intensity in NSO-TAg, CHO, and CHO-TAg by 3.0, 4.7, and 6.8-fold respectively (Table 4). Despite the additive effect of aphidicolin and butyrate in some cell lines, aphidicolin pre-treatment was not considered to be practical for use in large-scale electroporation experiments because of the extra time and manipulation involved. In subsequent large-scale experiments, only butyrate post-treatment was used.

Large-scale electroporation of CHO, CHO-TAg and 293-EBNA cells

Based on their high transfection efficiencies and rapid growth, CHO, CHO-TAg and 293-EBNA were selected for evaluation as large-scale protein expression vehicles. As prerequisites for analysis, the cell lines had doubling times of 24 hr or less and the ability to grow in suspension culture in Techne stir flasks under serum-free conditions. Large-scale electroporation experiments were carried out using a BTX 9000 Electro Flow Porator. After optimizing electroporation conditions for each cell line (see Section Materials and Methods), levels of SPAP expression were measured in mock-transfected (no DNA) control cells and SPAP-transfected cells, with and without post-electroporation butyrate treatment (Figure 1). Cell counts taken on Day 1 following electroporation with SPAP showed 73, 86, and 71% surviving cells in CHO, CHO-TAg and 293-EBNA respectively. SPAP-transfected CHO cells secreted very little SPAP, with butyrate treatment increasing the SPAP concentration 1.7-fold on Day 7 (Figure 1). SPAP production reached a plateau in CHO-TAg on Day 4 following electroporation (250 U/10^6 cells), while butyrate treatment increased the SPAP levels 2.5-fold (Figure 1). In 293-EBNA, the expression of SPAP in the media continued to rise throughout the seven day assay period, reaching a level of 854 U/10^6 cells on Day 7; butyrate increased this concentration by 2.2-fold to 1865 U/10^6 cells (Figure 1). These results correlate with our small-scale observations measuring the transfection efficiency and fluorescence intensity of GFP. As a follow-up to the SPAP expression studies, several excreted proteins have been expressed in the mg/liter (10^9 cells) range in 293-EBNA cells using this system (data not shown).

Conclusions

In summary, electroporation utililizing a flow chamber is a efficient and practical method for performing large-scale transfections. Although several cell lines were evaluated for growth rate, transfection efficiency, and GFP fluorescence intensity, 293-EBNA was found to produce the highest levels of SPAP protein in large-scale experiments. Butyrate post-treatment was found to enhance protein production, particularly in CHO-TAg and 293-EBNA, and is now routinely included in our large-scale transfection experiments.

Acknowledgements

We would like to acknowledge Jeff Northrop and Emily Tate of Affymax for kindly providing us with the CHO-TAg cells and the TAg plasmid. Also thanks to Jeff Robbins and Martin Rink for the large-scale preparation of high quality DNA, Eric Dixon for assistance with the SPAP assays, and Tom Kost for critical review of the manuscript.

References

Alwine J (1985) Transient gene expression control: effects of transfected DNA stability and trans-activation by viral early proteins. Mol Cell Biol 5: 1034–1042.

Aubin RA, Weinfeld M, Taghavi M, Mirzayans R and Paterson MC (1997) Highly effective delivery of foreign DNA to adherent cells via polybrene/DMSO-assisted gene transfer. Methods Mol Biol 62: 319–342.

Berger J, Hauber J, Hauber R, Geiger R and Cullen BR (1988) Secreted placental alkaline phosphatase: a powerful new quantitative indicator of gene expression in eukaryotic cells. Gene 66: 1–10.

Berglund P, Sjoberg M, Garoff H, Atkins GJ, Sheahan BJ and Liljestrom P (1993) Semliki Forest virus expression system: production of conditionally infectious recombinant particles. Biotechnol 11: 916–920.

Boletta A, Benigni A, Lutz J, Remuzzi G, Soria MR and Monaco L (1997) Nonviral gene delivery to the rat kidney with polyethylenimine. Human Gene Therapy 8: 1243–1251.

Chen WY, Bailey EC, McCune SL, Dong J-Y and Townes TM (1997) Reactivation of silenced, virally transduced genes by inhibitors of histone deacetylase. PNAS 94: 5798–5803.

Evans K, Hawley-Nelson P, Schifferli K and Jessee J (1996) pGreen Lantern^{TM}-1, a superior green fluorescent protein mammalian cell transfection reporter. Focus 18: 40–42.

Gorman CM and Howard BH (1983) Expression of recombinant plasmids in mammalian cells is enhanced by sodium butyrate. Nucleic Acids Research 11: 7631–7648.

Grossman SR and Laimins LA (1996) EBNA1 and E2: a new paradigm for origin-binding proteins? Trends in Microbiology 87: 87–89.

Harvey TJ, Macnaughton TB and Gowans EJ (1997) The development of a SV40 T-antigen positive cell line of hepatic origin. Journal of Virological Methods 65: 67–74.

Hawley-Nelson P and Ciccarone V (1996) Transient transfection efficiency of human keratinocytes in two serum-free media. Focus 18: 43–44.

Konesdi I, Brough DE, Bruder JT and Wickham TJ (1997) Adenoviral vectors for gene transfer. Current Opinions in Biotechnology 8: 583–588.

Kruh J (1982) Effects of sodium butyrate, a new pharmacological agent, on cells in culture. Molecular and Cellular Biochemistry 42: 65–82.

Lurquin PF (1997) Gene transfer by electroporation. Mol Biotechnol 7: 5–35.

Manfredi JJ and Prives C (1994) The transforming activity of simian virus 40 large tumor antigen. Biochimica et Biophysica Acta 1198: 65–83.

Nicolau C and Sene C (1982) Liposome-mediated transfer in eukaryotic cells: dependence of the transfer efficiency upon the type of liposome used and the host cell cycle stage. Biochimica et Biophysica Acta 721: 185–190.

Northrop JP, Ullman KS and Crabtree GR (1993) Characterization of the nuclear and cytoplasmic components of the lymphoid-specific nuclear factor of activated T Cells (NF-AT) complex. Journal of Biological Chemistry 268: 2917–2923.

Ogris E, Mudrak I and Wintersberger E (1993) Distinct amounts of polyoma-virus large T antigen are required for different functions of the protein. Oncogene 8: 1277–1283.

Pari GS and Keown WA (1997) Experimental strategies in efficient transfection of mammalian cells: calcium phosphate and DEAE-dextran. Methods in Molecular Biology 62: 301–316.

Parker GA, Crook T, Bain M, Sara EA, Farrell PJ and Allday MJ (1996) Epstein-Barr virus nuclear antigen (EBNA)3C is an immortalizing oncoprotein with similar properties to adenovirus E1A and papillomavirus E7. Oncogene 13: 2541–2549.

Possee RD (1997) Baculoviruses as expression vectors. Current Opinions in Biotechnology 8: 569–572.

Price TNC, Moorwood K, James MR, Burke JF and Mayne LV (1994) Cell cycle progression, morphology and contact inhibition are regulated by the amount of SV40 T antigen in immortal human cells. Oncogene 9: 2897–2904.

Rice PW and Cole CN (1993) Efficient transcriptional activation of many simple modular promoters by SV40 large T antigen. Journal of Virology 67: 6689–6697.

Schenborn E, Goiffon V and Oler J (1998) An efficient new transfection reagent for eukaryotic cells: TransFast^{TM} transfection reagent. Promega Notes 65: 2–6.

Strain AJ, Wallace WAH and Wyllie AH (1985) Enhancement of DNA-mediated gene transfer by high-M_r carrier DNA in synchronized CV-1 cells. Biochemical Journal 225: 529–533.

Su J-L, Kilpatrick KE, Champion BR, Morris DC, Lehmann JM and Kost TA (1997) Fluorescent microtiter screening assay for immunochemically reactive antibodies. Biotechniques 22: 320–324.

Tang DC, Johnston SA and Carbone DP (1994) Butyrate-inducible and tumor-restricted gene expression by adenovirus vectors. Cancer Gene Therapy 1: 15–20.

Yates JL, Warren N and Sugden B (1985) Stable replication of plasmids derived from Epstein-Barr virus in various mammalian cells. Nature 313: 812–815.

Yoshida M, Kijima M, AM and Beppu T (1990) Potent and specific inhibition of mammalian histone deacetylase both *in vivo* and *in vitro* by trichostatin A. Journal of Biological Chemistry 265: 17174–17179.

Zepter K, Haffner AC, Trefzer U and Elmets CA (1995) Reduced growth factor requirements and accelerated cell-cycle kinetics in adult human melanocytes transformed with SV40 large T antigen. Journal of Investigative Dermatology 104: 755–762.

Cytotechnology **28:** 157–162, 1998.
© 1998 *Kluwer Academic Publishers. Printed in the Netherlands.*

Population balance model of *in vivo* neutrophil formation following bone marrow rescue therapy

L. K. Nielsen[1]*, J. G. Bender[2], W. M. Miller[3] & E. T. Papoutsakis[3]
[1] *Department of Chemical Engineering, The University of Queensland, Brisbane, QLD 4072, Australia*
E-mail: LarsN@cheque.uq.edu.au
[2] *Nexell Therapeutics Inc, Irvine, CA, U.S.A.*
[3] *Department of Chemical Engineering, Northwestern University, IL, U.S.A.*

Received 26 August 1998; accepted 26 August 1998

Key words: bone marrow rescue therapy, *ex vivo* expansion, *in vivo* model, neutropenia

Abstract

In this paper, we develop a simple four parameter population balance model of *in vivo* neutrophil formation following bone marrow rescue therapy. The model is used to predict the number and type of neutrophil progenitors required to abrogate the period of severe neutropenia that normally follows a bone marrow transplant. The estimated total number of 5 billion neutrophil progenitors is consistent with the value extrapolated from a human trial. The model provides a basis for designing *ex vivo* expansion protocols.

Abbreviations: HDT – high dose therapy; BMT – bone marrow transplantation; G-CSF – Granulocyte-Colony Stimulating Factor.

Introduction

Traditional chemo- and radio-therapies are dose limited due to haematological toxicity (Maraninchi, 1993). Employing 'bone marrow' transplantation (BMT) to restore haematopoiesis (blood cell formation) after therapy enables the use of high dose therapies (HDTs), i.e., therapies using 2–10 times the normal dose.

The haematopoietic tissue used for BMTs is increasingly precursor cells collected from the patient prior to HDT through mobilisation and aphaeresis (Figure 1a). The precursor cells include so-called stem cells capable of fully restoring the haematopoietic system. Mobilisation is typically achieved using low dose chemotherapy followed by daily cytokine treatment with Granulocyte-Colony Stimulating Factor (G-CSF). After approximately 10 days, the blood is enriched in precursor cells, which are collected in the mononuclear cell fraction through aphaeresis. The collected cells can be further enriched for precursor cells through CD34 sorting (precursors contain the CD34 surface marker, while other mononuclear cells do not). The final product is frozen and stored. After recovery from the mobilisation process, patients undergo HDT followed by transplantation of the stored cells (Figure 1b). After transplantation it is now common to use cytokine treatment – in particular G-CSF – to accelerate engraftment.

Despite significant improvements in BMT, patients suffer a 5–10 day period with severe neutropenia (less than 500 neutrophils per μL) (Figure 2). Neutropenia is associated with a high risk of infection, a significant cause of acute morbidity and mortality after BMT. It has been speculated that additional transplantation of more mature neutrophil progenitors may abrogate this problem and that these progenitors could be produced by *ex vivo* expansion of (part of) the haematopoietic tissue used in the BMT.

Initial human trials using cells expanded under non-optimised conditions have not led to significant reduction in neutropenia (Brugger et al., 1995; Williams et al., 1996). These trials, however, did

* Author for all correspondence.

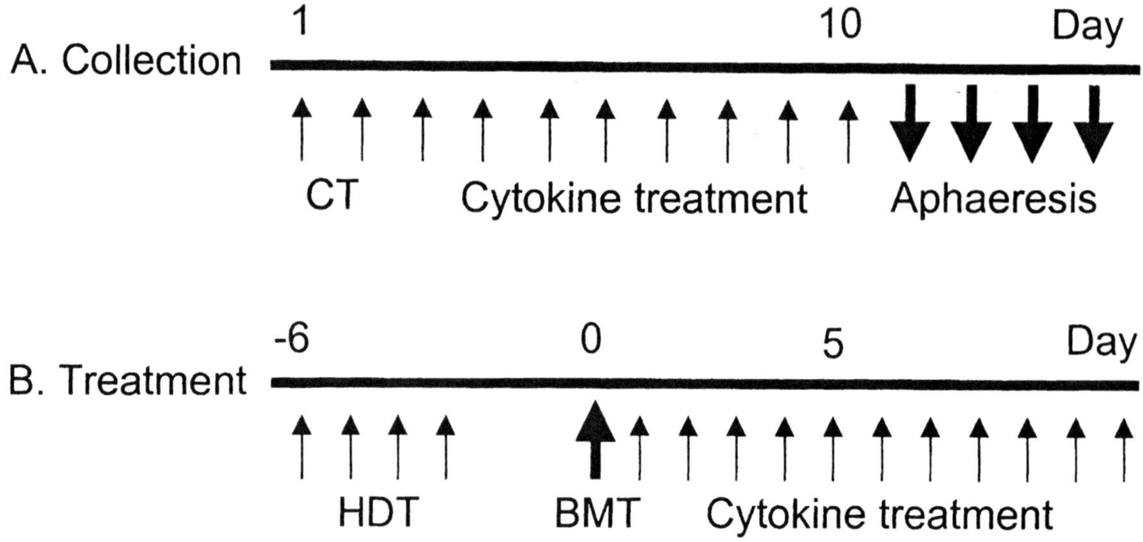

Figure 1. Collection of hematopoietic tissue (A) and treatment (B). See text for details. CT: Chemotherapy; HDT: High Dose Therapy; BMT: Bone Marrow Transplant.

not consider exactly what type of precursors and how many were required to abrogate neutropenia. Haematopoietic tissue was simply cultured for 10–12 days in a medium suitable for neutrophil formation.

Scheding et al. (1996) used a model of haematopoiesis to address the question of what type of precursors and how many are required to abrogate neutropenia. They concluded that ~40 billion neutrophil progenitors (see later) were required to abrogate neutropenia (in this case defined as less than 100 neutrophils per μL) in a normal patient. This estimate, however, is questionable.

Firstly, the basis for their model is an elaborate model of *homeostasis* in *mice*. Modelling the homeostatic response to a disturbance is very different to modelling engraftment following high dose chemotherapy. It corresponds to trying to model start up behaviour in a chemical plant with a model developed around steady state. The proposed pseudo-control loops are questionable, as is the proposed lumping of a naturally distributed process into 6 stages. These inadequacies of the model were highlighted by the need to assume 'reduced mitotic responsiveness' to make the model fit standard engraftment.

Secondly, the model was reparameterised for human neutrophil formation based on a very limited set of data, including some inaccurate data. For example, they use a neutrophil transit time of 5 hr based on data from the 60s, while a more recent publication – from the same group of people – suggests a transit time of ~24 hr and discusses the discrepancy (Steinbach et al., 1979). The parameters used by Scheding et al. led to the conclusion that under normal conditions neutrophil precursors alone amount to 800 billion cells (400 mL packed cell volume assuming a diameter of ~10 μm), which appears excessive given that the total system is estimated to contain 'only' 500–1000 billion cells (Koller and Palsson, 1993).

Finally, the simple initial condition used (a fixed reduction in all bone marrow populations) does not reflect the expected initial condition for a patient undergoing high dose therapy followed by BMT.

In this paper, no attempt is made to predict the full kinetics of engraftment. Rather, we use a simple population balance approach to estimate the potential benefit of transplanting additional *ex vivo* expanded neutrophil progenitor cells. The benefit of this approach is that the assumptions made become totally transparent, highlighting where additional information is required.

Theory

Model

Neutrophil development can be divided into a proliferative stage (P; CFU-G, blasts, promyelocytes, and proliferating myelocytes), a non-proliferative maturation stage (NP; non-proliferating myelocytes, metamyelocytes, and bands), and three stages for segment

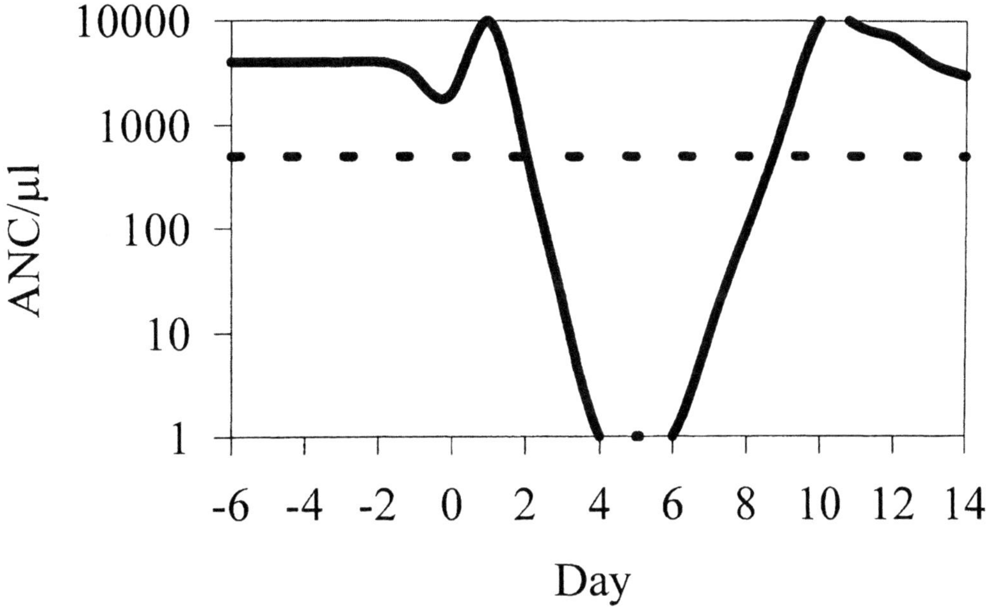

Figure 2. Idealised transplantation curve. The neutrophil count (ANC) typically drops slightly after therapy and before the BMT (day 0). The day after BMT – particularly when using G-CSF for recovery – the neutrophil number typically increases significantly, before dropping rapidly to below detection level. In a successful transplant, engraftment commences around day 6–10 and the period with severe neutropenia (dotted line) ends around day 8–12.

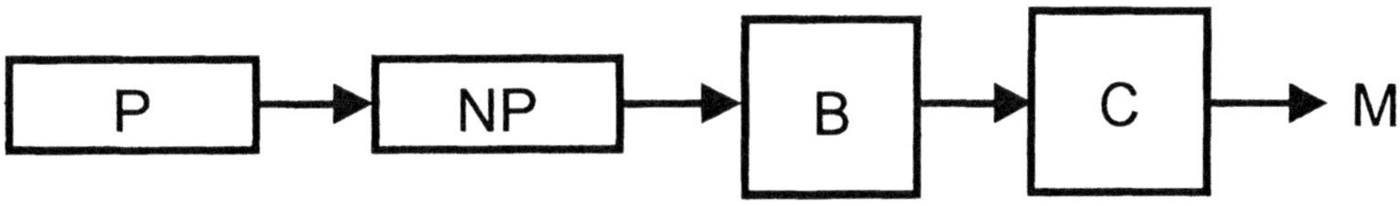

Figure 3. Model of neutrophil development (see text).

forms (neutrophils): bone marrow stage (B), circulation stage (C), and marginated stage (M) (Figure 3).

The P and NP stages are distributed stages, i.e., there is a deterministic progression through these stages. In contrast, labelling studies of the transit time of the circulating segment forms show the characteristic exponential decay (rather than linear) corresponding to an undistributed stage (Athens, 1969). In an undistributed stage, a cell has no 'memory' of how long it has been in a stage and can potentially leave the stage immediately after entering it. The three segment form stages will all be treated as undistributed.

The undistributed stages are described with ordinary differential equations

$$\frac{dB}{dt} = r_{NP}^{out} - \frac{B}{\tau_B} \tag{1}$$

$$\frac{dC}{dt} = \frac{B}{\tau_B} - \frac{C}{\tau_C} \tag{2}$$

where r_{NP}^{out} is the rate of cells leaving the non-proliferative precursor stage, τ_B is mean transit time for the bone marrow segment form stage, and τ_C is mean transit time for the circulating segment form stage. We do not consider the marginated stage as the kinetics are not well understood due to the inherent difficulty of sampling tissue.

The non-proliferative stage acts as a delay between the proliferative stage and the segment forms. If the transit time is τ_{NP}, we have that

$$r_{NP}^{out}(t) = r_{NP}^{in}(t - \tau_{NP}) \tag{3}$$

Progression through the proliferative stage will be defined in terms of residual number of divisions and the doubling time, T_2, which will be assumed constant throughout the stage. Thus, we can define relationships such as

Table 1. Parameters for myelopoiesis. Data are derived from Schmitz et al. (1993) and Price et al. (1996)

	Normal	G-CSF stimulated
	(Days)	
T_2	1	0.5
τ_{NP}	5	2.5
τ_B	3	0.5
τ_C	1	1

Table 2. Estimated number and type of precursors required to produce a flux of 2.5×10^9 cells day^{-1}

Division	2^n	Effective period	Total cells
0	1	1.0–2.5	3.8×10^9
1	2	2.5–3.0	6.3×10^8
2	4	3.0–3.5	3.1×10^8
3	8	3.5–4.0	1.6×10^8
4	16	4.0–4.5	7.8×10^7
5	32	4.5–5.0	3.9×10^7
6	64	5.0–5.5	2.0×10^7
7	128	5.5–6.0	9.8×10^6
8	256	6.0–6.5	4.9×10^6
9	512	6.5–7.0	2.4×10^6
10	1024	7.0–7.5	1.2×10^6
11	2048	7.5–8.0	6.1×10^5
Total			5.0×10^9

$$r_{NP}^{in}(t) = 2r_{P1}^{out}(t) = 2r_{P1}^{in}(t - T_2) = 4r_{P2}^{out}(t - T_2) = 4r_{P2}^{in}(t - 2T_2) = \ldots = 2^k r_{Pk}^{out}(t-[k-1]T_2) = 2^k r_{Pk}^{in}(t-kT_2) \quad (4)$$

where we have defined division as occurring upon exit of each substage, r_{Pk}^{out} is the rate of cells leaving stage k, and r_{Pk}^{in} is the rate of cells entering stage k.

Parameters

Under severe neutropenia and with co-stimulation from exogenous G-CSF, we expect neutrophil development to occur at near maximum rates. There are no data for the kinetics during engraftment, though it is well established that G-CSF accelerates engraftment (Price et al. 1996). We will use the kinetic data for healthy human adults stimulated with G-CSF (Table 1). The parameters have been rounded to nearest half-day, which is a reasonable indication of the variation between the data sets used. G-CSF causes both a decrease in doubling time and reduction in marrow transit time, while cycling time remains approximately constant. As a result, healthy individuals treated with high doses of G-CSF can have a circulating neutrophil count as much as 10 times greater than normal!

The fact that the G-CSF response is not counteracted by some negative regulation response lends credence to the use of parameters for healthy adults to describe neutrophil development during engraftment: the dominant G-CSF signal is the same in the two scenarios. One illustration that the G-CSF effect also dominates during engraftment was illustrated in Figure 2. Chemotherapy ablates the proliferative stages, but the neutrophil count is reasonably maintained during the 6 days up to the BMT drawing on the 8 days of neutrophil reserve in the non-proliferative and bone marrow segmented form stages. Immediately after BMT and onset of G-CSF treatment, the remaining reserve is purged producing neutrophil counts 2–3 times the normal level.

The use of an unchanged 1 day transit time for the circulating stage is more questionable. The kinetics of margination (neutrophils moving from circulation to tissue) is not well understood. One may speculate that after a period of neutropenia there could be a 'sponge effect' with tissue quickly 'absorbing' any circulating segment forms released from the bone marrow. A mechanism for this effect could be other cells (e.g., macrophages) signalling a need for neutrophils to fight infection. In any case, the effect of this parameter will be evaluated.

Results and discussion

Our objective is to transplant a sufficient number of proliferative and non-proliferative progenitors from the expanded graft to transiently maintain ANC>500 μL^{-1} until engraftment of the unexpanded graft. The neutrophil reserve will supply a sufficient number of neutrophils for one to two days after the BMT and the unexpanded graft is effective from around day 8. Thus, we need the expanded graft to supply an adequate flux out the non-proliferative stage from day 1 to day 8.

The required flux, r_{NP}^{out}, can be determined from the desired circulating cell count (500 μL^{-1}), the blood

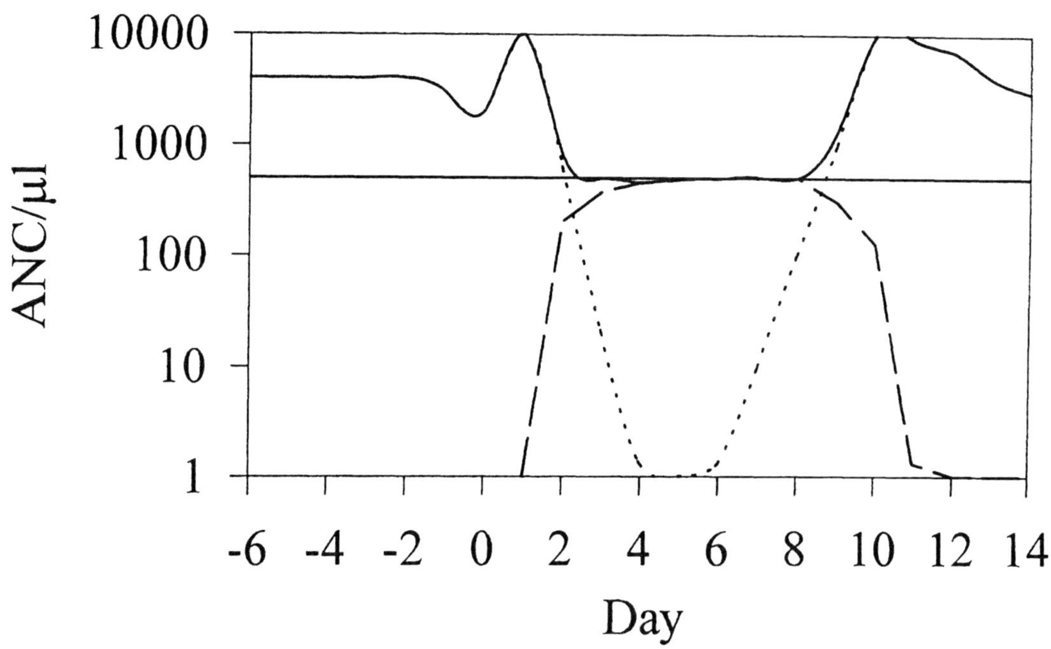

Figure 4. Simulated response with expanded cells added. Expanded cells (dashed), unexpanded cells (dotted), and total (solid).

volume (5×10^6 μL), and the blood transit time (1 day):

$$\begin{aligned} r_{NP}^{out} &= \text{desired count} \times \text{blood volume/transit time} \\ &= 500\,\mu L^{-1} \times 5 \times 10^6\,(1\text{ day})^{-1} \\ &= 2.5 \times 10^9 \text{ cells day}^{-1} \end{aligned} \quad (5)$$

A total of 5 billion precursors is required to generate this flux from day 1 to day 8 (Table 2). Of these cells, 2/3 are non-proliferative precursors needed to maintain the flux from day 1 to day 2.5. In estimating the numbers in Table 2, a fixed transit time was assumed for bone marrow segmented forms (τ_B = 0.5 day). Figure 4 shows the dynamics predicted using equations (1) and (2) (i.e., allowing for the natural variation around the mean, τ_B = 0.5 day, in transit time) together with the idealised normal response curve. The figure illustrates that by transplanting precursor cells that are 1 day away from reaching segmented form, a flux is generated that reaches the desired level of 2.5×10^9 new circulating segment forms per day approximately 2 days after transplantation, i.e., at the time where the neutrophil reserve is rapidly being depleted.

The estimates in Table 2 and Figure 4 are based on a transit time of one day for circulating neutrophils, which as mentioned may or may not reflect the situation during engraftment. The estimates are inversely related to the transit time, i.e., if the true transit time is 0.5 day the total number of expanded cells required is 10 billion. The true transit time can only be established through *in vivo* studies, e.g., using tritiated thymidine (Price et al., 1996).

The estimate in Table 2 is consistent with a cell number extrapolated (to be submitted) from the University of Chicago/Baxter trial with expanded cells (Williams et al., 1996). Although the trial failed to demonstrate total abrogation of neutropenia, the relative neutropenic risk (defined as the integral time under 500 neutrophils per μL) did decrease with increasing expanded cell dose. Linear extrapolation led to an estimated requirement of ~20 billion expanded cells to overcome neutropenia (the highest actual dose was 14 billion). Of these cells, approximately a third are of the progenitor type discussed in this paper, leading to an estimate of 6–7 billion neutrophil precursors required.

The model provides a basis for deciding what types of progenitors are required for abrogating neutropenia and hence a basis for optimising the *ex vivo* expansion cultures. It may also be useful for optimising the HDT/BMT/G-CSF regimen used for treatment. If we can delay the initial onset of neutropenia by 1–2 days (e.g., reducing the clearance period between HDT and BMT or by delaying G-CSF treatment), the need for non-proliferative progenitors in the expanded cell population disappears and thereby 2/3 of the cell requirements. Finally, the model highlights the effect of the four parameters introduced, thus identifying areas of further research.

Acknowledgments

This work is supported by National Institutes of Health (U.S.A.) Grant R01 HL48276.

References

Athens JW (1969) Granulocyte Kinetics in Health and Disease. In: Human Tumor Cell Kinetics, S Perry (ed.) Bethesda, MD: U.S. National Cancer Institute, 1969.

Brugger W, Heimfeld S, Berenson RJ, Mertelsmann R and Kanz L (1995) Reconstitution of hematopoiesis after high-dose chemotherapy by autologous progenitor cells generated *ex vivo*. N Engl J Med 333: 283–287.

Chatta GS, Price TH, Allen RC and Dale DC (1994) Effects of *in vivo* recombinant methionyl granulocyte colony-stimulating factor on the neutrophil response and peripheral blood colony-forming cells in healthy young and elderly adult volunteers. Blood 84: 2923–2929.

Koller MR and Palsson BO (1993) Tissue engineering: reconstitution of human hematopoiesis *ex vivo*. Biotech Bioeng 42: 909–930.

Maraninchi D (1993) The clinical consequences of haematological and non-haematological toxicity following bone marrow transplantation amd the possible impact of haematopoietic growth factors. Bone Marrow Transplantation, 11(2), 12–22.

Price TH, Chatta GS and Dale DC (1996) Effect of recombinant granulocyte colony-stimulating factor on neutrophil kinetics in normal young and elderly humans. Blood 88: 335–340.

Scheding S, Franke H, Brugger W, Kanz L and Schmitz S (1996) How many myeloid progenitors have to be transplanted to completely abrogate neutropenia after peripheral blood progenitor cell transplantation? Exp Hematol 24: 1044a.

Schmitz S, Franke H, Brusis J and Wichmann HE (1993) Quantification of the cell kinetic effects of G-CSF using a model of human granulopoiesis.

Steinbach KH, Schick P, Trepel F, Raffler H, Dohrmann J, Heilgeist G, Heltzel W, Li K, Past W, Van der Woerd-de Lange JA, Theml H, Fliedner TM and Begemann H (1979) Blut 39: 27–38.

Williams SF, Lee WJ, Bender JG, Zimmerman T, Swinney P, Blake M, Carreon J, Schilling M, Smith S, Williams DE, Oldham F and Van Epps D (1996) Selection and expansion of peripheral blood CD34+ cells in autologous stem cell transplantation for breast cancer. Blood 87: 1687–1691.

Cytotechnology **28:** 163–175, 1998.
© 1998 *Kluwer Academic Publishers. Printed in the Netherlands.*

Mammalian cell retention devices for stirred perfusion bioreactors

Steven M. Woodside[1], Bruce D. Bowen & James M. Piret[1]*
[1] *Biotechnology Laboratory and Department of Chemical and Bio-Resource Engineering, University of British Columbia, Vancouver, BC, V6T 1Z3 Canada*
E-mail: jpiret@chml.ubc.ca

Received 26 August 1998; accepted 26 August 1998

Key words: centrifuge, cross-flow filter, review, sedimentation, spin-filter, ultrasonic cell retention

Abstract

Within the spectrum of current applications for cell culture technologies, efficient large-scale mammalian cell production processes are typically carried out in stirred fed-batch or perfusion bioreactors. The specific aspects of each individual process that can be considered when determining the method of choice are presented. A major challenge for perfusion reactor design and operation is the reliability of the cell retention device. Current retention systems include cross-flow membrane filters, spin-filters, inclined settlers, continuous centrifuges and ultrasonic separators. The relative merits and limitations of these technologies for cell retention and their suitability for large-scale perfusion are discussed.

Introduction

Mammalian cell culture is important in biotechnology because many protein products require complex post-translational processing that is presently either difficult or impossible to achieve with bacteria, yeast or insect cells. The growing demand for these products has motivated the development of more efficient and reliable mammalian cell culture production technologies. Cell culture systems aim to provide optimal conditions for mammalian cell proliferation and protein production by controlling the cell environment (temperature, pH, osmolarity) and by supplying dissolved oxygen (DO), nutrients and growth factors at appropriate concentrations without exposing the cells to damaging conditions. Bioprocess selection is determined by economic as well as protein and cell line specific considerations, including the anticipated production needs and the dependence of product consistency on culture conditions. Changing culture conditions can influence the consistency of protein post-translational modifications (Maiorella et al., 1993; Marino et al., 1997). Thus, it is necessary to demonstrate consistent process performance and product quality for approval by regulatory agencies.

Packed-cell bioreactors such as hollow-fiber and ceramic-matrix reactors attain high cell densities ($>10^8$ cells mL^{-1}) as a result of continuous medium perfusion of the immobilized cells. In particular, hollow-fiber bioreactors are commonly used to produce gram quantities of MAbs (monoclonal antibodies) for diagnostic applications. High MAb concentrations ($\sim$1–10 g L^{-1}) are attained due to ultrafiltration retention of the product (Jackson et al., 1996). However, because of the spatial heterogeneity inherent in these plug-flow reactors, scale-up is limited (Piret et al., 1991).

Stirred tank culture is readily scaleable and its relatively uniform environment greatly facilitates monitoring and control. Anchorage dependent cells are often grown in stirred tanks on suspended microcarriers for large-scale production. Non-anchorage dependent cells grow directly in suspension, simplifying the process in a number of ways (Arathoon and Birch, 1986). In stirred batch reactors (Figure 1), the pH, DO and temperature are controlled, but nutrient and metabolite concentrations change dramatically with time, resulting in decreasing growth and protein pro-

* Author for all correspondence.

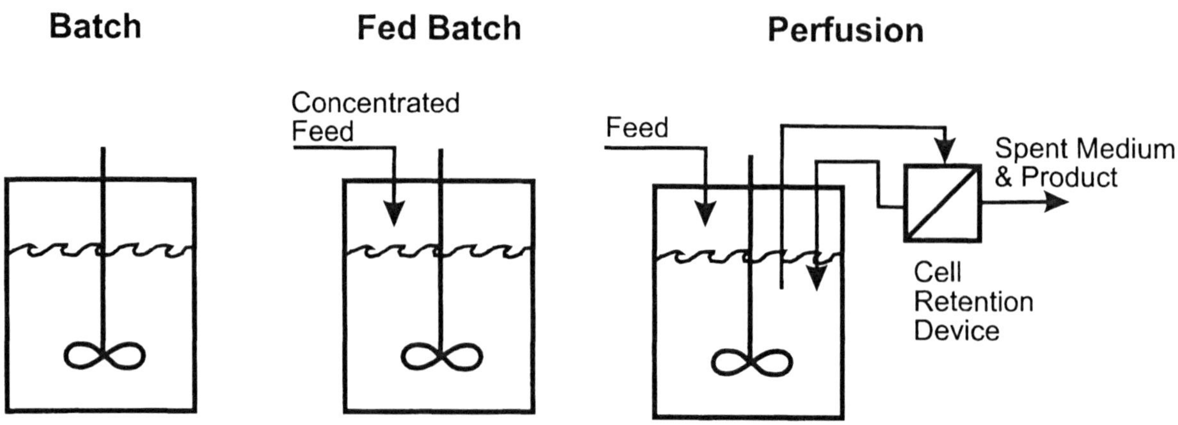

Figure 1. Alternate stirred bioreactor processes. In stirred mammalian cell culture the temperature, pH, and dissolved oxygen level are controlled. A defined concentrated feed stream is added to fed-batch culture in response to the depletion of one or more nutrients. In perfusion, nutrients are continuously supplied and metabolites diluted, while cells are separated from the product stream and retained in the reactor.

duction. The cell density in batch generally reaches a maximum of around 10^6 cells mL^{-1}, resulting in a much lower volumetric productivity than in a packed-cell bioreactor. In fed-batch, the initial batch culture is supplemented with fresh medium in response to the depletion of one or more nutrients, such as glucose and glutamine (Glacken et al., 1986; Bibila and Robinson, 1995). Careful optimization of the feed formulation can extend the culture duration to approximately 3 weeks from 1 week for simple batch, and can result in an order of magnitude increase in final product concentration.

In continuous stirred or chemostat bioreactors, nutrients are continuously replenished and metabolites diluted, minimizing time-varying culture conditions. In stirred perfusion cultures, higher cell densities (over 10^7 cells mL^{-1}) can be achieved by separating the cells from the outflow stream, and retaining them in the reactor (Figure 1). In these cultures, the volumetric productivity can be almost two orders of magnitude higher than in batch and in some cases product concentrations are increased up to 5-fold (Reuveny et al., 1986; Hiller et al., 1993; Bodeker et al., 1994; Trampler et al., 1994). Thus, the required reactor volume for a given production rate can be reduced by approximately 100-fold. Because perfusion cultures can last up to 6 months or more, further economic benefits are accrued due to reduced labor requirements for reactor inoculation and turnaround.

Fed-batch operation is frequently selected over perfusion for its shorter duration in cases where replicate processes must be validated in a minimum time. However, perfusion processes can be acceptable under such time-constrained circumstances by terminating the runs after as little as 1 month. Fed-batch is also often preferred to perfusion based on the potential for higher product concentration and yield on medium. This is especially true in cases where the cell line productivity is unstable, as fed-batch runs involve many fewer cell doublings. For these reasons, more effort has been directed toward fed-batch process development. However, perfusion may be preferred when temporal changes in culture conditions adversely affect product consistency or when the product is susceptible to degradation, because the mean product residence time is on the order of hours rather than days as in fed-batch (Kadouri and Spier, 1997). Perhaps most importantly, perfusion cultures are simpler to develop and operate than fed-batch in at least one respect: beyond the brief start-up period, only the single steady-state production phase of a perfusion process requires optimization, rather than the continuously changing dynamic culture and feeding of a fed-batch process. With the progression of the genomics initiative, increasing numbers of proteins will need to be produced rapidly, and will likely motivate the wide-spread use of simpler, higher productivity perfusion processing. The principal aspect of perfusion processing that can be more complex than fed-batch, particularly upon bioreactor scale-up, is the need for cell retention. This review will focus on the relative advantages and disadvantages of the various alternative technologies for perfusion bioreactor cell retention.

Cell retention devices

During a typical perfusion run inoculated at 10^5 cells mL^{-1}, the cell concentration will increase by two orders of magnitude or more over a span of approximately 10 days. As in any culture, some nonviable cells and debris will accumulate throughout the run due to apoptosis and necrosis (Mercille and Massie, 1994), and cell lysis will release intracellular macromolecules into the culture medium. These dynamic characteristics of long-term perfusion culture present a challenge to the robust operation of cell retention devices. Ideally, a cell retention device should operate satisfactorily for the required duration without replacement or maintenance in order to minimize additional risk of contamination. The device should not adversely affect cell viability or productivity and it should separate close to 100% of the viable cells from the effluent stream, regardless of the cell concentration, in order to maintain high viable cell concentration in the reactor. Finally, although the build-up of nonviable cells and debris can be mitigated by the use of a bleed stream (whereby the suspension is drawn directly from the reactor without cell separation), an ideal retention device should selectively retain viable cells while allowing the nonviable cells to pass through.

Table 1. Cell separation in retention devices. The cell attributes forming the basis for retention are indicated by check marks. Cell size only enhances the effect of one of the other factors when indicated by a bracketed check mark

Retention device	Basis for cell retention		
	Size	Density	Compressibility
Cross-flow filter	✓	×	×
Spin-filter	✓	×	×
Inclined settler	(✓)	✓	×
Centrifuge	(✓)	✓	×
Ultrasonic separation	(✓)	✓	✓

The majority of cell retention devices currently in use (Table 1) are adaptations of conventional particle-liquid separation methods and fall into two general categories: filtration and sedimentation devices. Spin-filters and cross-flow membrane filters retain cells on the basis of size. Sedimentation devices, such as inclined settlers and continuous centrifuges, take advantage of the difference in density between the cells and the medium to effect separation. Ultrasonic cell separation is a more recently developed approach that retains cells on the basis of differences in both density and compressibility between the cells and the medium. There is currently no consensus as to the best retention device for mammalian cell perfusion in stirred suspension reactors due to their different inherent strengths and weaknesses. The following discussion examines recent progress in the development of cell retention devices for mammalian cell perfusion in stirred suspension bioreactors.

Filtration

To sustain long-term perfusion culture performance, filter design and operation must mitigate fouling of the filtration surface by cells, cellular debris and macromolecules (Esclade et al., 1991; Bowen and Gan, 1992; Kelly and Zydney, 1997). Fouling occurs as these components are concentrated at the filter surface by the permeate flux (flow through the filter), a process known as concentration polarization. Filter fouling not only limits the filtration rate, but may result in ultrafiltration retention of high molecular weight products. In general, filtration systems are designed such that the accumulation of cells and cellular material is limited by flow tangential to the filter surface, with the extent of accumulation controlled in large part by the fluid shear rate (Frenander and Jonsson, 1996). The hydrodynamic lift force acting on cells in the shear field may also play a role in reducing filter clogging by cells and debris (Belfort, 1988; Yabannavar et al., 1992).

Cross-flow filters

Hollow-fiber and flat-plate cartridges are suitable as cross-flow filters for cell retention in mammalian cell perfusion applications (Table 2). In these systems, the suspension is pumped to the external cartridge and is concentrated as it flows across a membrane. The concentrated stream is recycled to the reactor, while the cell-free permeate forms the effluent stream. The pressure drop between the inlet and recycle streams drives the cross-flow and determines the surface shear rate. The resulting transmembrane pressure difference is thus highest near the inlet and may even be reversed at the outlet, generating a non-uniform permeate flux (Van Reis et al., 1991).

Permeate flux has been established to be the most important parameter in determining the time to detrimental fouling (clogging) of cross-flow filters (Maiorella et al., 1991). Mean fluxes in the range of 30 to 50 L m^{-2} h^{-1} are typically reported for the

Table 2. Cross-flow filter performance data in selected perfusion experiments

Membrane type, area and pore size	Reactor volume (L)	Maximum perfusion rate (L day^{-1})	Maximum permeate flux (L m^{-2} h^{-1})	Maximum total cell density (10^6 cells mL^{-1})	Viability	Comments	Study
Hollow-fiber 0.2 μm pore	1.0	1.3	0.5	18	>90%	perfusion lasted 32 days	Martin et al. (1989)
Flat plate 0.2 μm pore	1.1	1.2	7.7	25	84%	perfusion halted after 9 days	Velez et al. (1989)
Flat plate 1 10 μm pore	3.6	1.8	4.2	7	44%	filter replaced after 5 to 7 days	De la Broise et al. (1992)
Hollow-fiber 0.1 μm pore	2.0	8.8	8.1	34	80%	perfusion lasted 22 days, initial growth rate same as in batch	Shi et al. (1992)
Hollow-fiber 0.2 μm pore	1.0	4.0	2.8	14	70%	filter clogging at high flux, cell concentration	Zhang et al. (1993)
Flat plate 0.65 μm pore	0.5	1.45	2.5	30	(NA)	perfusion lasted 33 days, growth rate 66% of batch	Kawahara et al. (1994)
Flat-plate 5 μm pore	1.5	0.75	0.9	30	17%	filter replaced every 5 days	Mercille et al. (1994)

(NA) = Not available.

cross-flow filtration process during product harvest after batch culture (Maiorella et al., 1991; Van Reis et al., 1991). In successful perfusion experiments, reported fluxes were an order of magnitude lower due to the requirement for long-term operation (Table 2). Despite the lower fluxes, filter fouling remains a problem. The influence of membrane pore size, mean cell size, culture state, viability and surface shear rate on the maximum flux and on the operating life of the filter have been examined in order to optimize filter design and operation.

The majority of reported cross-flow filters use microporous membranes with 0.2 to 0.65 μm pores. Such relatively small pores might be expected to clog more readily than larger pore filters. However, a 5 μm pore flat-plate filter membrane required changing every 5 days during perfusion with a filter flux of 1 L m^{-2} h^{-1} (Mercille et al., 1994) while 2 and 10 μm diameter pore membranes were replaced every 5 to 7 days during perfusion with a filter flux of 4 L m^{-2} h^{-1} (De la Broise et al., 1992). Thus, there is no clear advantage to these larger pore filters, especially considering the poor cell retention efficiency (<70%) of the 10 μm membrane.

The maximum flux was shown to increase at higher surface shear rates, likely due to the reduced accumulation of cells and debris at the filter surface (Van Reis et al., 1991) However, the surface shear rate can be limited by the shear tolerance of the cells. Maiorella et al. (1991) reported that cells were damaged at wall shear rates over 6000 s^{-1}, while Zhang et al. (1993) observed a drop in culture viability at a shear rate of 1300 s^{-1} during perfusion using a hollow-fiber cartridge. Higher shear rates require higher inlet pressures and thus also increase the transmembrane pressure difference at the filter inlet. At sufficiently high transmembrane pressure differences, cell lysis has been shown to occur when cells occlude pores (Forstrom et al., 1975). Susceptibility to lysis was a function of transmembrane pressure difference, shear rate, pore size and cell membrane tension (Zydney and Colton, 1984). Increasing the shear rate to cell damaging levels can therefore ultimately reduce the maximum perfusion flux, because reduced culture viability is typically accompanied by a lower mean cell diameter and significant levels of debris, both of which decrease the maximum flux (Maiorella et al., 1991). Furthermore, the small pore radius (0.2 to 0.65 μm) of the majority

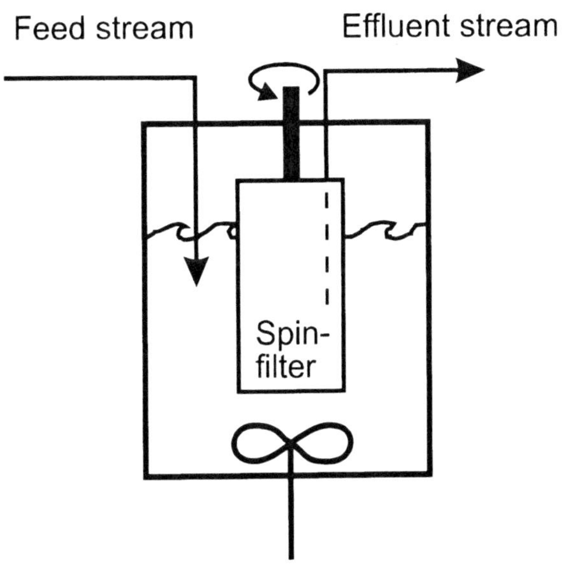

Figure 2. Spin-filter for cell retention in perfusion bioreactor. The effluent stream is withdrawn from the cell clarified interior of the cylindrical spin-filter. In this configuration, impeller and filter rotate independently.

of membranes results in the accumulation of nonviable cells and cell debris in the reactor, causing the maximum flux to decline with time during long-term operation.

One of the main advantages of cross-flow microfiltration for cell retention is that the cell free effluent facilitates integration with the purification process. Moreover, such external devices can be replaced or regenerated when fouling becomes a problem. However, replacement represents a contamination risk as well as additional expense and complications. Pumping of the cell suspension to the external device introduces further complexity and may also cause cell damage (Maiorella et al., 1991). There is little reported information about how cross-flow filters might be integrated into large-scale processes, but for scale-up it is clear that the mean perfusion flux should be kept constant by increasing the total membrane area in proportion to the required capacity.

Spin-filters

Himmelfarb et al. (1969) were the first to report the successful perfusion of mammalian cells in a stirred suspension bioreactor. Their cell retention device, referred to as a spin-filter (Figure 2), consisted of a cylindrical membrane rotating inside the bioreactor around the same centerline axis as the impeller. Spent culture medium was drawn from the culture volume into the cylinder and then out of the reactor, while the cells were retained by the membrane. Spin-filters are distinct from cross-flow filters in two major respects: fluid flow relative to the filtration surface is produced by the rotation of the cylinder, and the filtration surface is typically inside the reactor. As a result, the transmembrane pressure drop can be selected independently of the cross-flow velocity, the retentate cell concentration is more uniform over the whole filter surface, and the cells remain in a controlled environment.

High cell concentration and productivity were achieved in early spin-filter perfusion experiments using a variety of designs (Himmelfarb et al., 1969; Tolbert et al., 1981; Reuveny et al., 1986). However, perfusion was restricted to 10 days, with filter clogging leading in some cases to premature termination of the cultures. Because replacement of internal units is impractical, spin-filters should be designed and operated to avoid clogging during longer-term continuous operations. As with cross-flow filters, the rate of filter fouling increases with perfusion flux and cell concentration (Deo et al., 1996). However, the operating life and maximum perfusion flux are also functions of filter material, effective filter pore size and spin-filter rotational speed. The selection of these parameters for spin-filter optimization has been examined by a number of investigators.

By using larger-pore screens, the operating life of spin-filters has been prolonged (Table 3). While a 5 μm pore filter clogged within 7 days (Reuveny et al., 1986), continuous perfusion reactors were operated without clogging for 11 days using a 25 μm pore stainless steel mesh spin-filter (Yabannavar et al., 1992) and for 49 days using a 15 μm pore filter (Deo et al., 1996).

The upper limit to filter pore size is determined by the cell retention efficiency. Using a 53 μm mesh, Varecka and Scheirer (1987) found that 99% of an aggregating cell line was retained in the reactor during short-term separation experiments. During perfusion of the non-aggregating NS-1 cells (mean size 14 μm), the retention of a 25 μm pore screen varied between 60% at 100 rpm and 97% when the filter was at rest (Yabannavar et al., 1992). This was explained by the increase in the convective fluid exchange between the reactor volume and the volume inside the spin-filter that occurs at increased rotational speed. The retention efficiency was lower when the filter was rotated because cells were entrained with the exchange fluid flow. In perfusion experiments with a 50 μm mesh

Table 3. Spin-filter performance data in selected perfusion experiments

Filter pore size and area	Reactor working volume (L)	Maximum perfusion rate (L (L day^{-1})	Maximum permeate flux (L m^{-2} h^{-1})	Maximum total cell density (10^6 cells mL^{-1})	Viability	Comments	Study
5 μm pore (NA)	0.55	1.36	(NA)	22	70%	filter clogged every 7 days, culture moved to new vessel	Reuveny et al. (1986)
25 μm pore 470 cm^2	12	12	1.1	5.5	72%	perfusion lasted 14 days	Yabannavar et al. (1992)
Vortex flow filter 0.8 μm pore 200 cm^2	1.35	0.67	1.4	80	5%	filter replaced after 43 days, perfusion lasted 74 days	Mercille et al. (1994)
25 μm pore 2200 cm^2	175	87	1.6	4.4	65%	perfusion lasted 5 days (no fouling)	Yabannavar et al. (1994)
15 μm pore (NA)	500	500	(NA)	10 (viable)	(NA)	30 day duration	Deo et al. (1996)
Vortex flow filter 10 μm pore 200 cm^2	3	6	1.25	10	95%	perfusion lasted 60 days	Roth et al. (1997)

(NA) = Not available.

filter, there was no detectable retention of the NS-1 cells regardless of operating conditions (Yabannavar et al., 1992). The 50 μm pores were too large to retain the 14 μm cells on the basis of size, and other proposed mechanisms, such as the hydrodynamic lift and centrifugal forces, did not result in measurable retention.

Apart from extending operating life, large-pore spin-filters allow smaller, nonviable cells to pass more readily through the filter than viable cells, resulting in a selective retention of the viable population. A 25 μm pore spin-filter demonstrated modest selective viable cell retention for a range of rotational speeds (Yabannavar et al., 1992). The selectivity for viable cells varied, depending on rotation rate, agitation rate and perfusion rate. The same group found no selective retention by a 25 μm pore spin-filter in a scaled-up bioreactor, possibly due to the 2.5-fold higher permeate flux (Yabannavar et al., 1994).

The effect of filter material on fouling was examined by Avgerinos et al. (1990) for perfusion culture of an aggregating CHO cell line with aggregates between 200 and 600 μm in diameter. They found that stainless steel mesh filters ranging in nominal pore size from 44 to 105 μm became plugged within 11 to 21 days, with substantial cell attachment observed on the mesh. No fouling was apparent after more than 50 days of perfusion using ETFE (ethylene-tetrafluoroethylene) filters. Esclade et al. (1991) reported that clogging occurred after 7 days during hybridoma perfusion culture with a 10 μm pore stainless steel mesh filter, while a comparable 11 μm pore polyamide screen filter did not clog under similar conditions. The high surface charge density of stainless steel relative to polymer materials may be responsible for the increased fouling (Esclade et al., 1991). Nevertheless, stainless steel filters remain attractive because they are durable and reusable.

Increasing the filter rotation rate has a direct effect on the maximum perfusion rate and operating life. In experiments using a 15 μm pore stainless mesh spin-filter, the maximum perfusion rate before filter clogging was proportional to the rotational velocity squared (Deo et al., 1996). This dependence of clogging on rotational speed is also consistent with the observation that, in the event of mechanical failure, perfusion must be halted while repairs are made, or the stationary spin-filter will foul rapidly and irreversibly (Deo et al., 1996).

In typical spin-filters, fouling is reduced at high rpm due to the high shear rate generated by the filter motion relative to the bulk fluid volume. In external vortex flow filters (Rebsamen et al., 1987; Mercille et al., 1994; Roth et al., 1997), fouling is reduced by stable Taylor-Couette vortices generated in the gap between concentric cylindrical surfaces due to their relative rotation. A 0.8 μm pore vortex flow filter membrane was replaced after 45 days of perfusion culture (Mercille et al., 1994), while myeloma and CHO cell perfusion culture experiments lasted up to 60 days without clogging using a 10 μm pore steel mesh vortex flow filter (Roth et al., 1997). Compared to conventional spin-filters, vortex flow filters appear to be less susceptible to fouling (Table 3).

The scale-up of spin-filter perfusion devices is based on analyses of the mechanisms controlling the perfusion rate and fouling. Yabannavar et al. (1994) assumed that filter clogging due to the permeate flux was counteracted by a hydrodynamic lift force produced by the spinning filter, similar to the force observed on particles near a wall in a laminar flow field (Belfort, 1988; Drew et al., 1991). Keeping the ratio of the lift force to the permeation drag constant, they showed that the fluid exchange through the filter on a reactor volume basis was maintained in a scaled-up 175 L bioreactor and that the retention efficiency was similar to that in the 12 L reactor during a 10 day perfusion experiment. The importance of hydrodynamic lift in preventing clogging of large pore spin-filters is not clear, as there was no detectable contribution of such a force to the cell retention efficiency of a 50 μm pore filter. However, filter fouling will be reduced at higher shear rates due to the increased tangential drag forces preventing cell or debris deposition. Conserving the ratio of shear rate to permeation drag will result in an identical approach to scale-up because the shear rate at the filter surface has the same dependence on filter rpm as the hydrodynamic lift force (Favre and Thaler, 1992).

Deo et al. (1996) expressed the perfusion capacity as a function of rotation velocity, filter dimensions and cell concentration. At a constant rotation rate, the spin-filter perfusion capacity scaled with height and with diameter cubed. Using these relationships to determine dimensions and operating conditions, scale-up from 7 to 500 L was successful. Perfusion of the 500 L bioreactor, at a rate of one reactor volume per day, lasted 30 days at a cell concentration surpassing 10^7 cells mL^{-1}. The growth profile, cell concentration and antibody concentration were the same as in the 7 L bioreactor.

It should be noted that despite much recent progress in the design and operation of spin-filters, their reported permeate fluxes (Table 3) are of the same order as those in cross-flow filtration perfusion (Table 2). This is likely due to the more conservative approach required to ensure robust operation, since the filter cannot be easily replaced. In fact, the reported duration of most conventional spin-filter runs has been limited to one month.

Sedimentation

Sedimentation exploits the difference in density between the cells and the suspending medium by establishing a quiescent liquid volume in which the cells settle from suspension (Table 1). Failure due to clogging does not occur because the separation is not carried out by a physical barrier and selective retention of viable cells is possible because smaller nonviable cells have an approximately 2-fold lower sedimentation velocity relative to viable cells (Searles et al., 1994). The major challenge facing sedimentation-based separation is the slow sedimentation velocity of mammalian cells due to their small size (on the order of 10 μm) and low density (approximately 5% greater than that of the medium). This limits the maximum flow-rate and can result in long, potentially detrimental hold-up times in an unoxygenated, unmixed environment.

Vertical sedimentation devices (Figure 3A) have been used by a number of researchers for laboratory scale perfusion culture at cell concentrations up to 1.1×10^7 cells mL^{-1} (Takazawa and Tokashiki, 1989; Shitani et al., 1991). During perfusion at a cell density of up to 5×10^6 cells mL^{-1}, Hülscher et al. (1992) showed that 100% of the viable cells and 65% of the nonviable cells were retained by sedimentation at an upward flow velocity of 1.5 cm h^{-1}. Above 1.5 cm h^{-1}, the viable separation efficiency declined rapidly. This upper limit to effluent rate is overcome in two ways. With inclined sedimentation devices, the sedimentation path and effective upward fluid velocity are reduced by positioning narrow sedimentation channels at an angle to the vertical (Figure 3B). In continuous centrifuges the sedimentation velocity is increased by imposing a centrifugal acceleration many times greater than the gravitational acceleration.

Inclined sedimentation

Using an inclined settler for mammalian cell perfusion in a 1 L reactor, Batt et al. (1990) reached cell concentrations up to 1×10^7 cells mL^{-1} at a maximum

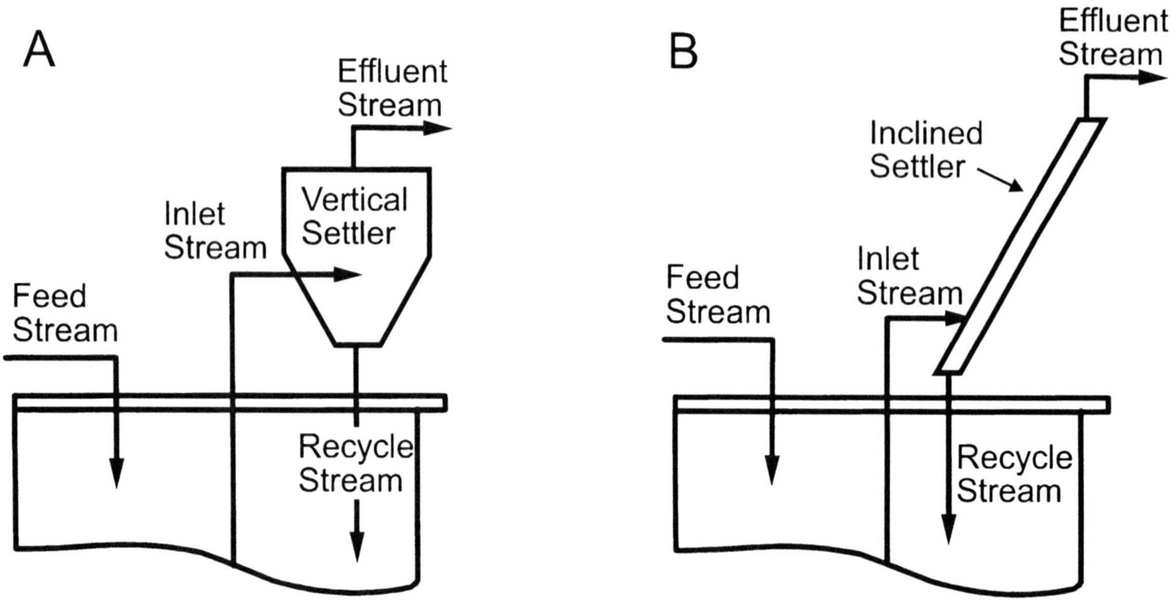

Figure 3. Sedimentation devices for mammalian cell retention in perfusion bioreactors. In vertical sedimentation (A), cells settle against the upward flow of the effluent stream. The settling path is shortened in inclined sedimentation devices (B) by angling a narrow channel from the vertical. Cells flow back to the bioreactor along the lower face of the channel.

perfusion rate of 2.5 L day^{-1} during a 12 day culture (Table 4). The retention efficiency of viable cells was greater than 99% throughout. The corresponding mean flow velocity of 28 cm h^{-1} was 20-fold higher than the limiting velocity for vertical sedimentation (Hülscher et al., 1992). At a mean flow velocity of 80 cm h^{-1}, over 90% separation efficiency was attained with a 50% longer settler (Batt et al., 1990).

The size selectivity of inclined sedimentation has been exploited to stably maintain a mixed culture of yeast and bacteria (Davison et al., 1985) and to preferentially retain a population of larger plasmid-bearing bacteria over a faster growing non-expressing reverted population (Henry et al., 1990). Selective retention of viable mammalian cells was demonstrated by Batt et al. (1990) as a function of flow-rate, with the difference between viable and nonviable cell retention being up to 50% within a range of intermediate flow rates.

Although there is an obvious improvement in maximum flow-rate and mean fluid hold-up time compared to vertical sedimentation, the mean cell residence time in the settler is not well defined because the cells form a sediment layer that must return to the bioreactor against the drag of the effluent flow (Davis and Acrivos, 1985). Also, fouling of the inclined surface due to the relative quiescence at the wall may result in a build-up of debris and attached cells over time, hindering the long-term operation of the separation device. Searles et al. (1994) counteracted this cell attachment by vibrating the sedimentation column and cooling it to 4 °C. Using labeled cells, they examined the mean cell residence time in the inclined settler. At two different concentrations (2.8 and 5.4 × 10^6 cells mL^{-1}), but comparable flow rates and settler configurations, the mean residence time was approximately 1.5 h for the cells that returned to the bioreactor. However, a large fraction of the labeled cells (36% of the total at the high cell concentration and 13% at the lower concentration), remained in the settler more than 4.3 h. When air was bubbled through the settler, these remaining cells were resuspended and the majority returned to the bioreactor within 30 min. There was no clear impact on cell viability from this prolonged period in unoxygenated medium, probably because the medium in the settler was cooled to 4 °C thereby slowing down the cell metabolism. For susceptible cells, prolonged residence in an anoxic environment and temperature cycling from 37 to 4 °C are causes for concern.

For design purposes, the maximum flow-rate for an inclined rectangular channel can be estimated as the product of the cell sedimentation velocity and the projected horizontal channel area (Davis and Acrivos, 1985). In practice the maximum measured velocity for total retention of viable cells was only approximately 50% of the predicted value (Batt et al., 1990) proba-

Table 4. Sedimentation, centrifugation and ultrasonic cell retention device performance data in selected perfusion experiments. Retention efficiency was greater than 95% in all cases

Type of retention device	Reactor working volume (L)	Maximum perfusion rate (L day^{-1})	Maximum total cell density (10^6 cells mL^{-1})	Viability	Comments	Study
Inclined settler	1.5	2.5	22	45%	selective retention of viable cells	Batt et al. (1990)
Continuous centrifuge mechanical seals	10	20	10 (viable)	(NA)	36 day perfusion	Tokashiki et al. (1990)
Vertical settler	5.4	6.8	6	50%	20 day perfusion	Hülscher et al. (1992)
Continuous centrifuge mechanical seals	6	33	20	88%	12 day perfusion growth rate and cell specific production rate same as in batch	Jäger et al. (1992)
Continuous centrifuge mechanical seals	21.5	63	10	90%	15 day perfusion	Jäger et al. (1992)
Inclined settler 7 channels	5.4	22	8	50%	30 day perfusion, recycle loop cooled to 20 °C, no pumping of cell suspension	Stevens et al. (1994)
Inclined settler multiple channels	25	50	10	47%	24 day perfusion	Thompson et al. (1994)
Ultrasonic separator	1	4.5	60	34%	29 day perfusion	Trampler et al. (1994)
Continuous centrifuge directly coupled	1.5	(NA)	13	35%	15 day perfusion cell specific productivity lower than in batch	Johnson et al. (1996)

(NA) = Not available.

bly because the prediction assumes a plug flow profile rather than the true parabolic velocity profile within each channel. Additional deviation from the ideal capacity can be attributed to disruptive convective flows due to non-uniform heating or cooling.

Scale-up of inclined sedimentation involves increasing the projected settling surface area. Because viscous damping of disruptive convective flows varies inversely with the smallest dimension of the channel, multiple shorter, shallower channels are favored over a single long, deep channel of the same width. A multiple channel settler with an internal volume of 2.2 L was operated at up to 50 L day^{-1} during a 24 day perfusion experiment (Thompson and Wilson, 1994). The cell concentration reached 5×10^6 cells mL^{-1} and viable cell retention was greater than 95%. Stevens et al. (1994) described a 7-channel settler operated at up to 22 L day^{-1} with a separation efficiency of greater than 95%. This settler was configured such that the recirculation flow could be established without pumping, by cooling the inlet stream to 20 °C. Despite the potential limitations of inclined sedimentation devices, they appear to have reached a reasonable degree of acceptance for large-scale perfusion operations.

Centrifugation

The challenge of implementing a continuous cell centrifuge for perfusion was initially overcome by using rotating mechanical seals between the rotor, and the inlet and outlet streams (Tokashiki et al., 1990). With such a system applied to perfusion (Table 4), virtually 100% separation was attained at a cell density of 2×10^7 cells mL^{-1} and a perfusion rate of 33 L day^{-1} for a 6 L working volume reactor (Jäger, 1992). Due to concerns about the effect of repeated centrifugation on cells, Tokashiki et al. (1990) demonstrated that two daily 10 min exposures over a one week period to centrifugal forces up to 500 times the force of gravity did not measurably affect the growth rate or antibody productivity of a hybridoma cell line. During the 12 day perfusion run of Jäger (1992), the growth rate of the cells was unchanged relative to batch, and culture viability remained high even though the cells passed through the centrifuge up to 6 times per hour.

The fallibility of rotating seals (Tokashiki et al., 1990) prompted the development of a rotor assembly that could be directly coupled to the bioreactor (Ito et al., 1975; Hodgson, 1991). This was ingeniously accomplished using a U-shaped tubing bundle, where the U arm at the outer edge of the centrifuge rotates at half the rate of the arm at the axis of rotation, canceling the twisting of the tubing that would otherwise result. Johnson et al. (1996) demonstrated that the viable hybridoma cell concentration (1×10^7 cells mL^{-1}) achieved in continuous perfusion with this novel centrifuge design was comparable to that using filtration-based cell retention under optimal conditions. However, the antibody productivity was in all cases approximately 30% lower. An analysis of the oxygen uptake rate and cell residence time in the centrifuge suggested that the pelleted cells experienced periodic oxygen starvation. Limiting nutrients were also likely depleted. As the cells passed through the centrifuge approximately once per hour, it appears that the drop in MAb production could indeed be attributed to oxygen and nutrient starvation.

Centrifuges are mechanically complex systems, and although recent improvements have been made to the disposable elements (Johnson et al., 1996), the separation insert required monthly aseptic replacement during continuous operation (or approximately every 20 million revolutions). That said, the manufacturer rated capacity of 2400 L day^{-1} for a production scale continuous centrifuge is the greatest of any retention device available for mammalian cell perfusion.

Ultrasonic cell retention

Ultrasonic separation is based on the acoustic forces that act on suspended cells in a standing wave field due to differences in density and compressibility between the cells and the medium (Gor'kov, 1962). When exposed in batch to an ultrasonic field, mammalian cells were shown to aggregate rapidly within planes corresponding to the pressure nodes of the standing wave field (Kilburn et al., 1989; Pui et al., 1995). When the field was extinguished, the aggregates quickly settled

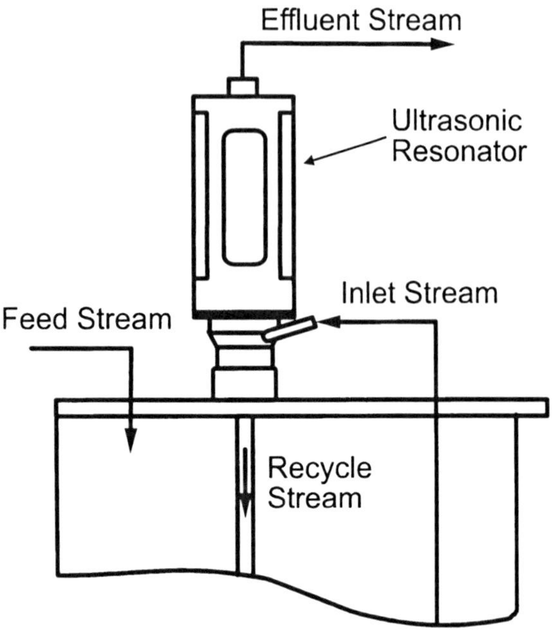

Figure 4. Ultrasonic cell separation device. The standing wave field in the resonator chamber aggregates and retains cells against the drag of the effluent flow. The aggregates settle out of the resonator, return to the bioreactor in the recycle stream where the cells are resuspended by the bioreactor agitator. Periodic shut-down of the power to the field and effluent pumps helps clarify the chamber.

out of suspension. Pui et al. (1995) demonstrated that there was no negative impact on cell viability, glucose uptake rate or antibody production resulting from exposure to ultrasonic standing waves at power inputs to the wave field up to 220 W L^{-1}. The viability was affected at 260 W L^{-1} likely due to cavitation in the standing wave. Cavitation is used in sonicators to disrupt microbial suspensions but is avoided in ultrasonic separators.

The application of ultrasonic cell separation to perfusion is attractive because it is not susceptible to fouling or mechanical failure, having no physical barrier or mechanical parts. Many implementations have been proposed for continuous ultrasonic separation where enriched and clarified streams are generated by concentrating the suspended particles on one side of the flow path (Benes et al., 1991; Whitworth et al., 1991; Frank et al., 1993). Trampler et al. (1994) successfully developed a cell retention device for perfusion culture by an alternate approach in which the cells were aggregated and retained in the standing wave field of an ultrasonic resonator, then returned to the bioreactor by sedimentation and dispersed by the impeller (Figure 4). Cell retention was greater than

97% throughout a one month culture, with total cell concentrations reaching 6×10^7 cells mL^{-1} (Table 4). There was no detectable impact on cell viability or evidence of cell lysis due to the retention device.

Factors that influence the maximum flow-rate in ultrasonic cell separation are the input power and cell concentration. Doblhoff-Dier et al. (1994) demonstrated that a higher input power was required to maintain a constant separation efficiency at high flow-rates. This trend is explained by the increased ultrasonic force required to retain the aggregated cells against the fluid drag, and by the lower mean fluid residence time. As demonstrated in batch, the formation of aggregates is time-dependent (Kilburn et al., 1989; Pui et al., 1995). The maximum flow-rate for 90% separation efficiency was almost 3-fold higher at 5.5×10^6 than at 1×10^6 cells mL^{-1} (Trampler et al., 1994). This effect is likely due to the increased total retention force on the larger aggregates formed at higher concentration (Woodside et al., 1997). Although the same trend was observed in batch (Kilburn et al., 1989), there is undoubtedly a separation efficiency maximum at some higher concentration.

Doblhoff-Dier et al. (1994) have reported the highest superficial velocities (720 cm h^{-1}) for ultrasonic cell separation, with greater than 90% separation during short-term retention experiments. This velocity is almost 500 times the sedimentation velocity and close to 10 times the highest reported rate for inclined sedimentation (Batt et al., 1990). Because ultrasonic separators are operated with periodic shut down of the power to the field and effluent pumps, cells and aggregates are allowed to settle back into the reactor, and mean cell hold-up times are greatly reduced relative to sedimentation devices. Selective retention of viable cells was investigated by Gaida et al. (1996), who demonstrated that the proportion of nonviable cells in the effluent was up to 30% greater than in the inlet stream.

To date, ultrasonic cell retention has been limited to small-scale perfusion. The recommended maximum flow rate for commercial ultrasonic separators is 50 L day^{-1}. Additional improvements in capacity are anticipated by further increasing resonator dimensions, and through optimization of the force distribution in the standing wave field (Woodside et al., 1998).

Conclusions

Both fed-batch and perfusion are capable of attaining high cell concentration, volumetric productivity and product concentration. Although fed-batch has been more widely adopted in recent years, there is no universal, optimum approach to efficient cell culture. Instead, particular aspects of each process need consideration to determine the method of choice. As outlined in this review, considerable progress has been made to overcome fouling, increase the separation capacity, and improve the reliability of cell retention devices, thus making large-scale perfusion an increasingly viable option for cell culture at any scale.

Acknowledgements

Steven Woodside was supported by Natural Sciences and Engineering Research Council (NSERC) and BC Science Council GREAT scholarships.

References

Arathoon WR and Birch JR (1986) Large-scale cell culture in biotechnology. Science 232: 1390–1395.

Avgerinos GC, Drapeau D, Socolow JS, Mao J, Hsiao K and Broeze RJ (1990) Spin filter perfusion system for high density cell culture: production of recombinant urinary type plasminogen activator in CHO cells. Bio/technol 8: 54–58.

Batt BC, Davis RH and Kompala DS (1990) Inclined sedimentation for selective retention of viable hybridomas in a continuous suspension bioreactor. Biotechnol Prog 6: 458–464.

Belfort G (1988) Membrane modules: Comparisons of different configurations using fluid mechanics. J Membr Sci 35: 245–270.

Benes E, Hager F, Bolek W and Gröschl M (1991) Separation of dispersed particles by drifting ultrasonic resonance fields. Ultrason. Int. Conf., Le Touquet, France, Butterworth-Heinemann, 167–170.

Bibila TA and Robinson DK (1995) In pursuit of the optimal fed-batch process for monoclonal antibody production. Biotechnol Prog 11: 1–13.

Bodeker BGD, Newcomb R, Yuan P, Braufman A and Kelsey W (1994) Production of recombinant factor VIII from perfusion cultures: I. Large-scale fermentation. In: Spier RE, Griffiths JB and Berthold W (eds.) Animal cell technology: Products of today, prospects for tomorrow, Butterworth-Heinemann, Oxford, pp. 580–583.

Bowen WR and Gan Q (1992) Properties of microfiltration membrane: the effects of adsorption and shear on the recovery of an enzyme. Biotechnol Bioeng 40: 491–497.

Davis RH and Acrivos A (1985) Sedimentation of noncolloidal particles at low Reynolds numbers. Ann Rev Fluid Mech 17: 91–118.

Davison BH, San K-Y and Stephanopoulos G (1985) Stable competitive coexistence in a continuous fermenter with size selective properties. Biotechnol Prog 1: 260–268.

De la Broise D, Noiseux M, Massie B and Lemieux R (1992) Hybridoma perfusion systems: A comparison study. Biotechnol Bioeng 40: 25–32.

Deo YM, Mahadevan MD and Fuchs R (1996) Practical considerations in operation and scale up of spin filter based bioreactors for monoclonal antibody production. Biotechnol Prog 12: 57–64.

Doblhoff-Dier O, Gaida T, Katinger H, Burger W, Gröschl M and Benes E (1994) A novel ultrasonic resonance field device for the retention of animal cells. Biotechnol Prog 10: 428–32.

Drew DA, Schonberg JA and Belfort G (1991) Lateral inertial migration of a small sphere in fast laminar flow through a small membrane duct. Chem Eng Sci 46: 3219–3224.

Esclade LRJ, Carrel S and Peringer P (1991) Influence of the screen material on the fouling of spin filters. Biotechnol Bioeng 38: 159–168.

Favre E and Thaler T (1992) An engineering analysis of rotating sieves for hybridoma cell retention in stirred tank bioreactors. Cytotechnol 9: 11–19.

Forstrom RJ, Bartlet K, Blackshear Jr, PL and Wood T (1975) Formed elements deposition onto filtering walls. Trans Am Soc Artif Intern Org 21: 602.

Frank A, Bolek W, Groeschl M, Burger W and Benes E (1993) Separation of suspended particles by use of the inclined resonator concept. Proc. Ultrason. Int. Conf., Vienna, Austria, Butterworth-Heinemann, Oxford, 519–522.

Frenander U and Jonsson AS (1996) Cell harvesting by crossflow microfiltration using a shear enhanced module. Biotechnol Bioeng 52: 397–403.

Gaida T, Doblhoff-Dier O, Strutzenberger K, Katinger H, Burger W, Gröschl M, Handl B and Benes E (1996) Selective retention of viable cells in ultrasonic resonance field devices. Biotechnol Prog 12: 73–76.

Glacken MW, Fleischaker RJ and Sinskey AJ (1986) Reduction of waste product excretion via nutrient control: Possible strategies for maximizing product and cell yields on serum in cultures of mammalian cells. Biotechnol Bioeng 28: 1376–1389.

Gor'kov LP (1962) On the forces acting on a small particle in an acoustical field in an ideal fluid. Sov Phys Dokl 6: 773–775.

Henry KL, Davis RH and Taylor AL (1990) Continuous recombinant bacterial fermentations utilizing selective flocculation and recycle. Biotechnol Prog 6: 7–12.

Hiller G, Clark D and Blanch H (1993) Cell retention chemostat studies of hybridoma cells. Analysis of hybridoma growth and metabolism in continuous suspension culture on serum free medium. Biotechnol Bioeng 42: 185–195.

Himmelfarb P, Thayer PS and Martin HE (1969) Spin-filter culture: the propagation of mammalian cells in suspension. Science 164: 555–557.

Hodgson (1991) Centrifugation takes a new turn. Bio/technol 9: 628–629.

Hülscher M, Scheibler U and Onken U (1992) Selective recycle of viable animal cells by coupling of airlift reactor and cell settler. Biotechnol Bioeng 39: 442–446.

Ito Y, Suaudeau J and Bowman RL (1975) New flow-through centrifuge without rotating seals applied to plasmapheresis. Science 189: 999–1000.

Jackson LR, Trudel LJ, Fox JG and Lipman NS (1996) Evaluation of hollow-fiber bioreactors as an alternative to murine ascites production for small scale monoclonal antibody production. J Immunol Methods 189: 217–231.

Jäger V (1992) A novel perfusion system for the large-scale cultivation of animal cells based on a continuous flow centrifuge. In: Spier RE, Griffiths JB and MacDonald C (eds.) Animal cell technology: Developments, processes, products, Butterworth-Hienemann, Oxford, pp. 397–402.

Johnson M, Lanthier S, Massie B, Lefebvre G and Kamen AA (1996) Use of the Centritech Lab centrifuge for perfusion culture of hybridoma cells in protein free medium. Biotechnol Prog 12: 855–864.

Kadouri A and Spier RE (1997) Some myths and messages concerning the batch and continuous culture of animal cells. Cytotechnol 24: 89–98.

Kawahara H, Mitsuda S, Kumazawa E and Takeshita Y (1994) High-density culture of FM-3A cells using a bioreactor with an external tangential filtration device. Cytotechnol 14: 61–66.

Kelly ST and Zydney AL (1997) Protein fouling during microfiltration: comparative behavior of different model proteins. Biotechnol Bioeng 55: 91–100.

Kilburn DG, Clarke DJ, Coakley WT and Bardsley DW (1989) Enhanced sedimentation of mammalian cells following acoustic aggregation. Biotechnol Bioeng 34: 559–562.

Maiorella B, Dorin G, Carion A and Harano D (1991) Crossflow microfiltration of animal cells. Biotechnol Bioeng 37: 121–126.

Maiorella BL, Winkelhake J, Young J, Moyer B, Bauer R, Hora M, Andya J, Thomson J, Patel T and Parekh R (1993) Effect of culture conditions on IgM antibody structure, pharmacokinetics and activity. Bio/Technol 11: 387–392.

Marino M, Corti A, Ippolito A, Cassani G and Fassina G (1997) Effect of bench-scale culture conditions on murine IgG heterogeneity. Biotechnol Bioeng 54: 17–25.

Martin N, Brennan A, Denome L and Shaevitz J (1989) High productivity in mammalian cell culture. Bio/technol 5: 838–840.

Mercille S, Johnson M, Lemieux R and Massie B (1994) Filtration-based perfusion of hybridoma cultures in protein free medium: reduction of membrane fouling by medium supplementation with DNase I. Biotechnology and Bioengineering 43: 833–846.

Mercille S and Massie B (1994) Induction of apoptosis in nutrient-deprived cultures of hybridoma and myeloma cells. Biotechnol Bioeng 44: 1140–1154.

Piret JM, Devens DA and Cooney CL (1991) Nutrient and metabolite gradients in mammalian hollow-fiber bioreactors. Can J Chem Eng 69: 421–428.

Pui PWS, Trampler F, Sonderhoff SA, Groeschl M, Kilburn DG and Piret JM (1995) Batch and semicontinuous aggregation and sedimentation of hybridoma cells by acoustic resonance fields. Biotechnol Prog 11: 146–52.

Rebsamen E, Goldinger W, Scheirer W, Merten O-W and Pálfi GE (1987) Use of a dynamic filtration method for separation of animal cells. Develop Biol Standard 66: 273–277.

Reuveny S, Velez D, Miller L and Macmillan JD (1986) Comparison of cell propagation methods for their effect on monoclonal antibody yield in fermenters. J Immunol Meth 86: 61–69.

Roth G, Smith CE, Schoofs GM, Montgomery JLA and Horwitz JI (1997) Using an external vortex flow filtration device for perfusion cell culture. BioPharm October: 30–35.

Searles JA, Todd P and Kompala DS (1994) Viable cell recycle with an inclined settler in the perfusion culture of suspended recombinant Chinese hamster ovary cells. Biotechnol Prog 10: 198–206.

Shi Y, Ryu DDY and Park SH (1992) Performance of mammalian cell culture bioreactor with a new impeller design. Biotechnol Bioeng 40: 260–270.

Shitani Y, Kohno Y-I, Sawada H and Kitano K (1991) Comparison of culture methods for human-human hybridomas secreting anti-HBsAg human monoclonal antibodies. Cytotechnol 6: 197–208.

Stevens J, Eickel S and Onken U (1994) Lamellar clarifier- a device for animal cell retention in perfusion culture systems. In:

Spier RE, Griffiths JB and Berthold W (eds.) Animal cell technology: Products of today, prospects for tomorrow, Butterworth-Heinemann, Oxford, pp. 234–239.

Takazawa Y and Tokashiki M (1989) High cell density perfusion culture of mouse-human hybridoma. Appl Microbiol Biotechnol 32: 280–284.

Thompson, KJ and Wilson JS (1994) A compact gravitational settling device for cell retention. In: Spier RE, Griffiths JB and Berthold W (eds.) Animal cell technology: Products of today, prospects for tomorrow, Butterworth-Heinemann, Oxford, pp. 227–229.

Tokashiki M, Arai T, Hamamoto K and Ishimaru K (1990) High density culture of hybridoma cells using a perfusion culture vessel with an external centrifuge. Cytotechnol 3: 239–244.

Tolbert WR, Feder J and Kimes RC (1981) Large-scale rotating filter perfusion system for high-density growth of mammalian suspension cultures. In Vitro 17: 885–890.

Trampler F, Sonderhoff SA, Pui PW, Kilburn DG and Piret JM (1994) Acoustic cell filter for high density perfusion culture of hybridoma cells. Bio/technol 12: 281–284.

Van Reis R, Leonard LC, Hsu CC and Builder SE (1991) Industrial scale harvest of proteins from mammalian cell culture by tangential flow filtration. Biotechnol Bioeng 38: 413–422.

Varecka R and Scheirer W (1987) Use of a rotating wire cage for retention of animal cells in a perfusion fermenter. Dev Biol Stand 66: 269–272.

Velez D, Miller L and Macmillan JD (1989) Use of tangential flow filtration in perfusion propagation of hybridoma cells for production of monoclonal antibodies. Biotechnol Bioeng 33: 938–940.

Whitworth G, Grundy MA and Coakley WT (1991) Transport and harvesting of suspended particles using modulated ultrasound. Ultrasonics 29: 439–44.

Woodside SM, Bowen BD and Piret JM (1997) Measurement of ultrasonic forces for particle-liquid separations. AIChE J 43: 1727–1736.

Woodside SM, Piret JM, Gröschl M, Benes E and Bowen B (1998), Acoustic force distribution in resonators for ultrasonic particle separation. AIChE J 44: 1976–1986.

Yabannavar VM, Singh V and Connelly NV (1992) Mammalian cell retention in a spinfilter perfusion bioreactor. Biotechnol Bioeng 40: 925–933.

Yabannavar VM, Singh V and Connelly NV (1994) Scaleup of spinfilter perfusion bioreactor for mammalian cell retention. Biotechnol Bioeng 43: 159–164.

Zhang S, Handa-Corrigan A and Spier RE (1993) A comparison of oxygenation methods for high-density perfusion cultures of animal cells. Biotechnol Bioeng 41: 685–692.

Zydney AL and Colton CK (1984) A red cell deformation model for hemolysis in cross flow membrane plasmapheresis. Chem Eng Commun 30: 191–207.

Cytotechnology **28:** 177–188, 1998.
© 1998 *Kluwer Academic Publishers. Printed in the Netherlands.*

Variable functions of bcl-2 in mediating bioreactor stress-induced apoptosis in hybridoma cells

A. Perani, R. P. Singh, R. Chauhan & M. Al-Rubeai*
Centre for Bioprocess Engineering, School of Chemical Engineering, University of Birmingham, Birmingham B15 2TT, United Kingdom
E-mail: m.al-rubeai@bham.ac.uk

Received 14 September 1998; accepted 14 September 1998

Key words: apoptosis, bcl-2, cell death, hybridoma, osmolarity, pH, shear, stress

Abstract

It has been demonstrated that the cell lines used for production of biopharmaceuticals are highly susceptible to apoptosis, and that over-expression of the bcl-2 oncogene can protect cells from death. Stress associated with the deprivation of nutrients has been shown to be the main cause of apoptosis in culture. We have extended these studies by investigating the mechanism of cell death under conditions of sub-optimal pH, shear stress and hyperosmolarity, and the protective action of bcl-2 over-expression. At pH 6, there was no clear evidence of protection from cell death. However, at pH 8, the viability of the bcl-2 transfected cells was about 20% higher relative to the control cells. Cultivation of control cells in a flat bottomed bioreactor with a magnetic stirrer bar without a pivot ring resulted in exposure of the cells to a high attrition effect. As a result, cell growth was retarded and a high level of cell death by apoptosis was observed. Under the same conditions, the bcl-2 transfected cell line exhibited a nearly five fold increase in viable cell number. This finding indicates that under apoptosis-suppressed conditions, shear stress can stimulate cell growth. Batch cultivation of both control and bcl-2 transfected cells in 350 and 400 mOsm media resulted in suppression of cell growth, athough the effect was most marked in the control cell line. Adaptation of control cells to 400 mOsm proved to be impossible to achieve. However, the bcl-2 transfected cells exhibited resistance to the osmotic stress resulting in long term adaptation to a high salt environment. Specific productivity of bcl-2 transfected cells grown in high osmolarity medium was 100% higher than that produced by non-adapted bcl-2 transfected cells grown in normal osmolarity medium. These results demonstrate that bcl-2 has a beneficial effect on hybridoma cultivation under a wide range of culture stresses.

Introduction

Apoptosis accounts for much of the cell death that occurs during the production of biopharmaceuticals in animal cell lines (Al-Rubeai et al., 1990; Franek and Dolnikova, 1991; Mercille and Massie, 1994; Singh et al., 1994; reviewed in: Al-Rubeai and Singh, 1998; Mastrangelo and Betenbaugh, 1998; Singh and Al-Rubeai, 1998). This cellular suicide programme has been extensively studied at the molecular and morphological level (for recent reviews see: Kroemer et al., 1998; King and Cidlowski, 1998; Thompson, 1998a and b; Kidd, 1998), which has aided its identification and characterisation in the bioreactor environment. The importance of apoptosis stems from its active and genetically defined nature. As a result, transfection of a range of cell types with genes which block the expression of this suicide pathway has led to the generation of cell lines which are able to tolerate some of the stresses associated with large scale production and intensification. This has resulted in an improvement in protein productivity.

The first and best studied of the anti-apoptosis genes is bcl-2, which encodes a 26 KDa membrane protein (Tsujimoto et al., 1985; Vaux et al., 1988; Chen-Levy et al., 1989). Over-expression of bcl-2

* Author for all correspondence.

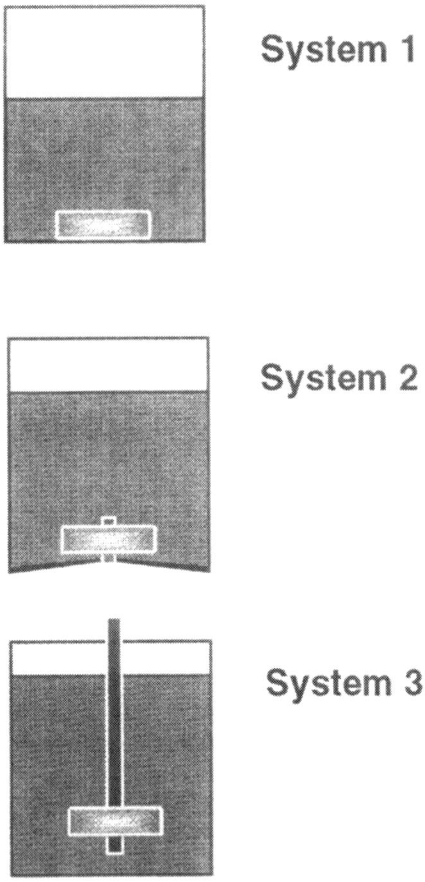

Figure 1. Schematic representation of the three bioreactor configurations used in the study.

in a range of cell types has been demonstrated to protect against numerous apoptosis-inducing insults (Yang and Korsemeyer, 1996). It has, therefore, been the primary target for studies which have attempted to block apoptosis in commercial cell cultures. For example, over-expression of bcl-2 in Burkitts Lymphoma and murine hybridoma cell lines resulted in the extension of batch culture duration by providing protection from amino acid and glucose starvation (Itoh et al., 1995; Singh et al., 1996; Simpson et al., 1997, 1998). Bcl-2 over-expression also aided cell adaptation to suspension culture and allowed rapid adaptation to serum free media. In cell lines which exhibit a negative correlation between growth rate and antibody productivity, cytostatic agents such as thymidine have been used to improve specific productivity (Fazekas, 1983; Boraston et al., 1984; Reuveny et al., 1986; Al-Rubeai et al., 1990, 1992). However, the main drawback with this strategy is the high level of apoptosis that results from the growth arrest (Singh et al., 1994). Over-expression of bcl-2 has been demonstrated to offer a high degree of protection under these conditions (Singh et al., 1996; Simpson et al., 1997). Transient virus-based expression systems, such as those employing baculovirus and insect cells, also result in the induction of apoptosis which limits the duration and therefore the productivity of the culture. Expression of bcl-2 results in a reduction in the rate of cell death under these conditions, thus leading to an increase in the duration of the protein productive phase of the cultures (Mastrangelo et al., 1996; Alnemri et al., 1992; Mitchell-Logean and Murhammer, 1997)

Despite the successes outlined above, numerous bioreactor stress factors remain to be assessed, both in terms of the level of apoptosis induced and the influence of bcl-2 expression on survivability. Most notable amongst these is the influence of high and low pH, which can have a significant affect on hybridoma growth and productivity even when the deviation from optimal conditions is very small. Indeed, Wayt et al., (1997) found that a pH difference of 0.1 unit resulted in a 50% reduction in maximum viable cell number but a 50% improvement in antibody titre in a murine hybridoma cell line. The physical culture environment is also a critical factor, and although there is evidence to suggest that hydrodynamic stress can induce an apoptotic response response (Al-Rubeai et al., 1995), the influence of bcl-2 expression under extreme levels of hydrodynamic stress has not been investigated.

In the present study, the effect of these two components of the culture environment on the cell death mechanism during hybridoma cultures was investigated, and the influence of bcl-2 over-expression on cell survival was examined. Furthermore, stressful bioreactor conditions, such as cell cultivation in hyperosmolarity medium, can result in substantial improvements in culture productivity (Oyaas et al., 1989; Ozturk and Pollson, 1991; Duncan et al., 1997; Oh et al. 1993). Clearly, however, such strategies are limited by the level of cell death which is triggered by the osmotic stress. The final objective of the present study was, therefore, to establish the effect of bcl-2 over-expression on cell adaptation and productivity under hyperosmolarity conditions.

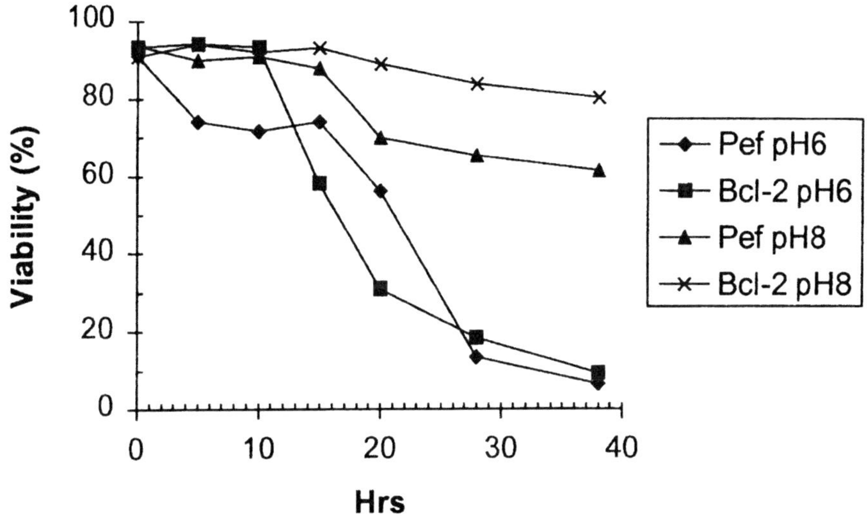

Figure 2. Influence of sub-optimal pH on hybridoma viability. (Pef: control transfected cell line; Bcl-2: Bcl-2 transfected cell line).

Materials and methods

Cell lines and culture medium

The NS1-derived murine hybridoma used in this study, TB/C3, produces Immunoglobulin G (IgG) monoclonal antibody specific to the hapten Cγ2 domain in the Fc region of human IgG. This cell line was transfected using the bcl-2 expression vector pEF bcl2-MCIneopA and control vector pEF-MC1neopA as previously described (Simpson et al., 1997).

Cells were maintained at 37 °C in RPMI 1640 medium (Gibco, U.K.) supplemented with 5% (v/v) Fetal Calf Serum (FCS) (Sigma, U.K.). Levels of bcl-2 expression were monitored at regular intervals by immunostaining of the bcl-2 protein followed by flow cytometric analysis as previously described (Simpson et al., 1997). Cells were grown in 50 and 250 mL T-flasks, and were removed from the mid-exponential phase of the culture for use in experimental work.

Effect of pH

RPMI 1640 culture medium was adjusted to pH 6 and inoculated with control and bcl-2 cells at a cell number of around 1.5×10^5 mL^{-1}. Samples were taken every 5 hr for evaluation of viability and viable cell number by the trypan blue exclusion method. The level of apoptosis was determined by fluorescence microscopic analysis of nuclear morphology as previously described (Simpson et al., 1997). At regular intervals, the pH was re-adjusted to the desired level by the manual addition of 1 mM HCl or NaOH as required.

Effect of hydrodynamic stress

Control and bcl-2 hybridoma cells were used to inoculate the following bioreactor configurations (Figure 1) at a cell number of around 2×10^5 cells mL^{-1}:

System 1. An Infors Sixfor flat-bottomed bioreactor with a 300 mL working volume was used. Agitation at 100 rpm was achieved using a flat magnetic stirrer bar which lacked a pivot ring. As a result, the cells were exposed to high levels of shear due to the grinding action of the magnetic bar on the bottom of the bioreactor. The culture was surface aerated with 5% CO_2/Air and temperature was controlled at 37 °C.

System 2. A 300 mL glass spinner flask was used. Agitation was achieved by means of a magnetic bar with a pivot ring rotating at 100 rpm. Together with the convex-shaped bottom of the flask, these features prevented the grinding action observed in system 1.

System 3. An Infors Sixfor bioreactor with a 500 mL working volume was used. Agitation at 100 rpm was achieved by means of a magnetic stirrer bar with stainless steel support and shaft. The culture was surface aerated with 5% CO_2/air.

Samples were removed daily from each culture in order to determine the viability, viable cell number and the level of apoptosis as described previously (Simpson et al., 1997).

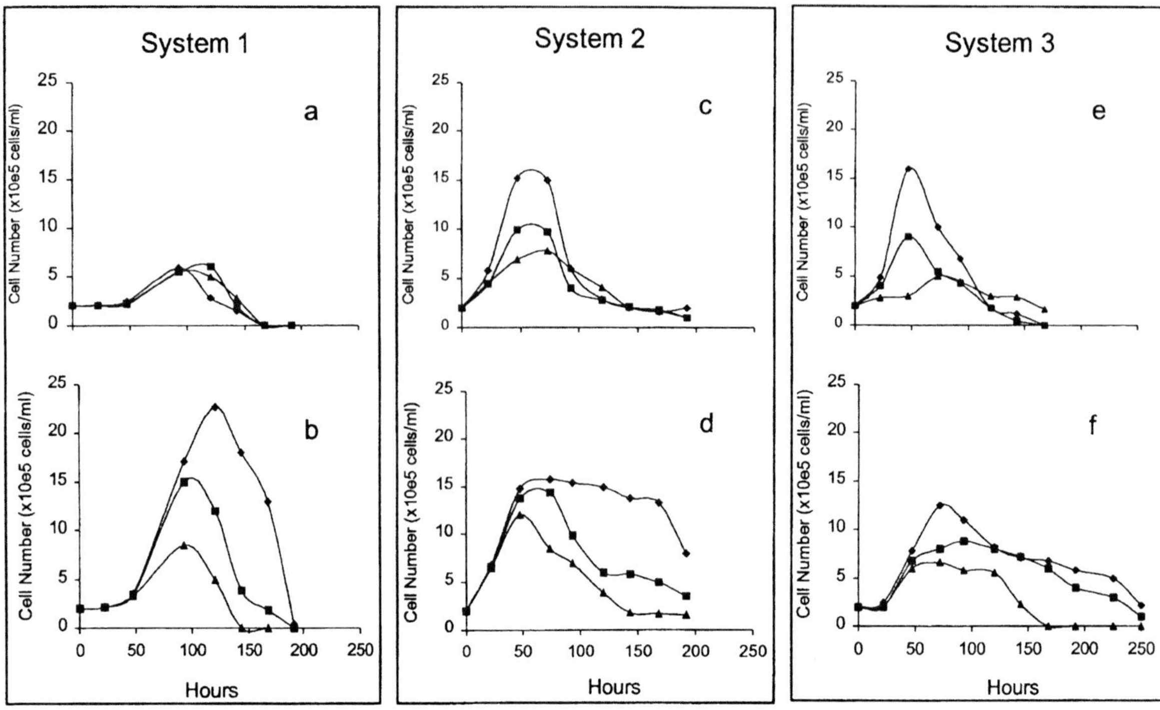

Figure 3. Viable cell numbers during batch cultures of control (a, c, and e) and bcl-2 transfected (b, d, and f) hybridoma cells in the three culture systems using 300, 350 and 400 mOsm culure medium. (Key: Diamonds: 300 mOsm; Squares: 350 mOsm; Triangles: 400 mOsm.)

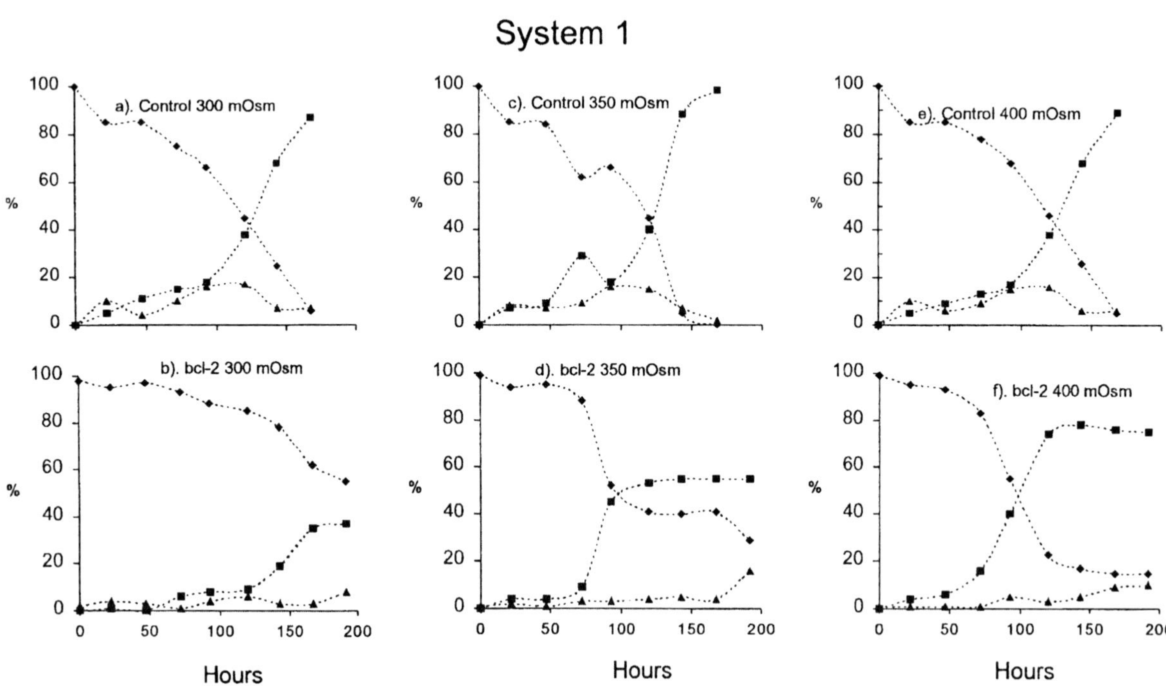

Figure 4. Level of viability, apoptosis and necrosis (each as a percentage of total cells) during batch cultures in culture system 1. (Key: Diamonds: viability; Squares: apoptosis; Triangles: necrosis).

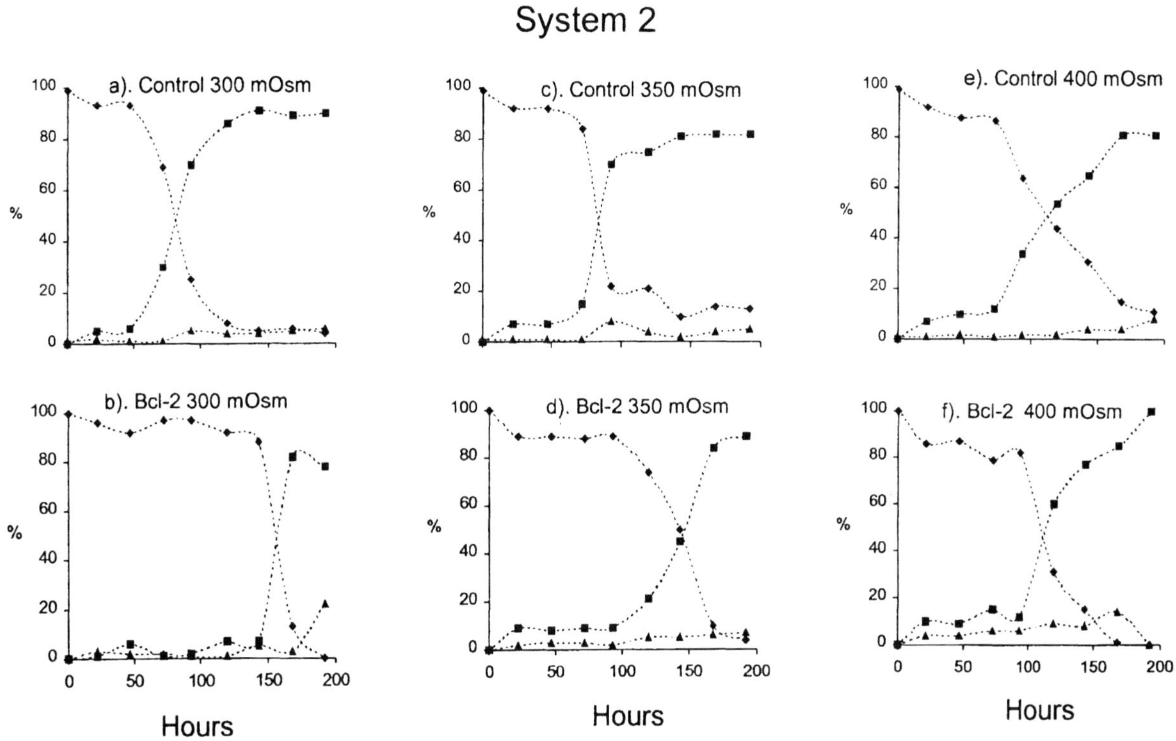

Figure 5. Level of viability, apoptosis and necrosis during batch cultures in system 2. (Key as for Figure 4.)

Influence of bcl-2 expression on adaptation to hyperosmolarity culture conditions

Batch cultivation in high osmolarity medium

Cells were inoculated at a cell number of around 2×10^5 mL^{-1}in reactor systems 1, 2 and 3 (described above) in medium at 300, 350 and 400 mOsm. Medium was prepared by the addition of a 0.1 g mL^{-1} sodium chloride solution to RPMI medium supplemented with 5% FCS. The cultures were sampled daily for the determination of viable cell number, viability and the level of apoptosis and necrosis.

Experiments on the effect of hydrodynamic forces and high osmolarity were repeated 3 times at different times using different inocula. The results of one typical experiment are reported.

Adaptation to high osmolarity medium

Cells were inoculated in reactor system 3 in 400 mOsm medium at a cell number of 2×10^5 mL^{-1}. The cells were subcultured every 50 hr by replacing 50% of the cell suspension with the same volume of fresh medium.

Spin filter cultivation

An LSL Biolafitte (LSL Group, Luton, U.K.) cell culture vessel (1.5 L working capacity), equipped with a four blade Rushton turbine (diameter: 4.4 cm), was used with a 25 μm spin filter mounted on the agitator shaft. Agitation speed was 110 rpm and dissolved oxygen was maintained at 50% of air saturation by a bubble-free oxygenation system using immersed silicone tubing. The tubing was 4.8 mm in diameter with a wall thickness of 0.6 mm and a total length of 3 m. pH was controlled by using sparged CO_2 and 0.1 M NaOH.

Samples were taken daily and viable and total cell numbers were determined. Monoclonal antibody was determined by a sandwich enzyme-linked immunosorbent assay (ELISA) using human IgG coated plates and sheep anti-mouse IgG peroxidase conjugate as the second antibody. The optical density at 490 nm was recorded by an ELISA reader after treating the plates with the peroxidase substrate, o-phenylenediamine dihydrochloride. All immunological reagents were obtained from Sigma (Poole, U.K.).

Results

Effect of sub-optimal pH

Figure 2 shows the effect of pH 6 and 8 on the viability of the two cell lines. Clearly, acidic pH values had the greatest affect on culture viability, with approximately 90% of cells having died within 40 hr. Generally, during the course of the culture, there was little difference in viability between the control and bcl-2 cell line. At pH 8, over the same time period, the viability had only declined by 20% and 40% in the bcl-2 and control cultures respectively, giving a clear indication of the activity of bcl-2 at this pH. Repeated experiments with different inocula confirmed this trend at both pH values and demonstrated the marginal protective effect of bcl-2 in an alkaline environment.

Effect of hydrodynamic stress and hyperosmolarity medium

When control cells were cultured in reactor System 1 using normal (300 mOsm) medium, there was an initial lag period of around 50 hr after which time the cell number increased from around 2×10^5 mL^{-1} to 6×10^5 mL^{-1} at 96 hr (Figure 3a). Fluorescence microscopic analysis of the dead cells indicated that over 10% of cells were apoptotic after only 24 hr, and this figure increased steadily such that after 168 hr about 96% of cells were apoptotic (Figure 4a). Cells overexpressing bcl-2 also exhibited a 50 hr lag phase after which cell number increased steadily, reaching a peak of 22.5×10^5 mL^{-1} after 120 hr (Figure 3b). The level of apoptosis remained well below 10% for in excess of 100 hr, clearly demonstrating the protective effect of bcl-2 (Figure 4b). When significant levels of cell death did occur, apoptosis was predominant. However, even by the end of the experiment, when no viable cells could be detected by means of trypan blue assay, analysis of nuclear morphology by fluorescence microscopy revealed that viability was still around 50% which, as discussed below, indicates the limitations of trypan blue staining.

The cultures described above were repeated in medium at 350 and 400 mOsm in order to compare batch growth under hyperosmolarity conditions. The control cell lines exhibited suppressed growth due to the extreme hydrodynamic conditions outlined above at all three osmolarity levels (Figure 3a). Fluorescence micrscopic analysis revealed apoptosis as the major form of cell death in all cases, with a similar rate of fall at each osmolarity level (Figure 4a, c, e). However, bcl-2 transfected cells exhibited significantly higher viable cell number and culture duration, although there was clear evidence that high osmolarity had suppressed cell growth relative to normal culture conditions (Figure 3b). Fluorescence miscroscopy indicated that the level of apoptosis remained low during the growth phase of the culture. However, once peak cell density had been attained, cell death by apoptosis ensued (Figure 4b, d and f). As in other experiments with the bcl-2 transfected cells in system 1, there was a slight disagreement in the viability data produced by fluorescence microscopy and trypan blue analysis.

The hydrodynamic environment of culture system 2 proved to be much less stressful to the control cell line, which reached a maximum cell number in excess of 15×10^5 cells mL^{-1} in 300 mOsm medium (Figure 3c). A similar maximum cell number was achieved by the bcl-2 transfected cell line. However, as expected, the death phase of the bcl-2 transfected cells was considerably longer with viable cell number falling to around 5×10^5 mL^{-1} after 200 hr (Figure 3d). By comparison, the control cell line had fallen to 5×10^5 mL^{-1} after only 100 hr. Substantial levels of apoptosis were seen in the bcl-2 cultures after 144 hr of cultivation (Figure 5b).

Because of the milder hydrodynamic environment in culture system 2, the effect of increased osmolarity on cell growth was clearly evident. Maximum viable cell number was suppressed by 33% and 50% at 350 and 400 mOsm respectively (Figure 3c). Notably, the high degree of growth suppression at 400 mOsm enabled the cells to exhibit a reduced rate of cell death, presumably due to the greater availability of nutrients at this stage of the culture. Maximum cell number for the bcl-2 transfected cell line was 50% and 100% higher than control cells at 350 and 400 mOsm respectively. Whilst a slight increase in culture duration for the bcl-2 cell line was evident in the 350 mOsm medium, no real difference could be detected at 400 mOsm (Figure 3d). In all experiments, analysis of nuclear morphology indicated that the majority of cell death occured by apoptosis (Figure 5). However, under high osmolarity conditions the rate of cell death in the control cell line cultures appeared to be lower than that seen in the bcl-2 transfected cell line.

The hydrodynamic environment of culture system 3 was also less stressful than that of system 1, although the growth profiles were significantly different to those seen for system 2. For the control cell line at 300 mOsm, the lag phase lasted less than 24 hr which was followed by an exponential growth phase during which cell number reached a maximum of 15.5×10^5 mL^{-1} after 50 hr (Figure 3e). The culture then went into death phase which lasted 120 hr. The bcl-2 cell line reached a maximum cell number of 12.5×10^5 cells mL^{-1} after 72 hr (Figure 3f). However, the death phase was considerably longer, with the viable cell number falling to only 2×10^5 mL^{-1} after around 250 hr. In comparison, the control cell line had completely lost viability by 170 hr. In both cases, cell death was predominantly by apoptosis (data not shown).

At 350 mOsm, viable cell numbers peaked at about 8×10^5 mL^{-1} for both control and bcl-2 cell lines, although the death phase was much longer in the bcl-2 culture (Figure 3e and f). At 400 mOsm, the control cells exhibited a highly extended lag phase, with only a slight, low rate increase in viable cell number, which peaked at around 5×10^5 mL^{-1}. Viable cell number then declined slowly, falling to just under 2×10^5 cells mL^{-1} after 170 hr in culture. The bcl-2 transfected cells, by contrast, exhibited a lag phase of only 24 hr, with cell number peaking at around 6×10^5 mL^{-1} after 50 hr. Following a stationary phase of approximately 75 hr, the cell number rapidly decreased, with complete loss of viability after 150 hr. As in all other experiments, fluorescence microscopic analysis of nuclear morphology revealed that apoptosis was the major form of cell death in each culture (data not shown).

Adaptation to high-osmolarity conditions

The adaptation of control and bcl-2 transfected cells to 400 mOsm medium is shown in Figure 6a and b. Despite initial growth of control cells, the viable cell number gradually declined over subsequent passages, and the culture was lost after the 5th passage. Repeated attempts to adapt the cells to these conditions failed. Bcl-2 expressing cells initially showed high viability, although after subsequent passages there was an increase in the proportion of dead cells. However, in contrast to the control cell line, the culture viability improved after the fourth passage, after which the cells exhibited a growth pattern which was similar to that seen in standard culture medium.

Cell cultivation under high-osmolarity conditions in spin filter perfusion culture

Having adapted the bcl-2 transfected cells to hyperosmolarity conditions, productivity was assessed in spin filter perfusion cultures. Comparison with control cells under hyperosmolarity conditions was not possible due to their failure to adapt to a high salt environment. Instead, a comparison is made with bcl-2 cells in normal osmolarity medium. Under such conditions, non-adapted bcl-2 cells reached a maximum of 1.6×10^7 cells mL^{-1} after around 250 hr in culture (Figure 7a). By contrast, at 400 mOsm, adapted bcl-2 transfected cells reached a maximum of only 8×10^6 cells mL^{-1}, with a consistently higher proportion of dead cells throughout the duration of the culture (Figure 7b). Specific antibody productivity during the course of the two cultures showed that the cells grown at 400 mOsm were approximately twice as productive as cells cultivated under standard conditions (Table 1). More importantly, variability of the specific productivity during the run was significantly lower than that obtained for the control run. The results clearly show a much higher consistence in productivity in the adapted cell line.

Table 1. Mean and standard deviation (σ) of several time-points of the specific antibody productivities during the cultivation of bcl-2 transfected hybridoma cells at 300 and 400 mOsm in spin filter perfusion cultures

	Antibody specific productivity (μg, 10^6 cells hr^{-1})	
	300 mOsm	400 mOsm
Mean	0.305	0.603
σ	0.356	0.227

Discussion

Influence of bcl-2 on growth at suboptimal pH

Acidification of culture medium due to lactate production can reduce the pH to around 6.6 by the end of the exponential phase of batch culture. This has been demonstrated to be cytotoxic to the cells (Harbour et al., 1989; Macmicheal, 1989), and also adversely affects antibody productivity (Harbour et al., 1989). Addition of sodium hydroxide to control pH can result in regions of high pH, particular in large scale bioreactors due to the increased mixing times (Wayte et al., 1997). Presumably, exposure of cells to large fluctuations in pH will also invoke a stress response and consequently will affect both culture viability and

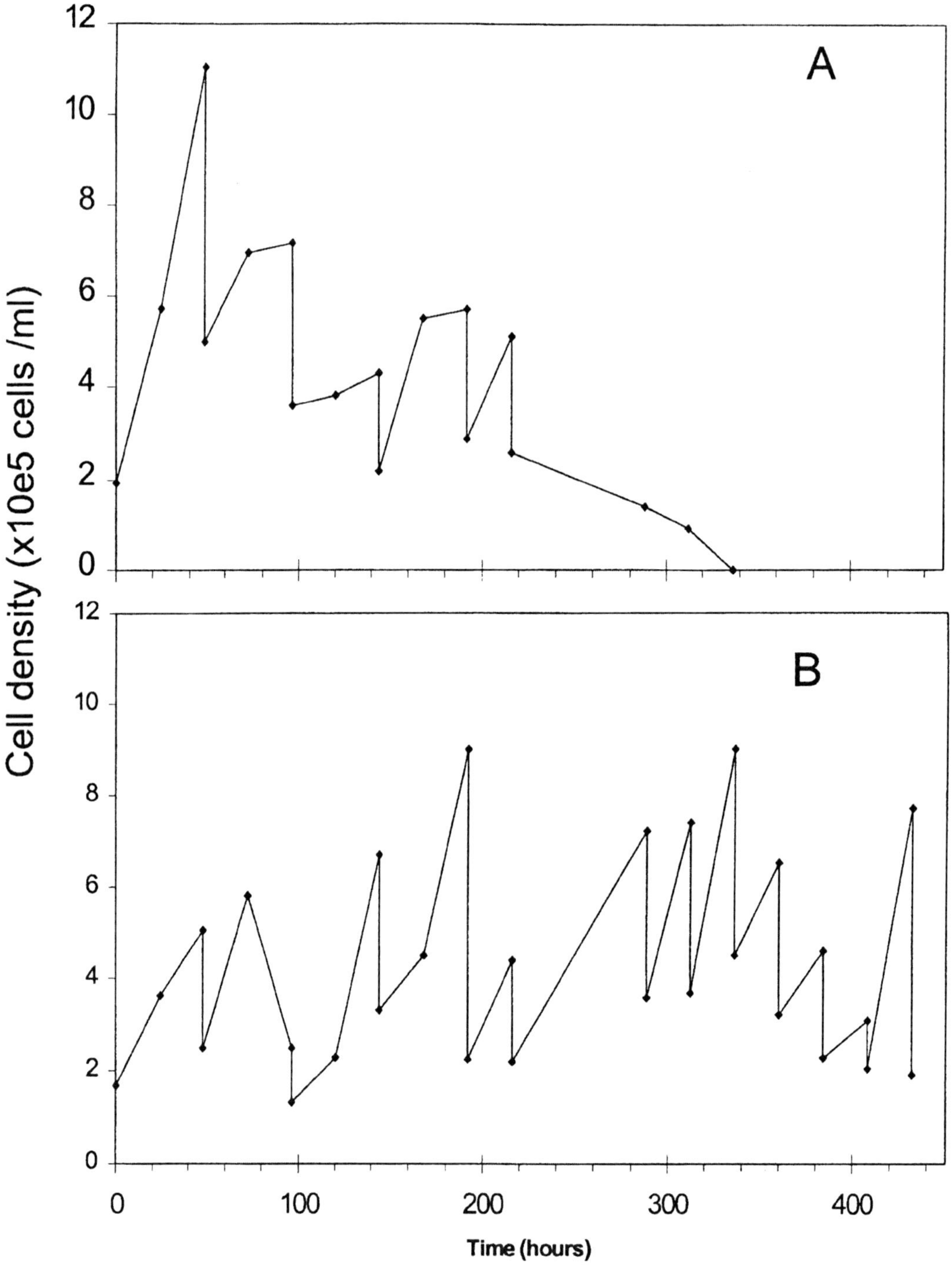

Figure 6. Viable cell number during adaptation to 400 mOsm medium for (A) control and (B) bcl-2 transfected cell lines.

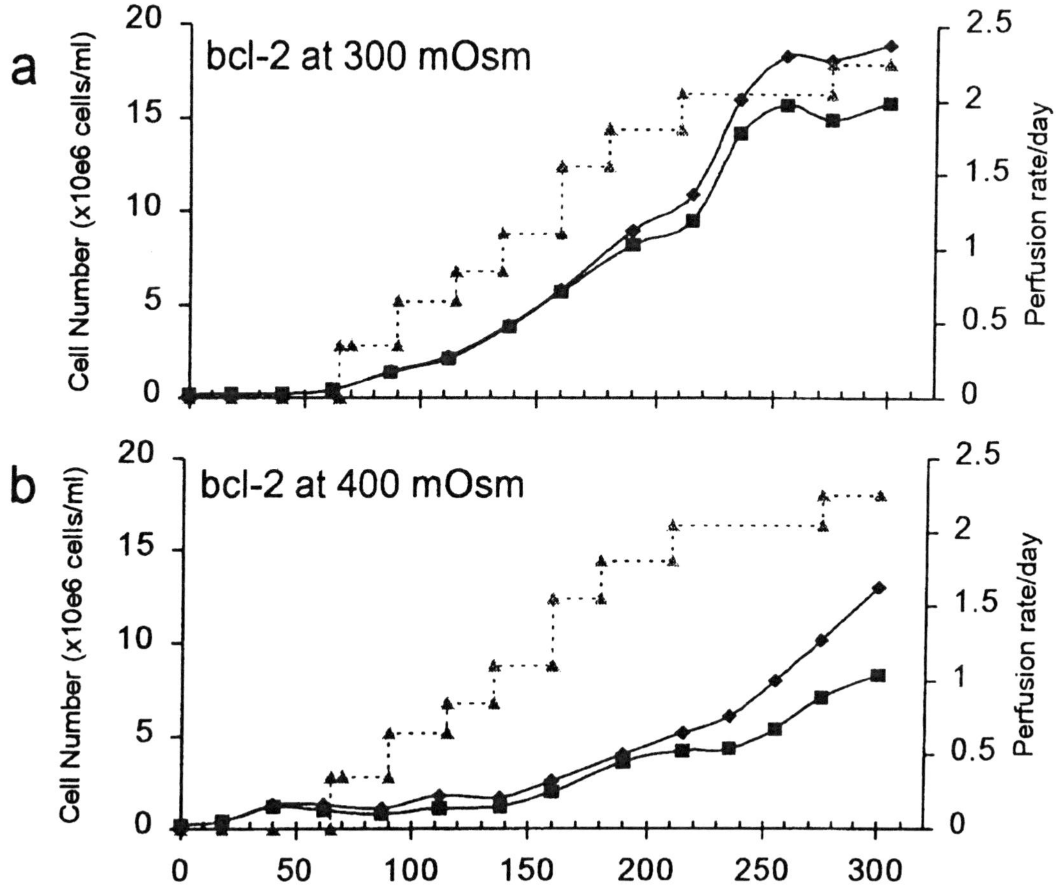

Figure 7. Spin filter perfusion culture using (a) non-adapted bcl-2 transfected cells at 300 mOsm and (b) adapted bcl-2 transfected cells at 400 mOsm. (Key: Diamonds: total cell number; Squares: viable cell number; Triangles: perfusion rate.)

antibody productivity. However, as confirmed in the present study, cells can tolerate a surprisingly large increase in pH, by as much as 1.1 units above neutral, whereas even slightly acidic conditions are more detrimental to culture viability. The results presented in Figure 2 suggest that cell tolerance to acidic conditions was not significantly influenced by bcl-2 expression. By contrast, there is a clear and consistent indication that bcl-2 expression did offer significant protection at pH 8. Further studies are required to assess the level of protection a) under more extreme culture conditions (i.e. higher than pH 8) and b) under both constant and fluctuating values of pH.

bcl-2 and the hydrodynamic environment

In recent years, two studies have investigated the role of hydrodynamic stress in modulating the apoptotic pathway. Initially, Al-Rubeai et al. (1995) reported that moderate levels of hydrodynamic stress resulted in the induction of significant levels of apoptotic cell death during murine B-cell hybridoma cultivation. However Dimmler et al. (1996) found that sub-lethal levels of shear stress resulted in suppression of apoptosis in endothelial cells. This different behaviour may indicate that the response to hydrodynamic stress is cell-type dependent and influenced by whether the cells are attached or in suspension when subjected to shear force.

In the present study, in reactor system 1, cells were exposed to a high shear environment generated by the action of the flat stirrer bar on the bottom of the glass culture vessel. In line with previous observations (Al-Rubeai et al., 1995), this resulted in the induction of high levels of apoptosis. As shown in Figure 3a, the expression of bcl-2 in the cells not only provided protection from apoptosis, but also allowed cell growth to levels not observed in standard suspension culture (reproducibility of results was obtained but not presented). Thus, it would appear that shear stress is extremely effective at stimulating cell growth, which is only evident when apoptosis is suppressed. Hydrodynamic stress has previously been reported to affect hybridoma cellular metabolic activity (Al-Rubeai et al., 1990), and a number of studies have identified genes which are inducible by shear stress in vascular tissue (JIN et al., 1996; Li et al., 1996; Hammond et al., 1997). Given the hydrodynamic environment of the B-cells from which hybridomas are derived, it is quite possible that the biochemical pathways which result in the induction of gene expression in response to physical stress are still active in hybridoma cells. This may explain the shear stress mediated stimulation of cell growth in the bcl-2 transfected cells reported in the present study.

Clearly, bcl-2 transfected cells exposed to the grinding action in system 1 will have undergone significant levels of physical damage and consequently will be extremely fragile. When cells in such a state are stained with trypan blue, they gradually become permeable to the stain. Thus, a culture which may initially have appeared to be highly viable, may exhibit a very large proportion of dead cells within a few minutes of staining. This may reflect the inability of stressed cells to tolerate the slight change in medium osmolarity which accompanies the addition of trypan blue solution. Such an effect, which has also been observed in previous unreported studies with stressed cells in our laboratory, would explain the difference in viability found between trypan blue analysis and fluorescence microscopic analysis. This clearly demonstrates the limitation of trypan blue staining for the determination of viability of highly stressed cells.

The difference in the growth profiles of the bcl-2 cultures in system 2 and 3 is also very interesting, as it suggests that the level of protection offered by bcl-2 expression varies depending on the exact configuration of the culture vessel. Thus, combining better bioreactor design with the use of apoptosis resistant cell lines may prove to be synergistic, and thereby further improve culture viability. Further studies are required to establish the basis of the differences observed in the present study.

bcl-2 and hyperosmolarity

Osmolarity of culture media typically ranges between 260 and 320 mOsm, values that were selected to mimic those conditions found *in vivo* (Ozturk and Palsson, 1991). Cell cultivation at osmolarity levels above this range has been reported to result in cellular growth arrest and an increase in specific antibody productivity (Oyaas et al., 1989; Ozturk and Palsson, 1991; Duncan et al., 1997). However, Oh et al. (1993) have found that when cells are initially passaged in 350 mOsm medium, growth arrest was clearly evident, although the specific productivity of the cells was not higher than that seen in control cells. Following a one week adaptation to these conditions, the cells exhibited a gradual increase in growth rate in addition to increased specific antibody productivity. This would indicate that the improved productivity of hybridoma cells under such conditions may not simply be a matter of growth arrest. It has been suggested that the high concentration of Sodium ions may enhance amino acid uptake, thereby stimulating the increased antibody productivity. However, this does not explain the increase in specific productivity seen following treatment with osmolytes such as sucrose and KCl.

The use of hyperosmolarity culture medium provides a simple and effective means of improving culture performance at large scale. However, when implementing such an approach, one has to consider the possibility that the resultant stress may lead to high levels of cell death. Clearly, under very high osmolarity conditions, the cells will simply undergo extensive damage resulting in passive cell death by necrosis. However, there will be a range of intermediate conditions in which cells will respond to the osmotic stress by undergoing apoptosis. In theory, the suppression of apoptosis by bcl-2 over-expression should allow cell cultivation at significantly higher stress levels than would otherwise be possible, thereby further stimulating antibody production. The results presented in this study support this prediction. As shown in Figure 6a and b, after the initial passage both cell lines exhibited a fall in viability. However, only the bcl-2 cell line recovered, and by the end of the adaptation period culture viability and cell growth were at levels similar to those observed in normal (300 mOsm) medium. Importantly, the cells adapted to 400 mOsm medium exhibited a doubling in specific antibody productivity.

Conclusion

An alkaline pH of 8 proved to be less cytoxic than an acidic pH of 6. Bcl-2 offers some degree of protection from apoptosis at the higher pH 8, an effect which is not so apparent at pH 6. High shear stress resulted in the induction of a high level of apoptosis in the control cell line. Expression of bcl-2 provided a high degree of protection under these conditions. Moreover, in the absence of apoptosis, the high shear stress environment stimulated a very high growth rate and resulted in a maximum cell number which was not achieved under standard culture conditions. The over-expression of bcl-2 also enabled adaptation of the cells to 400 mOsm medium, which was not possible in the control cell line. The adapted cell line then exhibited a 100% improvement in specific antibody productivity compared to the non-adapted bcl-2 over-expressing cells cultured in normal (300 mOsm) medium.

Together, these results demonstrate that bcl-2 over-expression results in the suppression of cell death induced by a wide range of stress factors which are not limited to nutrient and oxygen deprivation, growth arrest and viral infection.

Acknowledgements

This work is supported by the European Community Framework IV Programme.

References

Al-Rubeai M and Singh RP (1998) Apoptosis in cell culture. Current Opinion in Biotechnology 9: 152–156.

Al-Rubeai M, Oh SKW, Musaheb R and Emery AN (1990) Modified cellular metabolism in hybridomas subjected to hydrodynamic and other stresses. Biotechnology Letters 12: 323.

Al-Rubeai M and Emery AN (1990) Mechanisms and kinetics of monoclonal antibody synthesis and secretion in synchronous and asynchronous hybridoma cell cultures. Journal of Biotechnology 16: 67–86.

Al-Rubeai M, Emery AN, Chalder S and Jan, DC (1992) Specific antibody productivity and the cell cycle comparisons of batch, continuous and perfusion cultures. Cytotechnology 9: 85–97.

Al-Rubeai M, Mills D and Emery AN (1990) Electron Microscopy of hybridoma cells with special regard to monoclonal antibody production. Cytotechnology 4: 13–28.

Al-Rubeai M, Singh RP, Goldman MH and Emery AN (1995) Death mechanisms of animal cells in conditions of intensive agitation. Biotechnology and Bioengineering 45: 463–472.

Alnemri ES, Robertson NM, Fernandes TF, Croce CM and Litwack G (1992) Over-expressed full-length human bcl-2 extends the survival of baculovirus-infected sf9 insect cells. Proceedings of the National Acadamy of Sciences USA: 89: 7295–7299.

Boraston R, Thompson PW, Garland S and Birch JR (1984) Growth and oxygen requirements of antibody producing mouse hybridoma cells in suspension culture. Developments in Biological Standardization 55: 103–111.

Chen-Levy Z, Nourse M and Cleary ML (1989) The bcl-2 candidate proto-oncogene product is a 24 KDa integral-membrane protein highly expressed in lymphoid cell lines and lymphomas carrying the t(12:18). Molecular and Cellular Biology 9: 701–710.

Dimmeler S, Haendeler J, Rippmann V, Nehls M and Zeiher AM (1996) Shear Stress Inhibits Apoptosis of Human Endothelial Cells. FEBS Letters 399: 71–74.

Duncan PJ, Jenkins HA and Hobs G (1997) The effect of hyperosmotic conditions on growth and recombinant protein expression by NSO myeloma cells in culture. The Genetic Engineer and Biotechnologist 17:75-79.

Fazekas SG (1983) Automated production of monoclonal antibodies in a cytostat. Journal of Immunological Methods 57: 121–136.

Franek F and Dolnikova J (1991) Nucleosomes occuring in protein-free hybridoma cell cultures. Evidence for programmed cell death. FEBS Letters 248: 285–287.

Hammond TG, Goodwin TJ, Navar GL, Goda FO, Campbell WC, Ostberg JR, Verroust PJ and Kaysen JH (1997) Epithelial cell gene expression is shear stress responsive. Molecular Biology of the Cell 8: 2517.

Harbour C, Low KS, Marquis CP and Barford JP (1989) pH control options for hybridoma cultures. Biotechnology Techniques 3: 73–78.

Itoh Y, Ueda and Suzuki E (1995) Overexpession of bcl-2, apoptosis suppressing gene: prolonged viable culture period of hybridoma and enhanced antibody production. Biotechnology and Bioengineering 48: 118–122.

Jin G, Shyy Jy, Li YS and Chien S (1996) Effects of a dominant mutant of ras protein on shear-stress and cytokines induced gene-expression in the vascular endothelium. Circulation 94: 591.

Kidd VJ (1998) Proteolytic activities that mediate apoptosis. Annual Review of Physiology 60: 533–573.

King KL and Cidlowski JA (1998) Cell cycle regulation and apoptosis. Annual Review of Physiology 60: 601–617.

Kroemer G, Dallaporta B and Resche-Rigon M (1998) The mitochondrial death/life regulator in apoptosis and necrosis. Annual Review of Physiology 60: 619–642.

Li YS, Shyy JYJ, Li S, Lee Jd, Su B, Karin M and Chien S (1996) The ras-jnk pathway is involved in shear-induced gene-expression. Molecular and Cellular Biology 16: 5947–5954.

Macmicheal GJ (1989) The effectof pH and oxygen on cell growth, monoclonal antibody production, and metablism of a mouse hybridoma. International Biotechnology Laboratory, pp. 10–12.

Mastrangelo AJ and Betenbaugh MJ (1998) Overcoming apoptosis: New methods for improving protein-expression systems. Trends in Biotechnology 16: 88–95.

Mastrangelo AJ, Hardwick JM and Betenbaugh MJ (1996) Bcl-2 inhibits apoptosis and extends recombinant protein production in cells infected with Sindbis viral vectors. Cytotechnology 22: 169–178.

Mercille S and Massie B (1994) Induction of apoptosis in nutrient-deprived cultures of hybridoma and myeloma cells. Biotechnology and Bioengineering 44: 1140–1154.

Mitchell-Logean C and Murhammer DW (1997) bcl-2 Expression *in Spodoptera frugiperda* Sf 9 and trichoplusia ni bt-tn-5b1-4 insect cell: effect on recombinant protein expression and cell viability. Biotechnology and Bioengineering 56: 380–388.

Oh SKW, Vig P, Chua F, Teo WK and Yap MGS (1993) Substantial overproduction of antibodies by applying osmotic pressure and sodium buterate. Biotechnology and Bioengineering 42: 601–610.

Oyaas K, Berg TM, Bakke O and Levine DW (1989) Hybridoma growth and antibody productivity under conditions of hyperosmotic stress. In: RE Spier, JB Griffiths and PJ Crooy (eds.) Advances in animal cell biology and technology for bioprocesses. Butterworth, Essex, U.K., pp. 212–220.

Ozturk SS and Palsson BO (1991) Effect of medium osmolarity on hybridoma growth, metabolism and antibody production.

Biotechnology and Bioengineering 37: 989–993.

Reuveny S, Velez D, Miller L and MacMillan JD (1986) Comparison of cell propagation methods for their effect on monoclonal antibody yeild in fermentors. Journal of Immunological Methods 86: 61–69.

Simpson NH, Singh RP, Perani A, Goldenzon C and Al-Rubeai M (1998) In hybridoma cultures, deprivation of any single amino acid leads to the induction of apoptosis, which is suppressed by the *bcl-2* gene. Biotechnology and Bioengineering 59: 90–98.

Simpson N, Milner AN and Al-Rubeai M (1997) Prevention of hybridoma cell death by bcl-2 during sub-optimal culture conditions. Biotechnology and Bioengineering 54: 1–16.

Singh RP, Al-Rubeai M, Gregory CD and Emery AN (1994) Cell death in Bioreactors: A role for apoptosis. Biotechnology and Bioengineering 44: 720–726.

Singh RP, Emery AN and Al-Rubeai M (1996) Enhancement of survivability of mammalian cells by over-expression of the apoptosis supressor gene bcl-2. Biotechnology and Bioengineering 52: 166–175.

Singh RP and Al-Rubeai M (1998) Apoptosis and Bioprocess Technology. Advances in Biochemical Engineering/Biotechnology 62: 167–184.

Thompson EB (1998) Special topic: Apoptosis. Annual Review of Physiology 60: 525–532.

Thompson EB (1998) The many roles of c-Myc in apoptosis. Annual Review of Physiology 60: 575–600.

Tsujimoto Y, Cossman J, Jaff E and Croce CM (1985) Involvement of the bcl-2 gene in human follicular lymphoma. Science 228: 1440–1443.

Vaux DL, Cory S and Adam J (1988) bcl-2 gene promotes haemopoietic cell survival and co-operates with c-myc to immortilize pre-B-cells. Nature 335: 440–442.

Wayte R, Boraston H, Varley J and Brown M (1997) pH: effects on growth and productivity of cell line producing monoclonal antibodies: Control in large scale fermentors. The Genetic Engineer and Biotechnologist, 17: 125–133.

Yang E and Korsemeyer S (1996) Molecular Thanatopsis – a Discourse on the Bcl-2 Family and Cell-Death. Blood 88: 386–401.

Cytotechnology **28:** 189–203, 1998.
© 1998 *Kluwer Academic Publishers. Printed in the Netherlands.*

Apoptosis-resistant NS/0 E1B-19K myelomas exhibit increased viability and chimeric antibody productivity under cell cycle modulating conditions

Sylvain Mercille[1] & Bernard Massie[1,2,3]*

[1] *Animal Cell Engineering Group, Biotechnology Research Institute, National Research Council of Canada, 6100 Royalmount Ave, Montréal, P.Q., Canada H4P 2R2*

[2] *Centre de Recherche en Virologie, Institut Armand Frappier, Université du Québec, Laval, P.Q., Canada, H7N 4Z3*

[3] *Département de Microbiologie et Immunologie, Faculté de Médecine, Université de Montréal, C.P. 6128, Succursale Centre-Ville, Montréal, P.Q., Canada, H3C 3J7*

E-mail: bernard.massie@nrc.ca

Received 14 September 1998; accepted 14 September 1998

Key words: apoptosis, cell cycle, E1B-19K, hydroxyurea, hyperosmosis, hypertonic, monoclonal antibodies, OptiMAb^{TM}, thymidine

Abstract

Lymphoid cells expressing sufficient levels of Bcl-2 or E1B-19K are known to resist to induction of apoptosis in glutamine-free or nutrient-limited batch cultures. However, despite the increased viability and prolonged stationary phase achieved in batch culture, product yields are not necessarily improved. Here we have found that expression of E1B-19K in NS/0 myeloma cells cultivated in the presence of certain cell cycle modulators could result in a significant increase in MAb productivity as compared to untransfected control cells. The use of E1B-19K significantly enhanced cell survival in the presence of osmolytes (sorbitol, NaCl), DNA synthesis inhibitors (hydroxyurea, excess thymidine), and the cell culture additive OptiMAb^{TM}. E1B-19K myelomas cultivated in the presence of NaCl or OptiMAb^{TM} accumulated in the G_1 phase, while those arrested with excess thymidine were blocked in all phases. Interestingly, control NS/0 cells treated with these agents were found to die in a cell-cycle specific manner. Thus, while all G_1 and most S phase cells quickly underwent apoptosis, G_2/M cells remained alive and maintained MAb secretion for more than 10 days if supplied with adequate nutrients. For both control and E1B-19K cells, incubation with sorbitol or hydroxyurea was detrimental for MAb secretion, while addition of NaCl, excess thymidine and OptiMAb^{TM} resulted in an increased specific MAb productivity as compared to the batch culture. However, this increase resulted in an improvement of final MAb yields only in the case of OptiMAb^{TM}. The extension of viability conferred by E1B-19K allowed to further improve the final MAb yield obtained using OptiMAb^{TM} with a 3.3-fold increase for E1B-19K cells as compared to 1.8-fold for control NS/0 cells.

Abbreviations: HT – Hypertonic; HU – Hydroxyurea; IT – Isotonic; MAb – Monoclonal antibodies; PI – Propidium Iodide; q_{MAb} – Specific MAb production rate; TTP – Thymidine triphosphate; XST – Excess thymidine; X_t – Total cell density; X_v – Viable cell density; μ – Specific growth rate.

Introduction

In the last fifteen years, increasing demand for proteins derived from eukaryotic cells has led to many developments in the field of animal cell technology. The appreciation of the apoptotic nature of many cell types used in the production of biopharmaceuticals has stimulated efforts to generate genetically engineered apoptosis-resistant cell lines in an attempt to

* Author for all correspondence.

increase viability and productivity in culture processes (reviewed in Mosser and Massie, 1994; Mastrangelo and Betenbaugh, 1998). Genes such as Bcl-2 (Fassnacht et al., 1998; Itoh et al., 1995; Murray et al., 1996; Singh et al., 1996), Bcl-2 in combination with Bag-1 (Terada et al., 1997) and adenoviral E1B-19K (Mercille et al., submitted) have been tested into production cell lines. While some researchers have observed that Bcl-2 overexpression could result in improved MAb productivity in batch culture (Simpson et al., 1997; Suzuki et al., 1997), others have found no effect (Fassnacht et al., 1998). In a previous study, we found that the dose-dependent enhancement of viability conferred by E1B-19K in batch cultures of NS/0 cells did not lead to significant antibody productivity throughout the prolonged stationary phase (Mercille et al., submitted). We postulated that this could be explained by the early exhaustion of glutamine or other limiting nutrients essential for protein synthesis, and that cultivation of apoptosis-resistant cell lines in other modes of culture could result in both improved viability and increased MAb yields. Indeed, we have recently demonstrated that E1B-19K-expressing myeloma cells grown in perfusion culture exhibited a significant increase in MAb production (Mercille and Massie, submitted). Cultivation of apoptosis-resistant production cell lines in culture medium supplemented with cell cycle modulators such as osmolytes or inhibitors of DNA synthesis may also lead to significant increases in antibody concentration as compared to apoptosis-susceptible cell lines grown under similar conditions.

A number of studies have shown that hypertonic stress can increase the specific MAb productivity of hybridoma cells (Oh et al., 1993; 1995; Øyaas et al., 1994a; 1994b; Ozturk and Palsson, 1991; Reddy et al., 1992; Reddy and Miller, 1994; Yang et al., 1996). This has been attributed in part to increased amino acid uptake via the Na^+-dependent symport (Oh et al., 1995). The use of osmolytes such as NaCl, KCl, sucrose, choline chloride, and sorbitol usually results in apparent growth arrest and accumulation of cells in the G_1 phase of the cell cycle (Fuhrman Conti et al., 1990; Pellicciari et al., 1995). However, hypertonic conditions are a potent inducer of apoptosis in a variety of cell lines (Pandey et al., 1996; Qin et al., 1997; Matthews and Feldman, 1996; Mathews et al., 1997; Anderson, 1997; Ortiz et al., 1997) such as S49 lymphocytes, primary thymocytes (Bortner and Cidlowski,1996), SK-N-SH and HeLa cells (Galvan and Roizman, 1998). In neurons, the levels of Bcl-2 have been shown to modulate the susceptibility to induction of apoptosis in response to hypertonic conditions. Singleton et al. (1996) reported that osmotic stress decreases total Bcl-2 by 4-fold and correlates with the processing of neuronal ced-3/ICE-like proteases. In these cells, type I insulin-like growth factor receptor activation maintains normal Bcl-2 levels, enhances the negative death regulator Bcl-x_L and prevents apoptosis.

Another approach aiming at increasing the productivity of cultured cells has been through the inhibition of DNA replication with the use of hydroxyurea or excess thymidine (Terada et al., 1997; Suzuki and Ollis, 1990; Takahashi et al., 1994; Al-Rubeai et al., 1990). Inhibition of ribonucleotide reduction either by increasing dATP and dTTP pools by addition of deoxyadenosine and thymidine to culture medium, or by hydroxyurea, leads to a rapid cessation of DNA synthesis (Stubbe, 1990; Fontecave et al., 1992). However, stalled replication forks are unstable and DNA repair mechanisms become rapidly incapable of coping with long periods of DNA synthesis arrest. To avoid extensive DNA damage accumulation, apoptosis is rapidly induced following prolongation of deoxynucleotide depletion (Paulovich et al., 1997; Barbiero et al., 1995). Hydroxyurea has been extensively studied as a suitable chemotherapeutic agent capable of inhibiting tumor cell growth by induction of apoptosis (Schrell et al., 1997; Anand et al., 1995). However, since Bcl-2 and Bcl-X_L are expressed in a variety of tumors, their expression can modulate the sensitivity of tumor cells to chemotherapeutic agents such as hydroxyurea (Simonian et al., 1997). In an effort to increase MAb yields and prevent induction of apoptosis following addition of excess thymidine to culture medium, genes such as Bcl-2 (Simpson et al., 1997) and Bcl-2 in combination with Bag-1 (Terada et al., 1997) have been transfected into hybridoma cells. Although expression of those genes was successful in improving host cell survival (Terada et al., 1997) and increasing specific MAb productivity (Simpson et al., 1997), it still has not been demonstrated that this resistance can result in increased final MAb yields in growth-arrested conditions as compared to normal batch culture conditions.

The culture media additive known as OptiMAb™ (Gibco, Grand Island, NY) is a MAb production enhancer containing a cell cycle modulator and a nutrient supplement of proprietary composition. Enhancements in both specific MAb productivity and yields have been previously reported with the use of this agent (Fike, 1990; Fike et al., 1991a; 1991b; Jayme, 1991). In a previous study, we have shown that addition of OptiMAb™ to batch cultures of D5 hybridoma cells resulted in an increase in the G_1 phase population, in a 50% increase in specific MAb production rate and in an overall enhancement of MAb concentration (Mercille et al., 1994). However, this increase in productivity was only transient since cell viability declined after 2 days in culture. In the present paper, we have examined whether expression of E1B-19K in myeloma cells could prevent induction of apoptosis and increase final MAb yields in growth-modulating culture conditions achieved using agents such as sorbitol, NaCl, hydroxyurea, excess thymidine or OptiMAb™.

Materials and methods

Cell lines, media and additives

The NS/0 myeloma cell line used in this study was derived from a NS1/1.AG4.1 clone originally obtained from Caesar Milstein (Cambridge University, U.K.). The construction of the E1B-19K and chB43 chimeric monoclonal antibody expressor plasmids, the transfection protocol, and the establishment of stably transfected antibody-producing control NS/0 and E1B-19K cell lines have been described elsewhere (Mercille et al., submitted). Selection of the final sub-clone constitutively expressing both E1B-19K and chB43 was not only done on the basis of high levels of secretion, but also on the basis of high stability of chB43 expression and best apoptosis resistance. Although we initially tried to obtain control NS/0 and E1B-19K clones with similar MAb secretion levels in batch, we considered that stability of MAb and E1B-19K expression was paramount for long-term culture. Unfortunately, following the various rounds of cloning and selection of stable clones, the best E1B-19K clone selected for this study turned out to have a lower secretion in batch as compared to the control NS/0 clone. Both parental and E1B-19K myelomas used in the present study tested negative for mycoplasma contamination using a Hoechst staining and co-culturing detection assay (ATCC quality control methods for cell lines, 1st edition, 1985, pp. 12–15). Cells were cultivated in high glucose Dulbecco's modified Eagle's medium (DMEM) (Gibco, Grand Island, NY) supplemented with 5% fetal bovine serum (Hyclone, Logan, Utah), and maintained at 37 °C in a 5% CO_2 atmosphere. Spinner flask cultures were supplemented with 0.1% Pluronic F68 (Gibco) for protection against shear stress. Every culture was started from the same batch of inoculum vials kept in liquid nitrogen.

Cell cycle modulators

Hypertonic medium (HT) medium was prepared with sorbitol or NaCl as osmolytes. All osmolarity measurements were made using a micro-osmometer (Model 3MO+, Advanced Instruments Inc, Needham Heights, MA). The osmolarity of isotonic (IT) DMEM was 335 mOsm. Using sorbitol, 1400 mOsM medium was prepared by dissolving one mole of sorbitol into DMEM and filtering with a 0.22 μm filter (Millipore). Three different HT media with osmolarities of 385, 425, and 475 mOsM were also prepared by 1:150, 1:100 and 1:75 addition of a 5 M NaCl stock solution (prepared in water) into isotonic DMEM. Given that 110 mM NaCl is already present in DMEM, the final NaCl concentration of these three HT media was 143, 160 and 176 mM, respectively. In order to achieve inhibition of DNA synthesis, cells were grown in medium supplemented with hydroxyurea (HU) or excess thymidine (XST). Medium containing 10 mM HU (Bianchi et al., 1986) was prepared by 1:100 addition of a 1 M HU stock solution (prepared in water) into isotonic DMEM. Three different XST media were prepared by 1:50, 1:30 and 1:15 addition of a 150 mM thymidine stock solution (prepared in water) into DMEM. The final thymidine concentration of these three media was 3, 5 and 10 mM, respectively. In order to maintain the thymidine concentration, medium replacements were performed every 2 days by centrifuging the cells and resuspending them in fresh thymidine-supplemented medium. Both component A and B of the OptiMAb™ antibody production enhancer were added in concentrations prescribed by the manufacturer (Gibco). Batch controls and cultures supplemented with sorbitol, NaCl, HU and XST were performed in 6-well plates. OptiMAb™ experiments and their controls were performed in 100 mL spinner flasks.

Analytical methods

Viable and total cell concentrations were evaluated using erythrosin B (Sigma) dye exclusion tests and haemacytometer counts. The concentration of human immunoglobulin G_3 in culture supernatants was measured using a standard ELISA protocol described elsewhere (Gervais et al., 1998; Mercille et al., submitted).

Cell cycle distribution was assessed by measuring the DNA content of propidium iodide stained cells. Conditions of cell fixation, coloration, flow cytometric analyses, data collection and analysis and quantification of cell cycle phases are detailed elsewhere (Mercille and Massie, submitted; Mosser et al., 1997). All the samples selected for determination of cell cycle profiles have been analyzed during the same session. Appropriate controls such as DNA beads have been used before and after the analysis to ensure that no drift had occurred. Throughout the analysis run, G_1 and G_2/M peaks in all samples did not move, exhibiting constant PI staining intensity (DNA content). Two methods were used for quantification of apoptosis. In the first method, cell suspensions were mixed with fluorescent DNA-binding dyes and examined by fluorescence microscopy to visualise and count cells with apoptotic nuclear morphology (Duke and Cohen, 1992; Mercille and Massie, 1994). In the second method, apoptotic cells were quantified by determination of the sub-G_1 population following cytometric cell cycle analysis. Cell fixation in ethanol followed by rinsing in aqueous buffers (Hotz et al., 1994; Telford et al., 1992) leads to extraction of the mono- and oligonucleosomes produced consequent to the action of endonucleases (Sokolova et al., 1992). Subsequent staining with DNA fluorochromes such as propidium iodide (PI) results in a discrete peak with reduced fluorescence as compared to G_1 non-apoptotic cells of the same ploidy. Since it can be difficult to discriminate between apoptotic cells and debris in the sub-G_1 population, we have used a logarithmic instead of a linear scale on the x-axis (DNA content/intensity of fluorescence from PI) to achieve a better discrimination between debris and apoptotic cells. This slight modification to the established protocol is documented in Mercille and Massie (submitted). Similar results were obtained with these two methods.

Computational methods

Specific growth rate (μ) was defined as $(dX_t/dt)(1/X_v)$ where X_t ($\times 10^8$ cells L^{-1}) is the total cell density (including both viable and dead cells) and X_v ($\times 10^8$ cells L^{-1}) is the viable cell density. Considering the fact that the control and test clones did not secrete equivalent levels of MAbs in batch culture, we did not make direct comparisons between control NS/0 and E1B-19K cells under arrested conditions. The levels of secreted MAbs and specific MAb productivity (q_{MAb}) of control NS/0 and E1B-19K cells were compared on the basis of ratios obtained by dividing the MAb concentrations or q_{MAb} value of a given clone under cell cycle modulating conditions by the MAb or q_{MAb} value obtained with the same clone in batch culture in isotonic medium. Specific MAb production rate (q_{MAb}) was defined as $(dMAb/dt)(1/X_v)$ where MAb is the antibody concentration (mg L^{-1}). For both μ and q_{MAb}, rates were evaluated at each data point by taking 3-point slopes as a function of time (days). For example, in the case of μ: $dX_t/dt = \Delta X_t/\Delta t$. This slope was then divided by the corresponding average value of X_v (middle point) to obtain the specific rates.

Results

Effect of E1B-19K on viability and MAb productivity in hypertonic medium

In order to assess whether E1B-19K could protect NS/0 myeloma cells from induction of apoptosis in hypertonic (HT) medium and could allow for increased MAb concentration, both control NS/0 and E1B-19K-expressing cells were cultivated in medium supplemented with sorbitol or NaCl osmolytes. Sorbitol was used to temporarily increase the osmolarity of the culture medium to 1400 mOsm. Cells seeded at 8×10^5 cells mL^{-1} were incubated for 2 hr in medium containing 1 M sorbitol and then reincubated in isotonic (IT) medium for 18 hr. Following the 18-hour incubation period, control NS/0 cells exhibited a viability of 15% while E1B-19K cells had a viability of 68%. Various clones containing different levels of E1B-19K were also tested similarly. The resistance to induction of apoptosis following a 2-hour sorbitol osmotic shock and 18-hour recovery was found to be proportional to the amount of E1B-19K expressed in a given clone (data not shown). For both cell lines, dead cells were found to exhibit apoptotic morphology. E1B-19K therefore partially protected the cells from apoptotic death following sorbitol osmotic shock. In order to achieve an HT medium more suitable for long term cell cultivation and antibody production, the osmolyte NaCl was used instead of sorbitol. Concentrations of 33, 50 and 66 mM were tested on the cells but only the latter was sufficient to stop the increase in viable cell density (data not shown). Supplementation with 66 mM NaCl resulted in an increase in the osmolarity from 335 to 475 mOsm. For both control NS/0 and E1B-19K cells, a batch culture performed in IT medium was included in the experiment. This batch

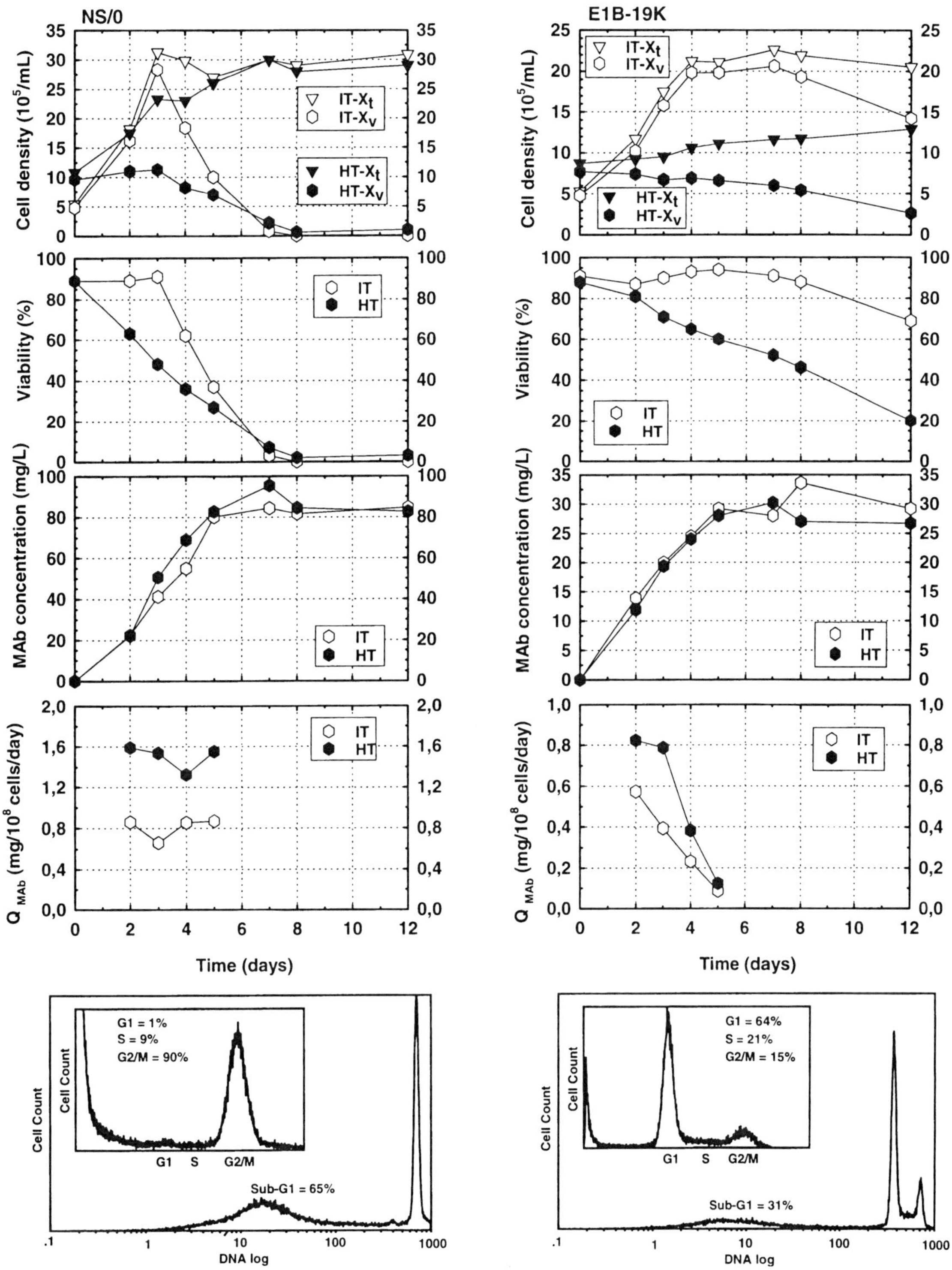

Figure 1. Effect of E1B-19K on cell survival, MAb productivity and cell cycle distribution of NS/0 myelomas grown in hypertonic medium (supplemented with 66 mM NaCl). This figure presents the evolution of viable (X_v) and total (X_t) cell density, viability (%), MAb concentration and specific MAb productivity (q_{MAb}) of control NS/0 (left panels) and E1B-19K-expressing cells (right panels) cultivated throughout a 12-day period in isotonic (IT) and hypertonic (HT) medium (supplemented with 66 mM NaCl). Bottom panels present cell cycle profiles of control NS/0 (left panel) and E1B-19K cells (right panel) after 5 days in HT medium. Each lower panel presents both log and linear (inset) scales of the same sample. Log scale profiles allow to distinguish the sub-G_1 apoptotic peak while linear scale insets present the cell cycle distribution into G_1, S and G_2/M phases of non-apoptotic cells, excluding the sub-G_1 cells which make up for a large fraction of the sample, especially in the case of control NS/0 cells.

control was important since large variations are often observed in the specific growth rates, maximal viable cell densities, and specific MAb productivities of different clones. The viable density of control NS/0 cells exposed to NaCl leveled off initially around 10×10^5 cells mL^{-1} for 3 days and then decreased steadily until day 8 (Figure 1, left panels). Immediately after exposing the cells to the HT medium, a rapid and linear decrease in viability was observed concomitant with the accumulation of sub-G_1 apoptotic cells (Figure 1, bottom left panel). Although viable cell density quickly leveled off and declined, the growth of control NS/0 myeloma cells was not arrested. Indeed, the accumulation of total cells up to 28×10^5 cells mL^{-1} throughout the culture indicated that cells were still dividing. A constant specific growth rate (μ) of 0.28 day^{-1} was measured for control NS/0 cells during the first 5 days of culture in HT medium, as compared to a μ of 0.51 day^{-1} for exponentially growing control NS/0 cells in IT medium. Therefore, a leveling-off or decrease in the viable cell density in HT medium does not necessarily imply that cells are undergoing growth arrest, in contrast to what is sometimes reported in the literature (Oh et al., 1993; 1995). At day 5, control NS/0 cells in HT medium were collected for cell cycle analysis. The sub-G_1 cells made up for 65% of the population (Figure 1, bottom left panel) and the resulting viability estimate of 35% obtained with this cytometric technique was close to the 27% measurement made using the standard dye exclusion test. The remaining viable control NS/0 cell population at day 5 (excluding sub-G_1 cells) was composed of 1% G_1 , 9% S and 90% G_2/M phase cells. At the beginning of the experiment (day 0), exponentially growing control NS/0 cells used as inoculum were composed of 32% G_1, 52% S and 16% G_2/M phase cells. Thus at day 5, all G_1 and most S phase cells had disappeared, concomitant with a large increase in the sub-G_1 cells, whereas the G_2/M phase cells had increased from 16 to 32% of the total population. Indeed, on day 5, only 35% of the total population was viable and 90% of these viable cells were composed of G_2/M, making up for a resulting percentage of 32%. For control NS/0 cells, the evolution of the MAb concentration followed the same trend in both HT and IT media. As is typically observed in batch culture of lymphoid cells, MAbs were produced in IT medium throughout the exponential growth phase and further accumulation could be observed in the 2 to 3 days that followed the onset of the decline phase. In HT conditions, MAbs were produced continuously until day 5 when the viable cell number started to decrease drastically. A 2-fold increase in specific MAb productivity (q_{MAb}), from 0.8 mg/10^8 cells day^{-1} in IT conditions as compared to 1.5 mg/10^8 cells day^{-1} in HT conditions, was concomitant with the 2-fold decrease in specific growth rate described above. However, due to the continuous production of a very large number of dead cells in the HT conditions, a significant contribution from the passive release of antibodies from dying cells may have led to an overestimation of q_{MAb}. In both IT and HT media, the MAb concentration reached a maximum of 80 to 85 μg mL^{-1} at day 5 and no real gain was observed.

E1B-19K-expressing NS/0 cells exhibited very different growth kinetics in HT medium. Indeed, upon incubation with the osmolyte, the viable density of E1B-19K cells leveled off before decreasing gradually (Figure 1, right panels), but at a much lower rate than the control NS/0 culture. Thus, after 7 days in HT medium, 52% of the E1B-19K cells were still viable, as compared to only 7% in the control NS/0 culture. Furthermore, in contrast to control cells, a greater reduction in μ from IT to HT conditions was observed for E1B-19K-expressing cells. Indeed, a constant μ of 0.07 day^{-1} was measured for E1B-19K cells during the first 8 days of culture in HT medium, as compared to a μ of 0.40 day^{-1} (measured at day 2) for the same cells in IT medium. The cell cycle profile of E1B-19K cells in HT medium at day 5 was markedly different from the control NS/0 cells; only 31% of the population appeared as apoptotic sub-G_1 cells while the viable cell population (excluding sub-G_1 cells) was composed of 64% G_1cells, 21% S phase cells and 15% G_2/M cells. Cells transfected with the E1B-19K gene product and exposed to HT medium were therefore able to accumulate in the G_1 phase without extensive apoptosis. Fluorescent microscopic evaluation of the culture revealed that non-viable E1B-19K cells nonetheless exhibited apoptotic morphology (data not shown), as confirmed by the presence of the sub-G_1 peak. The evolution of the MAb concentration of E1B-19K cells followed similar trends in both IT and HT conditions with a similar maximal MAb concentration of 28 μg mL^{-1} at day 5. As for control NS/0 cells, an increase in q_{MAb} was observed for E1B-19K cells in HT conditions. However, the greater reduction in specific growth rate achieved under HT conditions with the E1B-19K cell (6-fold reduction for E1B-19K cells versus 2-fold reduction for control NS/0 cells) did not result in a greater increase in q_{MAb}, and was not translated into a greater antibody concentration. In fact, although the E1B-19K cells retained a much higher viability in HT conditions throughout the culture, they stopped producing antibodies at day 5, at which point they were still 60% viable.

Effect of E1B-19K on viability and MAb productivity under DNA synthesis inhibition

The effect of E1B-19K on apoptosis protection and MAb productivity under conditions of DNA synthesis inhibition was tested using 10 mM HU or 10 mM XST. Immediately after exposing control NS/0 cells to HU, the viable cell density decreased rapidly and most cells died from apoptosis within 2 days (Figure 2, left panels). The negligible increase in the total cell number throughout the culture indicated that HU inhibited DNA replication and prevented cell division. As for control NS/0 cells, incubation of E1B-19K cells in HU resulted in total growth arrest (Figure 2, right panels). In this case, the decrease in viability was significantly slower than in the control NS/0 culture. Thus, after 4 days in HU, 65% of the E1B-19K cells were still viable, whereas the control NS/0 cells were all dead. E1B-19K cells therefore exhibited resistance to apoptosis induced by inhibition of DNA replication. No cell cycle profiles are presented in Figure 2 for HU conditions as all control NS/0 cells were dead when the analysis was performed on day 5. MAb concentrations measured at day 9 were as follows: 81.6 and 25.1 μg mL^{-1} for control NS/0 cells in batch and HU conditions respectively; 21.1 and 5.6 μg mL^{-1} for E1B-19K cells in batch and HU conditions respectively. Since MAb productivity in the presence of HU was considerably reduced for both cell lines, the use of this cell cycle modulator presented little advantage.

In an effort to find more physiological conditions under which long-term DNA synthesis inhibition would not result in inhibition of MAb synthesis, another agent, XST, was subsequently tested at three different concentrations: 3, 5 and 10 mM. Concentrations of 3 and 5 mM were not sufficient to inhibit DNA synthesis or prevent accumulation of viable cells (data not shown). Experiments were therefore performed in medium supplemented with 10 mM thymidine with medium replacements every 2 days to insure that thymidine remained in excess. As usual, a batch control culture was included in the experiment for both control NS/0 and E1B-19K cells. The viable density of control NS/0 cells exposed to XST conditions increased during the first 2 days, decreased from day 2 to day 5 where it reached 5×10^5 cells mL^{-1} and only marginally increased afterwards (Figure 3, left panels). Immediately after exposing the cells to XST medium, a rapid decrease in viability could be observed associated with the accumulation of apoptotic sub-G_1 cells (Figure 3, bottom left panel). This decrease in viability lasted for 5 days where it reached 36% and leveled-off afterwards until the experiment was terminated. The small increase observed between day 8 and day 10 was probably artefactual and can be explained by the loss of a fraction of the dead cells upon medium changes every two days. The negligible increase in the total cell number throughout the culture indicated that XST inhibited DNA replication and prevented cell division. Indeed, in contrast to HT conditions (NaCl), the specific growth rate of control NS/0 cells in XST conditions was considerably reduced. A constant μ of 0.13 day^{-1} was measured for control NS/0 cells during the first 5 days of culture in XST medium, as compared to of 0.61 day^{-1} for exponentially growing control NS/0 cells. At day 5, the sub-G_1 cells made up 68% of the control NS/0 cell population in XST and the remaining viable cell population (excluding sub-G_1 cells) was composed of 2% G_1, 42% S and 56% G_2/M phase cells (Figure 3, left bottom panel). The G_1 cell population that made 32% of total cells at day 0 had disappeared almost totally after 5 days of exposure to XST, concomitant with the increase in the sub-G_1 cells. The S phase cells also decreased from 55% at day 0 to 13% of the total population at day 5. Indeed, on day 5, only 32% of the total population was viable and 42% of these viable cells were composed of S phase cells, making up for a resulting percentage of 13%. The percentage of G_2/M cells present at day 0 (16%) remained stable at 18% on day 5. In control batch conditions, MAbs reached a maximum of 98 μg mL^{-1} on day 5. In XST conditions, the cumulative evolution of the MAb concentration was computed by adding the MAb concentration found in the successive medium replacements performed at 48-hour intervals. The evolution of the cumulative MAb concentration followed the same trend as in standard medium until day 5. However, the S and G_2/M phase cells that remained viable after day 5 kept on secreting antibodies until day 10, accumulating up to 155 μg mL^{-1}. A 3-fold increase in q_{MAb} from 0.75 mg/10^8 cells day^{-1} in standard batch conditions to 2.25 mg/10^8 cells day^{-1} in XST conditions was concomitant with the 4-fold decrease in growth rate described above.

Myeloma cells expressing E1B-19K exhibited very different growth kinetics in XST medium. While incu-

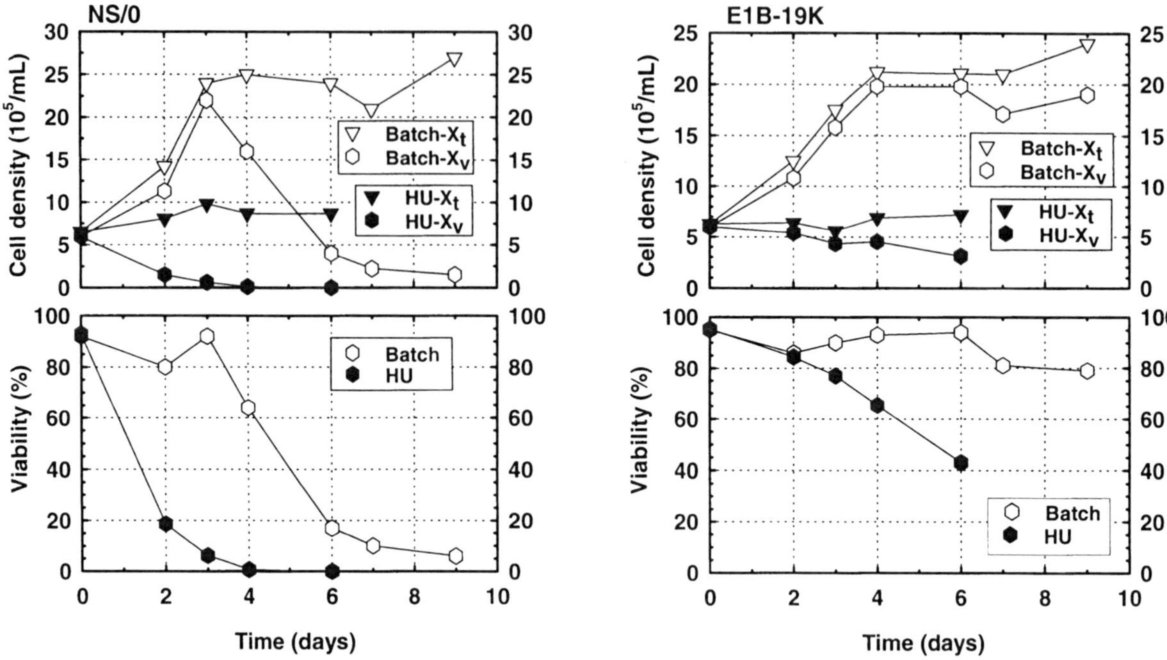

Figure 2. Effect of E1B-19K on cell survival of NS/0 myelomas grown in the presence of the DNA synthesis inhibitor hydroxyurea. This figure presents the evolution of viable (X_v) and total (X_t) cell density and viability of control NS/0 (left panels) and E1B-19K cells (right panels) cultivated throughout a 9-day period in standard (batch) and 10 mM hydroxyurea (HU) medium.

bation of E1B-19K cells in XST medium also resulted in an arrest in growth rate (Figure 3, right panels), the viability remained between 80–90% throughout the culture. Thus, after 5 days in XST medium, the viability of the E1B-19K cells measured by dye exclusion test was 89%, as compared to 36% for the control NS/0 cells at day 5. Similarly as for the HT (NaCl) condition, the cell cycle profile of E1B-19K cells at day 5 was totally different from the control NS/0 cells (Figure 3, right bottom panel). Indeed, only 16% of the population appeared as sub-G_1 apoptotic cells and the cell cycle profile of viable cells (excluding sub-G_1 cells) was similar to what was found for exponentially growing cells with 32% G_1, 51% S and 17% G_2/M phase cells. The cell cycle profile of E1B-19K cells seeded at day 0 in XST medium therefore remained unchanged for 5 days while G_1 cells and S phase cells were protected from apoptosis. Such an arrest in all phases of the cell cycle is consistent with what others have found following a single prolonged treatment with excess thymidine (Becker and Bartholomew, 1979; Cos et al., 1996).

The evolution of the cumulative MAb concentration of E1B-19K cells in the XST medium followed the same trend as for control NS/0 cells, with continuous production until day 10. A final concentration of 35 μg mL^{-1} was found for XST medium as compared to 20 μg mL^{-1} for standard batch condition. Similarly as for control NS/0 cells, an increase in the q_{MAb} of E1B-19K cells was observed in XST conditions as compared to the control batch culture.

Effect of E1B-19kD on viability and MAb productivity in medium supplemented with the monoclonal antibody production enhancer OptiMAb™

As can be observed on Figure 4 (left panels), the viable density of control NS/0 cells exposed to OptiMAb™ slowly increased on the first 2 days in culture and then decreased steadily until day 8. Two days after exposing these cells to OptiMAb™, a rapid decrease in viability could be observed concomitant with the accumulation of sub-G_1 apoptotic cells (Figure 4, bottom left panel). Although viable cell density quickly declined, the growth of control NS/0 myeloma cells was not arrested. Indeed, as for cells exposed to HT (NaCl) conditions, the significant accumulation of total cells throughout the culture up to 20×10^5 cells mL^{-1} indicated that cells were still dividing. A μ of 0.24 day^{-1} was measured for control NS/0 cells exposed to OptiMAb™ at day 3 (at which point the viable cell number was steadily declining), as compared

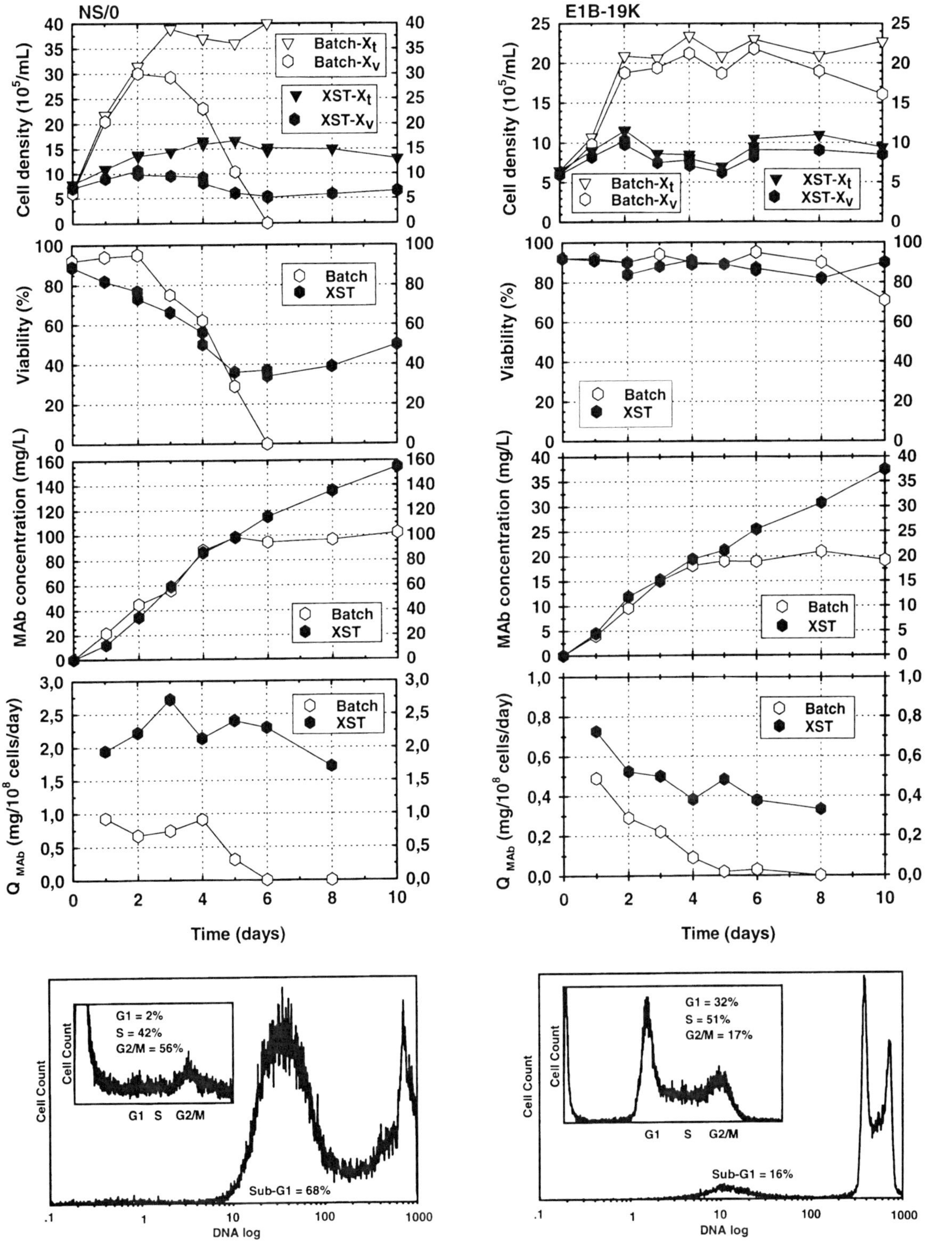

Figure 3. Effect of E1B-19K on cell survival, MAb productivity and cell cycle distribution of NS/0 myelomas grown in the presence of excess thymidine (DNA synthesis inhibitor). This figure presents the evolution of viable (X_v) and total (X_t) cell density, viability (%), MAb concentration and specific MAb productivity (q_{MAb}) of control NS/0 (left panels) and E1B-19K-expressing cells (right panels) cultivated throughout a 10-day period in standard (batch) and excess thymidine (XST) medium. Bottom panels present cell cycle profiles of control NS/0 (left panel) and E1B-19K cells (right panel) after 5 days in XST medium. Each lower panel presents both log and linear (inset) scales of the same sample. Log scale profiles allow to distinguish the sub-G_1 apoptotic peak while linear scale insets present the cell cycle distribution into G_1, S and G_2/M phases of non-apoptotic cells only. The evolution of the MAb concentration in XST conditions is cumulative and has been computed by addition of the MAb concentration found in the successive medium replacements performed at 2-day intervals.

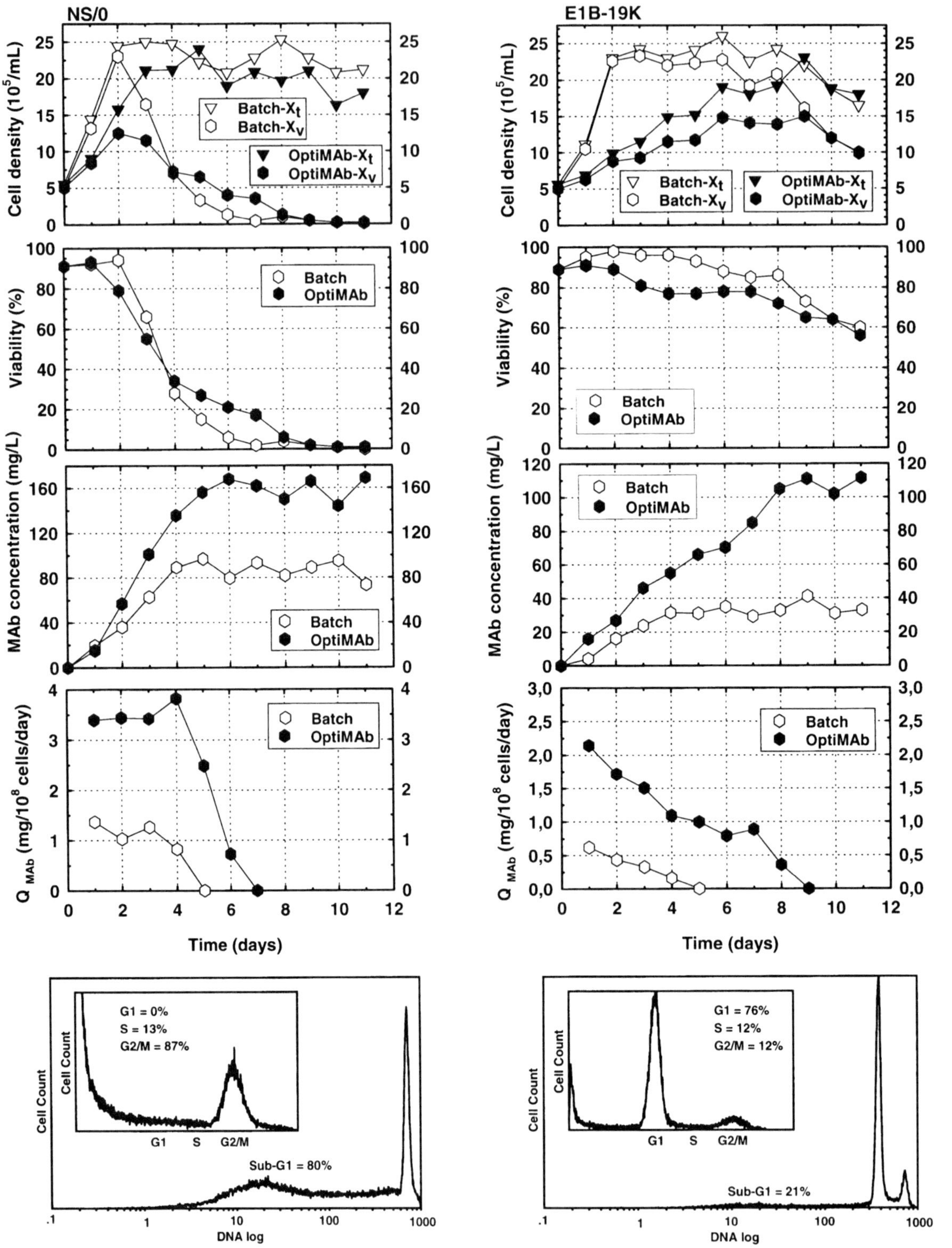

Figure 4. Effect of E1B-19K on cell survival, MAb productivity and cell cycle distribution of NS/0 myelomas grown in medium supplemented with the monoclonal antibody production enhancer OptiMAb^{TM}. This figure presents the evolution of viable (X_v) and total (X_t) cell density, viability (%), MAb concentration and specific MAb productivity (q_{MAb}) of control NS/0 (left panels) and E1B-19K-expressing cells (right panels) cultivated throughout a 11-day period in standard (batch) and OptiMAb^{TM}-supplemented medium. Bottom panels present cell cycle profiles of control NS/0 (left panel) and E1B-19K cells (right panel) after 5 days in OptiMAb^{TM}-supplemented medium. Each lower panel presents both log and linear (inset) scales of the same sample. Log scale profiles allow to distinguish the sub-G_1 apoptotic peak while linear scale insets present the cell cycle distribution into G_1, S and G_2/M phases of non-apoptotic cells only.

to a μ of 0.72 day^{-1} for exponentially growing control NS/0 cells in standard medium. At day 5, the sub-G_1 cells made up 80% of the control NS/0 population and the remaining viable cell population (excluding sub-G_1 cells) was almost entirely composed of G_2/M cells (87%) with some remaining S phase cells (13%). The G_1 phase cells present at inoculation totally disappeared concomitantly with the increase in the sub-G_1 cells. The S phase cells decreased from 52% at day 0 to 3% of the total population at day 5. Similarly as for XST conditions, the percentage of G_2/M cells present at day 0 (16%) remained stable at 17% on day 5. In OptiMAb™ conditions, MAbs were produced continuously until day 5 when the viable cell number started to decrease drastically, as compared to normal batch culture where MAb accumulation was observed for 3 to 4 days. Antibody concentration was higher in the OptiMAb™ condition with an average final MAb concentration of 156 μg mL^{-1} as compared to a maximum of 87 μg mL^{-1} for the batch condition, a 1.8-fold increase in antibody concentration. Interestingly, a 3-fold increase in q_{MAb} from 1.22 mg/10^8 cells day^{-1} in batch to 3.52 mg/10^8 cells day^{-1} in OptiMAb™ was concomitant with the 3-fold decrease in μ described above.

Like for control NS/0 cells, incubation of E1B-19K cells in OptiMAb™ medium resulted in a much slower accumulation of viable cells (Figure 4, right panels). However, viability decreased at a lower rate than in the control NS/0 culture. Indeed, after 7 days in OptiMAb™ medium, 78% of the E1B-19K cells were still viable, as compared to only 2% in the control culture. As observed in HT and XST conditions, the cell cycle profile of E1B-19K cells in OptiMAb™ at day 5 was totally different from the control NS/0 cells. Thus, only 21% of the population appeared as sub-G_1 apoptotic cells whereas the viable cell population (excluding sub-G_1 cells) was composed of 76% G_1, 12% S and 12% G_2/M phase cells. E1B-19K cells exposed to OptiMAb™ could therefore accumulate into G_1 presumably because they underwent apoptosis at a much reduced rate. As in all other conditions, the non-viable E1B-19K cells (sub-G_1) exhibited morphology typical of apoptosis (data not shown). A 4-fold increase in q_{MAb} was obtained through the use of OptiMAb™ for E1B-19K cells. MAbs were produced continuously until day 8 when the viable cell number and viability started to decrease. At day 5, the gain in antibody concentration of E1B-19K cells (1.8-fold increase from to 33 μg mL^{-1} in batch to 60 μg mL^{-1} in OptiMAb™) was similar to that of control NS/0 cells (1.8-fold increase from to 87 μg mL^{-1} in batch to 156 μg mL^{-1} in OptiMAb™). However, the extension of viability beyond day 5 resulting from expression of E1B-19K allowed to further extend the gain of productivity obtained. Thus, the final antibody concentration of E1B-19K cells at day 11 was even higher at 107 μg mL^{-1} as compared to 33 μg mL^{-1} for the batch condition, accounting for the final 3.3-fold increase in antibody concentration compared to only 1.8-fold for control NS/0 cells.

Discussion

The results presented above indicate clearly that antibody-producing E1B-19K myeloma cells are more resistant than control NS/0 cells against apoptosis induced by cell cycle modulating conditions such as incubation in HT medium (achieved with sorbitol or NaCl osmolytes), inhibition of DNA synthesis (achieved by supplementation with HU or XST), or use of the production enhancer OptiMAb™. The dose-dependent apoptosis-protective effect of E1B-19K observed in the presence of some of these cell cycle modulators is consistent with the dose-effect observed previously in glutamine-deprived cultures (Mercille et al., submitted). While supplementation with sorbitol and HU were detrimental in terms of their effect on MAb secretion, the use of NaCl, XST and OptiMAb™ resulted in an increase of q_{MAb} as compared to the batch culture in both control NS/0 and E1B-19K-expressing cells. However, this increase in q_{MAb} was translated in significantly higher MAb yields only in the case of OptiMAb™, and that, for both cell lines. Moreover, the inhibition of apoptosis conferred by the protective action of E1B-19K allowed to further extend the production period, and that resulted in an additional enhancement in productivity.

Cell cycle kinetics of control NS/0 cells cultivated in the presence of cell cycle modulators showed that non-resistant G_1 and S phase cells were much more susceptible than G_2/M cells to induction of apoptosis. After 5 days in the presence of NaCl and OptiMAb™, all G_1 and most S phase cells had disappeared, while the G_2/M phase cells had either increased in number (in the case of NaCl) or remained stable (in the case of OptiMAb™). These results, along with the progressive increase in the total cell density, indicate that under these conditions: (i) S phase cells retained the ability to progress into the G_2/M phase, showing that these agents were not inhibitory for DNA synthesis,

(ii) G_2/M cells could still enter mitosis and produce G_1 cells, and (iii) G_1 cells were unable to initiate another cycle and died by apoptosis. This confirms previous observations which showed that HT conditions arrest the cells and induce apoptosis in G_1 (Pellicciari et al., 1995; Pandey et al., 1996; Qin et al., 1997; Matthews and Feldman, 1996; Matthews et al., 1997; Anderson, 1997; Ortiz et al., 1997) and is in agreement with the 'grow or die' cell cycle model of De la Broise et al. (1991). In sharp contrast with the cell cycle profile of control NS/0 cells at day 5, a high percentage of apoptosis-resistant E1B-19K cells was found in the G_1 phase. Again, this is consistent with the notion that the apoptotic death of control NS/0 cells occurs preferentially in the G_1 phase of the cell cycle.

In the case of control NS/0 cells supplemented with XST, most of the G_1 cells disappeared, S phase cells decreased from 55% to 13% of the total population and G_2/M cells remained stable. Considering that the evolution of total cells was negligible in this case, these results demonstrate that upon addition of XST to non-protected cells: (i) most S phase cells died by apoptosis, (ii) G_2/M and a small fraction of S phase cells remained viable without undergoing apoptosis for more than 10 days if adequately fed; (iii) G_1 cells were unable to initiate another cycle and died by apoptosis. We can therefore conclude that, for all three conditions where cell cycle distribution was analyzed (66 mM NaCl, OptiMAb^{TM}, XST), the apoptosis susceptibility of non-protected lymphoid cells exposed to cell cycle modulators was cell cycle dependent; while G_1 and most S phase cells readily died by apoptosis, G_2/M phase cells could remain viable for more than 10 days. In contrast to cells arrested using cell cycle modulators, there was no selective survival of G_2/M cells in the control batch cultures. Thus, the cell cycle profile of the remaining viable control NS/0 cells at day 5 in a standard batch culture consisted of 55% G_1, 40% S and 5% G_2/M phase cells while for E1B-19K cells, the distribution was 51% G_1, 42% S and 7% G_2/M phase cells (data not shown).

Curiously, we observed that the increase in the total cell concentration of NaCl- and OptiMAb^{TM}-treated E1B-19K cells was slower than in the control NS/0 cells, in contrast to exponential batch cultures where both cell lines exhibited similar growth rates. This indicates that, in addition to protection against apoptosis, expression of E1B-19K altered the growth kinetics of NS/0 cells under cell cycle modulating conditions. This behavior of E1B-19K cells is consistent with what has been recently reported by O'Reilly et al. (1996) who showed that E1B-19K, and it's functional homologues Bcl-2 and Bcl-x_L, could exert a growth-inhibitory effect predominantly during transition from the quiescent G_0/G_1 state into the initial S phase, while having little or no effect once cells are continuously cycling.

As reported in other studies (Oh et al., 1993; Ozturk and Palsson, 1991; Reddy et al., 1992, Ryu and Lee, 1997), the increase in q_{MAb} observed in HT conditions did not result in an increase in antibody concentration. Indeed, despite the fact that E1B-19K cells retained a much higher viability in HT conditions throughout the culture, they stopped producing antibodies on day 5 in the culture, at which point they were still 60% viable. It has been reported that cells grown under stressful conditions can exhibit increased nutrient consumption rate (Oh et al., 1995). In such a case, nutrient exhaustion could have impaired the protein synthesis machinery by limiting the availability of some essential amino acids. While supplementation with nutrients throughout the culture could prevent nutrient limitation, increased accumulation of lactate and ammonia may in turn inhibit secretion. Increased antibody concentration may also be achieved through gradual adaptation of the cells to HT conditions (Oh et al. 1993, 1995) or by supplementation with osmoprotective agents such as glycine betaine (Øyaas et al., 1994a, b).

As mentioned above, E1B-19K cells exposed to OptiMAb^{TM} accumulated into the G_1 phase and exhibited a 4-fold increase in q_{MAb}. In a previous study we have shown that treatment of D5 hybridoma cells with OptiMAb^{TM} had also resulted in increased q_{MAb} and final MAb concentration, associated with prolongation of the G_1 phase (Mercille et al., 1994). Although we did not analyze the cell cycle distribution of control NS/0 cells in OptiMAb^{TM} conditions during the peak production period (day 2 to day 4), it is also likely that the increased q_{MAb} of control NS/0 cells (3-fold) was associated with enrichment in G_1 phase cells. This is in agreement with a number of studies reporting that synthesis and secretion of immunoglobulins are maximal in late G_1/early S phase (Byars and Kidson, 1970; Garatun-tjeldsto et al., 1976; Suzuki and Ollis, 1989; Al-Rubeai et al., 1990; Ramirez and Mutharasan, 1990; Richieri et al., 1991; Cazzador and Mariani, 1993). With the use of OptiMAb^{TM}, it was possible to establish that through an appropriate combination of cell cycle modulator and nutrient supplements, one can benefit from the use of an apoptosis-resistant cell line via an extension of the productive period, thus leading to significant gains in antibody concentration. Is it also possible that the prolongation of the production period from day 5 to day 8 (Figure 4, right panels) achieved here through the use of OptiMAb^{TM} could be have been extended further with appropriate feeding strategies. Indeed, we have recently shown that, under slow growth conditions in perfusion culture, E1B-19K cells exposed to a constant supply of nutrients exhibited an increased productivity that could be maintained for an extended period of time (Mercille and Massie, submitted).

In the case of DNA synthesis inhibition with XST, although cumulative productivity was greater than in batch for both control NS/0 (155 vs 98 μg mL^{-1}) and E1B-19K cells (35 vs 20 μg mL^{-1}), it is difficult to compare XST and batch control conditions as medium replacements were only performed in the XST condition. However, a comparison of the q_{MAb} and cell cycle profiles of control NS/0 and E1B-19K cell lines in XST medium leads to interesting speculations. While G_1 and a fraction of S phase control NS/0 cells died from apoptosis within the first 5 days (Figure 3, left panels), nutrient-fed G_2/M and a fraction of S phase cells remained viable for up to 10 days, secreting MAbs with a q_{MAb} 3-fold higher than in standard batch culture. These observations are consistent with studies reporting that a peak in synthetic activity can not only be associated with the late G_1/early S phase but also with the G_2 phase (Watanabe et al., 1973; Cherlet et al., 1995; Kronemaker and Srienc, 1991). These studies also point to the fact that both the rate of synthesis and secretion are at the lowest level in the S phase. This is also substantiated by our results since, for E1B-19K cells, the cell cycle profile of viable cells after 5 days in XST was similar to exponentially growing cells with more than 50% cells in S phase (Figure 3, right bottom panel), and the q_{MAb} from day 4 to day 8 was similar to what was found during the exponential phase of the batch culture. Thus, whereas G_2/M-enriched control NS/0 cells kept secreting MAbs in XST at a q_{MAb} 3-fold higher than in standard batch culture, S phase-enriched E1B-19K cells showed no increase in q_{MAb}. Since non-protected control NS/0 myeloma cells in G_2/M can remain viable under apoptosis-inducing conditions and exhibit such an elevated q_{MAb}, strategies could be developed to specifically block the cells in the G_2/M phase using cytostatic agents such as mitomycin C (Herrera et al., 1991) or gene products such as vpr from HIV (He et al., 1995). In order to avoid possible nutrient limitations, G_2/M cell cycle blocking agents and/or genes could be used in conjunction with feeding strategies such as perfusion cultures (Fong et al., 1997; Mercille and Massie, submitted).

Conclusions

Although E1B-19K protected against apoptosis induced by all cell cycle modulators tested in this study, only in one condition did we see a significant improvement in antibody productivity. This could be explained by pleitropic effects of these modulators which could not only induce apoptosis but also interfere with MAb synthesis through unknown mechanisms. Therefore, in order to exploit the extended viability conferred by anti-apoptotic genes, a careful optimization of all culture conditions (concentration of cell cycle modulators, medium formulation, etc.) has to be achieved. Alternatively, cell engineering targeting modulation of the cell cycle (Fussennegger et al., 1997) could alleviate the putative pleitropic effects of chemical modulators. Work is in progress to explore this avenue.

Acknowledgements

We would like to thank Lucie Bourget for technical assistance on the cytometer, Yves Langelier for background information on DNA synthesis inhibitors, Maria Koutroumanis and Dick D. Mosser for reviewing the manuscript. This manuscript is NRC #41439.

References

Al Rubeai M and Emery AN (1990) Mechanisms and kinetics of monoclonal antibody synthesis and secretion in synchronous and asynchronous hybridoma cell cultures. J Biotechnol 16: 67–85.

Anand S, Verma H, Kumar L and Singh N (1995) Induction of apoptosis in chronic myelogenous leukemia lymphocytes by hydroxyurea and adriamycin. Cancer Lett 88: 101–105.

Anderson P (1997) Kinase cascades regulating entry into apoptosis. Microbiol Mol Biol Rev 61: 33–46.

Barbiero G, Duranti F, Bonelli G, Amenta JS and Baccino FM (1995) Intracellular ionic variations in the apoptotic death of L cells by inhibitors of cell cycle progression. Exp Cell Res 217: 410–418.

Becker JF and Bartholomew JC (1979) Aryl hydrocarbon hydroxylase induction in mouse liver cells–relationship to position in the cell cycle. Chem Biol Interact 26: 257–266.

Bianchi V, Pontis E and Reichard P (1986) Changes of deoxyribonucleoside triphosphate pools induced by hydroxyurea and their relation to DNA synthesis. J Biol Chem 261: 16037–16042.

Bortner CD and Cidlowski JA (1996) Absence of volume regulatory mechanisms contributes to the rapid activation of apoptosis in thymocytes. Am J Physiol 271: C950–961.

Byars N and Kidson C (1970) Programmed synthesis and export of immunoglobulin by synchronized myeloma cells. Nature 226: 648–650.

Cazzador L and Mariani L (1993) Growth and production modeling in hybridoma continuous cultures. Biotechnol Bioeng 42: 1322–1330.

Cherlet M, Kromenaker SJ and Srienc F (1995) Surface IgG content of murine hybridomas: Direct evidence for variation of antibody secretion rates during the cell cycle. Biotechnol Bioeng 47: 535–540.

Cos S, Verduga R, Fernandez-Viadero C, Megias M and Crespo D (1996) Effects of melatonin on the proliferation and differentiation of human neuroblastoma cells in culture. Neurosci Lett 216: 113–116.

De la Broise D, Noiseux M, Lemieux R and Massie B (1991) Long-term perfusion culture of hybridoma: a 'grow or die' cell cycle system. Biotechnol Bioeng 38: 781–787.

Duke RC and Cohen JJ (1992) Morphological and biochemical assays of apoptosis. In: K Janssen (ed.) Current protocols in immunology. Wiley, New York, pp. 3.17.1–3.17.16.

Fassnacht D, Rossing S, Franek F, Al-Rubeai M and Pörtner R (1998) Effect of Bcl-2 expression on hybridoma cell growth in serum-supplemented, protein-free and diluted media. Cytotechnology 26: 219–225.

Fike RM (1990) Enhancement of monoclonal antibody production *in vitro*. In Vitro cell Devel Biol 26: 54A.

Fike RM, Jayme DW and Weiss SA (1991a) Monoclonal antibody enhancement in protein-free and serum-supplemented hybridoma culture. Am Biotechnol Lab 9: 40.

Fike RM, Pfohl JL, Epstein DA, Jayme DW and Weiss SA (1991b) Hybridoma growth and monoclonal antibody production in protein-free hybridoma medium. Biopharm 4: 26–29.

Fong W, Zhang Y and Yung P (1997) Optimization of monoclonal antibody production: combined effects of potassium acetate and perfusion in a stirred tank bioreactor. Cytotechnol 24: 47–54.

Fontecave M, Nordlund P, Eklund H and Reichard P (1992) The redox centers of ribonucleotide reductase of Escherichia coli. Adv Enzymol Relat Areas Mol Biol 65: 147–183.

Fuhrman Conti AM, Tori R, Ronchetti E, De Grada L, Pellicciari C and Manfredi Romanini MG (1990) Cell kinetics of PHA-activated lymphocytes are slowed by prolonged hypertonic stress. Basic Appl Histochem 34: 269–279.

Fussenegger M, Mazur X and Bailey JE (1997) A novel cytostatic process enhances the productivity of chinese hamster ovary cells. Biotechnol Bioeng 55: 927–939.

Galvan V and Roizman B (1998) Herpes simplex virus 1 induces and blocks apoptosis at multiple steps during infection and protects cells from exogenous inducers in a cell-type-dependent manner. Proc Natl Acad Sci USA 95: 3931–3936.

Garatun-Tjeldsto, Pryme IF, Weltman, JK and Dowben RM (1976) Synthesis and secretion of light-chain immunoglobulin in two successive cycles of synchronized plasmacytoma cells. J Cell Biol 68: 232–239.

Gervais C, Paquette D, Burns-Tardif A, Martin L and Massie B (1998) Development of high output expression vectors for antibody production in mammalian cells. In: K Nagai and M Wachi (eds.) Animal Cell Technology: Basic and Applied aspects. Vol. 9. Kluwer, The Netherlands, pp. 349–354.

He J, Choe J, Walker R, Di Marzio P, Morgan DO and Landau NR (1995) Human immunodeficiency virus type 1 viral protein R (Vpr) arrests cells in the G_2 phase of the cell cycle by inhibiting $p34^{cdc2}$. J Virol 69: 6705–6711.

Herrera LA, Tittelbach H, Gebhart E and Ostrosky-Wegman P (1991) Changes in the proliferation of human lymphocytes induced by several cytostatics and revealed by the premature chromosome condensation technique. Mutat Res 263: 101–106.

Hotz MA, Gong J, Traganos F and Darzynkiewicz Z (1994) Flow cytometric detection of apoptosis: comparison of the assays of in situ DNA degradation and chromatin changes. Cytometry 15: 237–244.

Itoh Y, Ueda E and Suzuki E (1995) Overexpression of Bcl-2, apoptosis-suppressing gene: Prolonged viable culture period of hybridoma and enhanced antibody production. Biotechnol Bioeng 48: 118–122.

Jayme DW (1991) Nutrient optimization for high density biological production applications. Cytotechnology 5: 15–30.

Kromenaker SJ and Srienc F (1991) Cell cycle-dependant protein accumulation by producer and non-producer murine hybridoma cell lines: A population analysis. Biotechnol Bioeng 38: 665–677.

Mastrangelo AJ and Betenbaugh MJ (1998) Overcoming apoptosis: new methods for improving protein expression systems. TIBTECH 16: 88–95.

Matthews CC and Feldman EL (1996) Insulin-like growth factor I rescues SH-SY5Y human neuroblastoma cells from hypertonic induced programmed cell death. J Cell Physiol 166: 323–331.

Matthews CC, Odeh HM and Feldman EL (1997) Insulin-like growth factor-I is an osmoprotectant in human neuroblastoma cells. Neuroscience 79: 525–534.

Mercille S and Massie B (1994) Induction of apoptosis in nutrient-deprived cultures of hybridoma and myeloma cells. Biotechnol Bioeng 44: 1140–1154.

Mercille S and Massie B Apoptosis-resistant NS/0-E1B-19K myeloma cells exhibit increased viability and chB43 monoclonal antibody productivity under perfusion culture conditions. Biotechnology and BioEngineering (accepted).

Mercille S, Johnson M, Lanthier S, Bourget L and Massie B (1994) Kinetic analysis of the 'Grow or Die' cell cycle system in perfusion culture of hybridomas. Abstracts of the 44th Canadian Chemical Engineering Conference, Calgary, Alberta, Canada, october 3–7.

Mercille S, Jolicoeur P, Gervais C, Paquette D, Mosser DD and Massie B Dose-dependent reduction of apoptosis in nutrient-limited cultures of NS/0 myeloma cells transfected with the E1B-19K adenoviral gene. Biotechnology and BioEngineering (accepted).

Mosser DD and Massie B (1994) Genetically engineering mammalian cell lines for increased viability and productivity. Biotech Adv 12: 253–277.

Mosser DD, Caron AW, Bourget L, Jolicoeur P and Massie B (1997) Use of a dicistronic expression cassette encoding the green fluorescent protein for the screening and selection of cells expressing inducible gene products. Biotechniques 22: 150–61.

Murray K, Ang C-E, Gill K, Hickman JA and Dickson AJ (1996) NS0 myeloma cell death: influence of bcl-2 overexpression. Biotechnol Bioeng 51: 298–304.

Oh SKW, Chua FKF and Choo ABH (1995) Intracellular responses of productive hybridomas subjected to high osmotic pressure. Biotechnol Bioeng 46: 525–535.

Oh SKW, Vig P, Chua FKF, Teo WK and Yap MSG (1993) Substantial overproduction of antibodies by applying osmotic pressure and sodium butyrate. Biotechnol Bioeng 46: 525–535.

Ortiz A, Ziyadeh FN and Neilson EG (1997) Expression of apoptosis-regulatory genes in renal proximal tubular epithelial cells exposed to high ambient glucose and in diabetic kidneys. J Investig Med Feb 45: 50–56.

Øyaas K, Ellingsen TE, Dyrset N and Levine DW (1994a) Utilization of osmoprotective compounds by hybridoma cells exposed to hyperosmotic stress. Biotechnol Bioeng 43: 77–89.

Øyaas K, Ellingsen TE, Dyrset N and Levine DW (1994b) Hyperosmotic hybridoma cell cultures: Increased monoclonal antibody production with addition of glycine betaine. Biotechnol Bioeng 44: 991–998.

Ozturk SS and Palsson BO (1991) Effect of medium osmolarity on hybridoma growth, metabolism and antibody production. Biotechnol Bioeng 37: 989–993.

Pandey P, Raingeaud J, Kaneki M, Weichselbaum R, Davis RJ, Kufe D and Kharbanda S (1996) Activation of p38 mitogen-activated protein kinase by c-Abl-dependent and -independent mechanisms. J Biol Chem 271: 23775–23779.

Paulovich AG, Toczyski DP and Hartwell LH (1997) When checkpoints fail. Cell 88: 315–321.

Pellicciari C, Filippini C, De Grada L, Fuhrman Conti AM and Manfredi Romanini MG (1995) Cell cycle effects of hypertonic stress on various human cells in culture. Cell Biochem Funct 13: 1–8.

Qin S, Minami Y, Kurosaki T and Yamamura H (1997) Distinctive functions of Syk and Lyn in mediating osmotic stress- and ultraviolet C irradiation-induced apoptosis in chicken B cells. J Biol Chem 272: 17994–17999.

Ramirez OT and Mutharasan R (1990) Cell cycle and growth-phase dependent variations in size distribution, antibody productivity, and oxygen demand in hybridoma cultures. Biotechnol Bioeng 36: 839–848.

Reddy S and Miller WM (1994) Effects of abrupt and gradual osmotic stress on antibody production and content in hybridoma cells that differ in production kinetics. Biotechnol Prog 10: 165–173.

Reddy S, Bauer KD and Miller WM (1992) Determination of antibody content in live versus dead hybridoma cells: analysis of antibody production in osmotically stressed cultures. Biotechnol Bioeng 40: 947–964.

Richieri RA, Williams LS and Chau PC (1991) Cell cycle dependancy of monoclonal antibody production in asynchronous serum-free hybridoma cultures. Cytotechnology 5: 243–254.

Ryu JS and Lee GM (1997) Effect of hypoosmotic stress on hybridoma cell growth and antibody production. Biotechnol Bioeng 55: 563–570.

Schrell UM, Rittig MG, Anders M, Kiesewetter F, Marschalek R, Koch UH and Fahlbusch R (1997) Hydroxyurea for treatment of unresectable and recurrent meningiomas. I. Inhibition of primary human meningioma cells in culture and in meningioma transplants by induction of the apoptotic pathway. J Neurosurg 86: 845–852.

Simonian PL, Grillot DA and Nunez G (1997) Bcl-2 and Bcl-XL can differentially block chemotherapy-induced cell death. Blood 90: 1208–1216.

Simpson NH, Milner AE and Al-Rubeai M (1997) Prevention of hybridoma cell death by Bcl-2 during sub-optimal culture conditions. Biotechnol Bioeng 54: 1–16.

Singh RP, Emery N and Al-Rubeai M (1996) Enhancement of survivability of mammalian cells by overexpression of the apoptosis-supressor gene Bcl-2. Biotechnol Bioeng 52: 166–175.

Singleton JR, Dixit VM and Feldman EL (1996) Type I insulin-like growth factor receptor activation regulates apoptotic proteins. J Biol Chem 271: 31791–31794.

Sokolova IA, Volgin AU, Makarova NV, Volgina VV, Shishkin SS and Khodarev NN (1992) Internucleasomal chromatin degradation in myeloma and B-hybridoma cell cultures. FEBS Lett 313: 295–299.

Stubbe J (1990) Ribonucleotide reductases: amazing and confusing. J Biol Chem 265: 5329–5332.

Suzuki E and Ollis DF (1989) Cell cycle model for antibody production kinetics. Biotechnol Bioeng 34: 1398–1402.

Suzuki E and Ollis DF (1990) Enhanced antibody production at slow growth rates: Experimental demonstration and a simple structured model. Biotechnol Prog. 6: 231–236.

Suzuki E, Terada S, Ueda H, Fujita T, Komatsu T, Takayama S and Reed J (1997) Establishing apoptosis resistant cell lines for improving protein productivity of cell culture. Cytotechnology 23: 55–59.

Takahashi K, Terada S, Ueda H, Makishima F and Suzuki E (1994) Growth rate suppression of cultures mammalian cells enhances protein productivity. Cytotechnology 12: 1–8.

Telford WG, King LE and Fraker PJ (1992) Comparative evaluation of several DNA binding dyes in the detection of apoptosis-associated chromatin degradation by flow cytometry. Cytometry 13: 137–143.

Terada S, Fukuoka K, Fujita T, Komatsu T, Takayama S, Reed JC and Suzuki E (1997b) Anti-apoptotic genes, bag-1 and bcl-2, enabled hybridoma cells to survive under treatment for arresting cell cycle. Cytotechnology 25: 17–23.

Watanabe S, Yagi Y and Pressman D (1973) Immunoglobulin production in synchronized cultures of human hematopoietic cell lines. Variation of synthetic and secretion activities during the cell cycle. J Immunology 111: 797–804.

Yang X, Oehlert GW and Flickinger MC (1996) Use of the weighted Jackknife method to calculate the variance in cellular-specific protein secretion rate: Application to monoclonal antibody secretion rate kinetics in response to osmotic stress. Biotechnol Bioeng 50: 184–196.

Cytotechnology **28:** 205–211, 1998.
© 1998 Kluwer Academic Publishers. Printed in the Netherlands.

Effect of temperature on recombinant protein expression in Semliki Forest virus infected mammalian cell lines growing in serum-free suspension cultures

Ernst-Jürgen Schlaeger* & Kenneth Lundstrom
F. Hoffmann-La Roche Ltd., Research Laboratories, CH-4070 Basel, Switzerland
E-mail: ernst-j.schlaeger@roche.com

Received 25 August 1998; accepted 25 August 1998

Key words: expression levels, low temperature, Semliki Forest virus, serum-free

Abstract

The firefly luciferase gene was introduced into the Semliki Forest virus (SFV) vector and high titer recombinant SFV particles generated. The broad host range of SFV allowed efficient infection and high level expression of four mammalian cell lines growing in serum-free suspension cultures. The incubation temperature had dramatic effects on the level and duration of recombinant protein expression. For example, the luciferase activity was significantly higher in the rodent BHK and CHO cell lines incubated at 33 °C compared to 37 °C when harvested 19 h post-infection. At 33 °C the specific expression levels increased 10–20 fold during prolongation of the post-infection time up to 50 h. In contrast, a significant decrease in luciferase activity was observed from 26 h post-infection for cell cultures incubated at 37 °C. Only a slight temperature effect on luciferase expression was seen in the human cell line HEK293 and no effect was observed for the subclone293(EBNA).

Introduction

Recombinant protein expression is an essential part of modern molecular biology. Studies on gene function and regulation have been significantly facilitated by application of appropriate expression techniques to mammalian cells. Many different vectors and transfection methods have been developed to create both transient (Fong et al., 1992) and stable expression (Nakajima et al., 1992) systems. For large scale expression of recombinant proteins in mammalian cells it has been essential to establish the methodology for efficient and cost-effective production (Blasey et al., 1996). In these cases, much attention has been focused on cell culture growth conditions and viability. Cell death or apoptosis has become an increasingly important factor in these bioprocesses. To prevent cell death, or at least prolong the viability of the culture, two approaches have been taken. Stable expression of anti-apoptotic genes have shown a significant effect on cell viability. For example, the introduction of the Bcl-2 gene into RIN cells resulted in strong decrease of cell death after infection with Semliki Forest virus (SFV) (Lundstrom et al., 1997). Another approach has been to enhance the cell survival by addition of anti-apoptotic chemicals to the cell culture. It has recently been reported that the addition of Z-VADfmk (N-benzyoxycarbonyl-Val-Ala-Asp-fluoromethylketone) to Sindbis virus infected AT3-*neo* cells prolonged the CAT (chloramphenicol acetyltransferase) expression for at least 24 h and maintained cell viability for 48 h longer (Mastrangelo and Betenbaugh, 1998).

The incubation temperature of cell cultures might also influence cell behaviour. When adherent BHK cells were grown at a lower temperature (30 °C) slower growth rate and glucose consumption rate were observed, resulting in reduced lactate production (Weidemann et al., 1994). Maximum cell density and productivity were not affected. Similar results were obtained for suspension cultures of CHO cells, where

* Author for all correspondence.

the lower temperatures promoted the resistance of the cells to shear stress (Kretzmer et al., 1997). There was therefore an indication that lowering of the temperature might be a feasible parameter for process optimization.

Here we have investigated the effect of cell culture temperature on the expression levels of a reporter gene, namely firefly luciferase. We have applied the highly efficient SFV vectors to generate high titer recombinant SFV particles expressing the luciferase gene. Due to the broad host range of SFV several mammalian cell lines growing in serum-free suspension cultures could be efficiently infected. The expression levels of luciferase were monitored in the different host cells at time points from 19 to 65 h post-infection after incubation at the normal growth temperature of 37 °C and the lowered temperature of 33 °C, respectively.

Materials and methods

Cell culture

The rodent cell lines BHK-21 C13-2P (ECACC No 84111301) and CHO-dhfr$^-$ (ATCC CRL9096) were adapted to serum-free growth in DHI medium (Schlaeger, 1996) and the human HEK293 cells (ATCC CRL1573) and the subclone 293(EBNA) (Invitrogen, Leek, The Netherlands) were adapted to serum-free growth in HL medium (Schumpp and Schlaeger, 1990). Both fortified powder media, calcium-free DHI and calcium-free HL, were developed in-house and purchased from Life Technologies (Basel, Switzerland). The cells were routinely cultured in spinner flasks (Bellco, Inotech AG, Dottikon, Switzerland) with 95–105 rpm using about 70% of the recommended working volume. If 33 °C incubation were used for SFV infection BHK and CHO cells were adapted for two passages at lower temperature. HEK293 and 293(EBNA) cells which do not grow well at 33 °C were used from a 37 °C preculture for such an experiment.

Expression of luciferase from SFV vectors

The coding region of the firefly luciferase gene was cloned into the *Sma*I site of the pSFV1 vector resulting in pSFV1-Lux. *In vitro* transcribed RNA from pSFV1-Lux and pSFV-Helper2 (Berglund et al., 1993) were coelectroporated into BHK cells and virus-containing medium collected from the cells 24 h later as earlier described (Lundstrom et al., 1994). The recombinant SFV particles were activated with α-chymotrypsin (500 μg mL^{-1}) (Boehringer Mannheim) for 15 min at 25 °C and the reaction terminated with Aprotinin (250 μg mL^{-1}) (Sigma) prior to infection of mammalian cell lines. The expression of recombinant luciferase was verified by metabolic labeling with [^{35}S] methionine followed by SDS-PAGE and autoradiography.

Luciferase assay

The enzymatic activity of luciferase was measured at the different time points post-infection as follows. 2 × 1ml samples were centrifuged for 5 min at 2000 rpm at 4 °C and the supernatants carefully removed. The cell pellets were resuspended mechanically and 1 mL ice cold lysis buffer (Boehringer Mannheim) added. The cell lysates were mixed, incubated on ice for 13 min and centrifuged at 12 000 rpm for 10 min at 4 °C to remove cell debris. The supernatants were stored frozen at –20 °C until analyzed. After thawing, the samples were diluted in lysis buffer and assayed in the Lumat LB9501 (Berthold AG, Regensdorf, Switzerland) as earlier described (Legendre et al., 1996). The protein concentration of the cell extracts was determined using the BCA protein assay reagent (Pierce, Rockford, IL, USA).

Results and discussions

The introduction of the coding region of the luciferase gene into the pSFV1 vector made it possible to *in vivo* package recombinant SFV-Lux particles with an approximate titer of 5×10^8 infective units mL^{-1}. The titer was indirectly estimated by determination of the titer of SFV-LacZ virus *in vivo* packaged in parallel with SFV-Lux. The luciferase expression was verified by infection of BHK cells growing in monolayer cultures on 6 well plates. The SFV-infected cells were subjected to metabolic labeling 18 h post-infection, which resulted in high-level expression of luciferase as identified by the presence of a strongly labeled 66 kD band in SDS-PAGE (not detectable in uninfected control cells [data not shown]). The morphological changes in the cells were also dramatic and made it possible to microscopically distinguish infected cells from uninfected ones.

BHK, CHO, HEK293 and 293(EBNA) cells grown in serum-free suspension cultures achieving cell densities of 3.5–5.0 × 10^6 cells mL^{-1} were subjected to

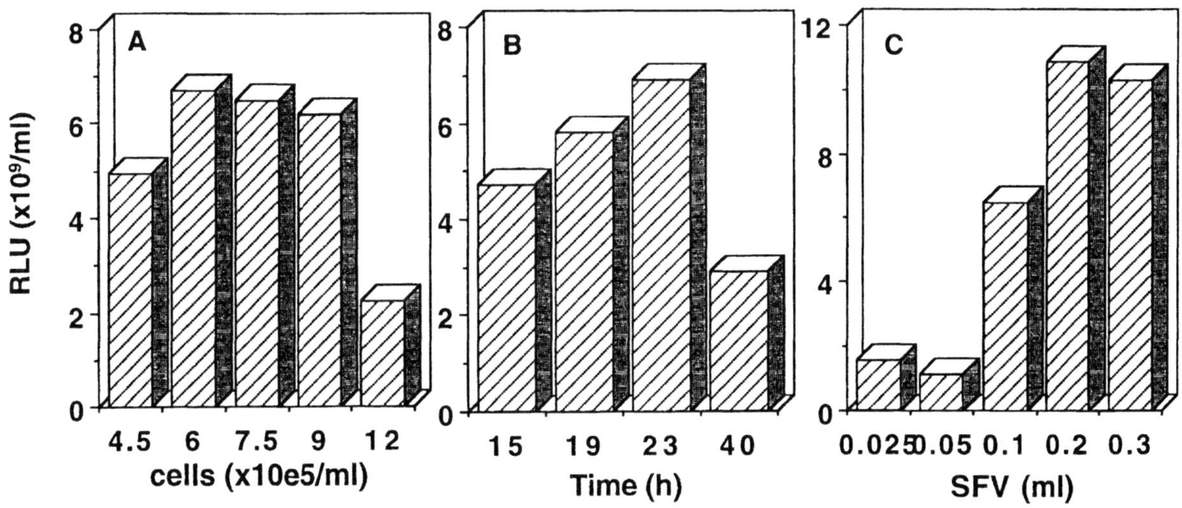

Figure 1. SFV-infection of BHK cells grown in suspension cultures at 37 °C. A) Cells adjusted to various concentrations in 15 mL volume, infected with 0.1 mL SFV-Lux and the luciferase activity determined at 19 h post-infection. B) 15 mL spinner cultures with 7.5×10^5 cells mL^{-1} infected with 0.1 mL SFV-Lux virus were assayed at different post-infectional time points. C) 7.5×10^5 cells mL^{-1} infected with different SFV-Lux concentrations and assayed at 19 h post-infection.

infection studies with recombinant SFV-Lux virus. The BHK cells were chosen to study the effect of the cell density and virus concentration on expression levels. This was because the BHK cells were used for *in vivo* packaging of SFV-Lux particles and were known to be efficiently infected by SFV in suspension cultures from previous studies (Blasey et al., 1996). A relatively broad cell density, from $6.0–9.0 \times 10^5$ cells mL^{-1}, resulted in high luciferase expression (Figure 1A). For all further studies, cell densities of $7.0–7.5 \times 10^5$ cells mL^{-1} were used. Next the optimal post-infection time for luciferase expression was determined. Our results indicated that the best harvest time would be 19 to 23 h post-infection (Figure 1B). The effect of virus concentration was studied by infection of 15 mL spinner cultures with different virus concentrations (0.025–0.3 mL). These amounts correspond to estimated multiplicity of infection (MOI) values from 1 to 15. The maximal luciferase activity was obtained with a MOI of 10 with no further increase in expression levels with higher MOI values (Figure 1C). For practical reasons all further experiments were performed with 0.1 mL virus/15 mL culture (MOI 5).

Using the conditions described above we compared the expression levels of luciferase in different host cells by infection of two rodent cell lines, Baby Hamster Kidney (BHK) and Chinese Hamster Ovary (CHO), with SFV-Lux virus. Previous studies had shown that CHO cells were infected with SFV as efficiently as BHK cells (Lundstrom et al., 1994). We have also demonstrated the high efficiency of HEK293 and 293(EBNA) cells by SFV-lacZ infection and by X-gal staining (Lundstrom, unpublished results). The infected spinner cultures were incubated at 33 and 37 °C, respectively. The 33 °C incubation temperature was chosen because it was shown recently that elevated production was observed with a recombinant CHO cell line at lower temperature (Kretzmer et al., 1997). The expression levels of luciferase were much higher in both BHK and CHO cells cultured at 33 °C than at 37 °C (Figure 2A and B). Additionally, while the luciferase activity rapidly decreased from 26 h post-infection onwards at 37 °C there was a significant increase in expression levels at 33 °C that continued until 65 h post-infection. As shown below no difference in viable cell numbers was observed during post-infection at 33 °C (Table 1). The increase in luciferase activity at the different time points for cells cultured at 33 °C in comparison to 37 °C is shown in Figure 2C. At 50 h post-infection more than 10-fold increase in luciferase levels was measured in CHO cells and almost 20-fold in BHK cells. The influence of other temperatures on the luciferase expression was not analyzed.

Likewise, we studied the temperature effect on the human HEK293 cell line and its subclone 293(EBNA). Surprisingly, we could not detect a significant difference in the expression levels of luciferase at the two temperatures studied (Figure 3). Even though the ex-

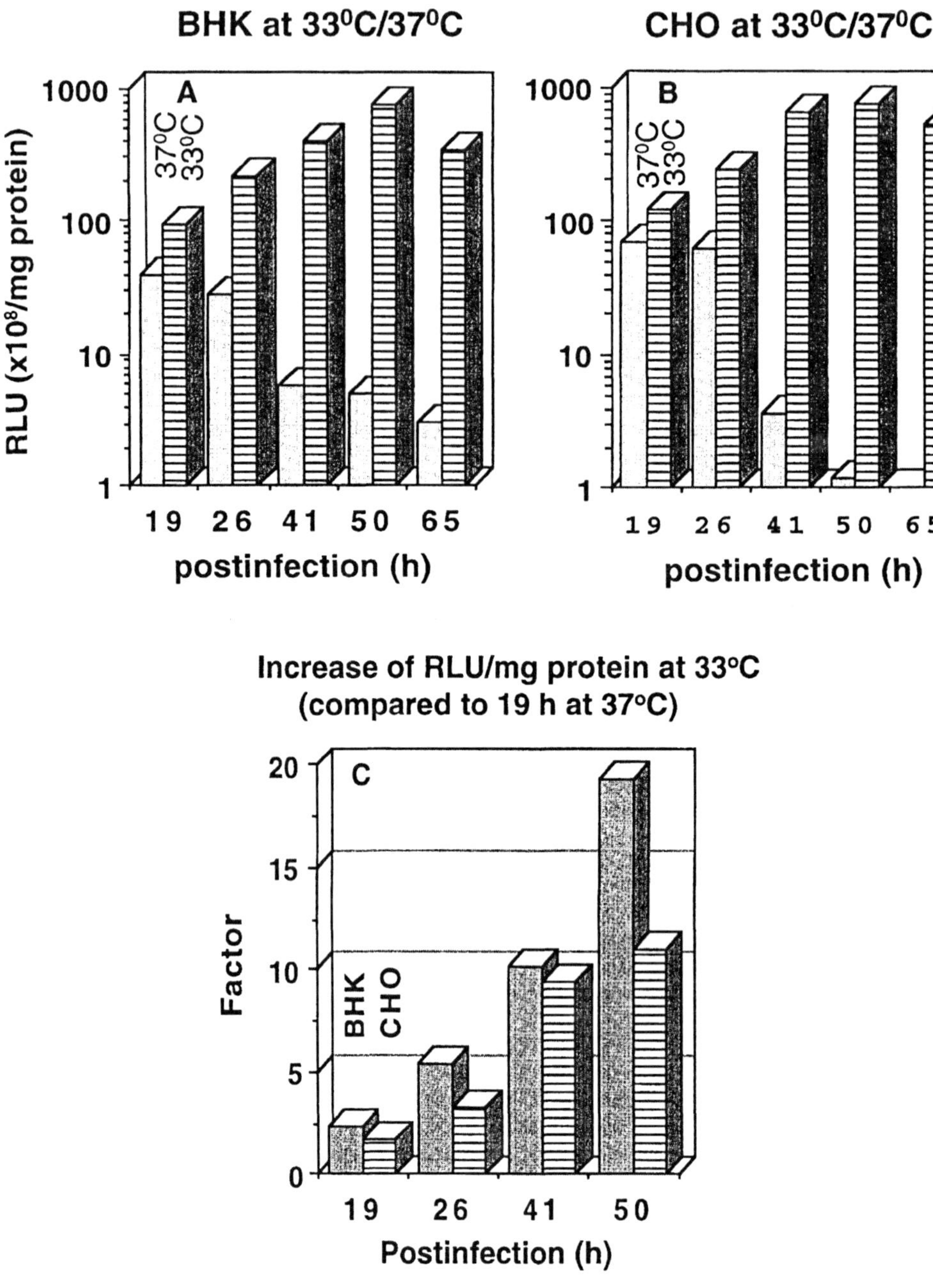

Figure 2. Kinetics of SFV-Lux infected BHK and CHO cells. 20 mL spinner cultures (7.0×10^5 cells mL^{-1}, from a 37 °C culture) infected with 0.135 mL of SFV-Lux virus were incubated at 33 and 37 °C, respectively for 65 h. 2×1 mL aliquots were taken at 19, 26, 41, 50 and 65 h post-infection and assayed for luciferase activity (expressed as relative light units/mg protein of the soluble fraction of the cell lysate). A) BHK cells; B) CHO cells. C) The relative increase of luciferase activity at different time points for SFV-Lux infected cells incubated at 33 °C compared to 19 h at 37 °C.

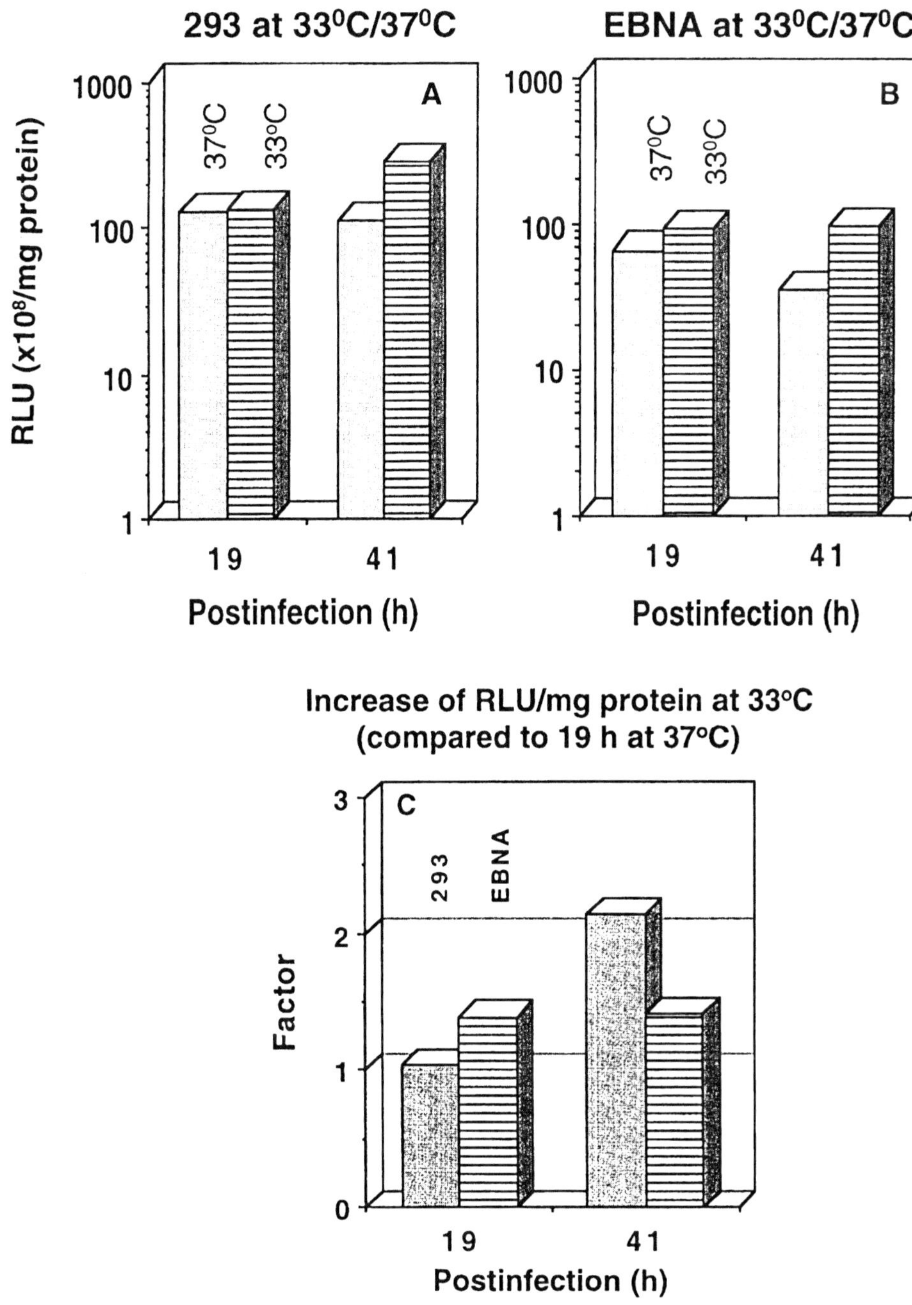

Figure 3. Kinetics of SFV-Lux infected HEK293 and 293(EBNA) cells. 20 mL spinner cultures (7.0×10^5 cells mL^{-1}) infected with 0.135 mL of SFV-Lux virus were incubated at 33 and 37 °C, respectively for 65 h. 2 × 1 mL aliquots were taken at 19, 26, 41, 50 and 65 h post-infection and assayed for luciferase activity (expressed as relative light units/mg protein of the soluble fraction of the cell lysate). A) HEK 293 cells; B) EBNA cells. C) The relative increase of luciferase activity at different time points for SFV-Lux infected cells incubated at 33 °C compared to 19 h at 37 °C.

Table 1. Viable cell density during SFV post-infection. The viable cell counts of the time kinetics of SFV infection were measured microscopically. Non-infected cells in brackets. Values are in 10^5 cells mL^{-1}.

Cell line		BHK	CHO	HEK293	293(EBNA)
0 h	33 °C	7			
	37 °C	7			
19 h	33 °C	14 (12.6)	8 (14.4)	8 (7.6)	7 (7.0)
	37 °C	22 (19.0)	10 (16.7)	8 (8.2)	8 (9.2)
41 h	33 °C	16 (40)	10 (42.0)	6 (10.0)	8 (10.0)
	37 °C	22 (40)	23 (52.0)	8 (17.8)	9 (17.0)
65 h	33 °C	11 (40)	9 (48.0)	4 (10.3)	4 (9.5)
	37 °C	21 (38)	28 (44.0)	7 (20.0)	7 (21.0)

pression levels were already relatively high at 19 h post-infection (almost 5-fold higher than for BHK cells) a 2-fold increase in luciferase activity levels was detected in HEK293 cells at 41 h post-infection at 33 °C. There was almost no difference in the expression levels seen at 33 °C for the 293(EBNA) cells. Thus, no improvement in luciferase expression levels could be achieved with human cell lines at the lower temperature (Figure 3C).

To analyze the differences in cell growth and viability properties for cultures incubated at 33 and 37 °C, we determined the number of viable cells in the SFV-Lux infected cultures and in non-infected samples. The viable cell numbers of infected BHK and CHO cells were lower at 33 °C but no significant difference was observed between 19 and 65 h (Table 1). Incubation at 37 °C showed some differences between BHK and CHO cells. Whereas the BHK cell numbers remained constant during 19 and 65 h post-infection, the cell density of CHO continued to increase. We could therefore conclude that the observed temperature effect on the expression level in both rodent cell lines was independent of the reduced cell growth rate at 33 °C compared to 37 °C. Additionally, the experiments also demonstrated that the infection of the cells in the spinner cultures was highly efficient since the cells divided poorly during the entire post-infection period. In contrast uninfected BHK and CHO cells grew within 40 h to cell densities of 4–5 × 10^6 cells mL^{-1}.

Infected HEK293 and 293(EBNA) cells did not show differences in viable cell numbers during two days incubation at 33 and 37 °C. Further growth of both cell lines was immediately inhibited after SFV infection.

In this study we have been able to show a dramatic increase in luciferase expression levels by simply lowering the growth temperature of SFV-Lux infected cell cultures of rodent BHK and CHO cell lines from 37 to 33 °C. This procedure resulted in up to 20-fold higher expression levels and, moreover, the expression phase could be prolonged to about 50 h. However, only a minor effect was seen for the human cell line HEK293 and no effect was detected for the subclone 293(EBNA). Both HEK cell lines grow poorly at low temperatures, in contrast to the rodent cells which grow normally. We therefore postulate that our finding could be important for further development of recombinant protein expression techniques to increase yields, especially for poorly expressed proteins. Obviously, it will be crucial to follow up studies on expression of some more complex mammalian recombinant proteins under these same conditions to evaluate if similar improvements in quantity can be achieved without loss in the quality of the product. Furthermore, it could be interesting to measure other (lower) temperatures for the expression of recombinant proteins in rodent cells using the SFV vector system.

Acknowledgements

We are thankful to Ms. Catherine Diener for subcloning the luciferase gene into the pSFV1 vector. Mr. Andreas Kunz is acknowledged for excellent assistant in SFV-Lux virus production and Mr. Marcello Foggetta for his help with the cell cultures.

References

Berglund P, Sjöberg M, Garoff H, Atkins GJ, Sheahan BJ and Liljeström P (1993) Bio/Technology 11: 916–920.

Blasey HD, Aubry J-P, Mazzei G and Bernard AR (1996) Cytotechnology 18: 183–192.

Blasey HD, Lundstrom K, Tate S and Bernard AR (1996) Cytotechnology 24: 65–72.

Fong TM, Anderson SA, Yu H, Huang RC and Strader C (1992) Mol Pharmacol 41: 24–30.

Kretzmer G, Rössler B, Hüners M, Lübben H and Reich T (1997) In: Animal Cell Technology. Carrondo MJT (ed.) Kluwer Academic Press The Netherlands, pp. 319–323.

Legendre JY, Trzeciak A, Bohrmann B, Deuschle U, Kitas EA and Supersaxo A (1996) Bioconjugate Chem 8: 57–63.
Lundstrom K, Mills A, Buell G, Allet E, Adami N and Liljestrom P (1994) Eur J Biochem 224: 917–921.
Lundstrom K, Pralong W and Martinou J-C (1997) Apoptosis 2: 189–191.
Mastrangelo AJ and Betenbaugh MJ (1998) Tibtech 16: 88–95.
Nakajima Y, Tsuchida K, Negishi M, Ito S and Naganishi S (1992) J Biol Chem 267: 2437–2442.
Schlaeger E-J (1996) J Immunol Meth 194: 191–199.
Schumpp B and Schlaeger E-J (1990) J Cell Sci 97: 639–647.
Weidemann R, Ludwig A and Kretzmer G (1994) Cytotechnology 15: 111–116.

Cytotechnology 28: 213–227, 1998.
© 1998 Kluwer Academic Publishers. Printed in the Netherlands.

Effects of CO_2 and osmolality on hybridoma cells: growth, metabolism and monoclonal antibody production

Vivian M. deZengotita*, Roy Kimura*,† & William M. Miller‡
Chemical Engineering Department, Northwestern University, Evanston, IL 60208-3120, U.S.A.
E-mail: wmmiller@nwu.edu

Received 14 September 1998; accepted 14 September 1998

Key words: antibody production, carbon dioxide, cell metabolism, continuous culture, inhibition

Abstract

CO_2 partial pressure (pCO_2) in industrial cell culture reactors may reach 150–200 mm Hg, which can significantly inhibit cell growth and recombinant protein production. Due to equilibrium with bicarbonate, increased pCO_2 at constant pH results in a proportional increase in osmolality. Hybridoma AB2-143.2 cell growth rate decreased with increasing pCO_2 in well-plate culture, with a 45% decrease at 195 mm Hg with partial osmolality compensation (to 361 mOsm kg^{-1}). Inhibition was more extensive without osmolality compensation, with a 63% decrease in growth rate at 195 mm Hg and 415 mOsm kg^{-1}. Also, the hybridoma death rate increased with increasing pCO_2, with 31- and 64-fold increases at 250 mm Hg pCO_2 for 401 and 469 mOsm kg^{-1}, respectively. The specific glucose consumption and lactate production rates were 40–50% lower at 140 mm Hg pCO_2. However, there was little further inhibition of glycolysis at higher pCO_2. The specific antibody production rate was not significantly affected by pCO_2 or osmolality within the range tested. Hybridomas were also exposed to elevated pCO_2 in continuous culture. The viable cell density decreased by 25–40% at 140 mm Hg. In contrast to the well-plate cultures, the death rate was lower at the new steady state at 140 mm Hg. This was probably due to higher residual nutrient and lower byproduct levels at the lower cell density (at the same dilution rate), and was associated with increased cell-specific glucose and oxygen uptake. Thus, the apparent effects of pCO_2 may vary with the culture system.

Abbreviations: CC – continuous culture; D – dilution rate; f_v – fraction of viable cells; k_d – specific death rate; pCO_2 – carbon dioxide partial pressure; q_{glc} – specific glucose consumption rate; q_{lac} – specific lactate production rate; q_{O_2} – specific oxygen consumption rate; q_P – specific protein production rate; RQ – respiratory quotient; X_d – dead cell density; X_t – total cell density; X_v – viable cell density; $Y_{lac,glc}$ – apparent yield of lactate from glucose; $Y_{n,glc}$ – apparent yield of cells from glucose; $Y_{O_2,glc}$ – ratio of oxygen to glucose consumption; μ – specific growth rate.

Introduction

CO_2 is produced via catabolic reactions, and is required for the synthesis of pyrimidines, purines, and fatty acids. Dissolved CO_2 is hydrated and is in equilibrium with H^+ and HCO_3^-:

$$CO_2 + H_2O \leftrightarrow H_2CO_3 \leftrightarrow H^+ + HCO_3^- \quad (1)$$

Inhibition by CO_2 is due in part to medium acidification if pH is not controlled. If pH is controlled by base addition, the equilibrium in reaction (1) is driven farther to the right, such that osmolality increases with pCO_2. High osmolality by itself may have detrimental effects on cells (Kurano et al., 1990; Ozturk and Palsson, 1991).

* VMdZ and RK contributed equally to the results in this article.
† Present address: MedImmune, Inc., 35 W. Watkins Mill Rd., Gaithersburg, MD 20878, U.S.A.
‡ Author for all correspondence.

Information on CO_2 production by cultured cells is still relatively scarce. Reported values for the respiratory quotient (RQ; mol CO_2 produced per mol O_2 consumed) range from 0.6 to 1.2 (Aunins and Henzler, 1993; Lovrecz and Gray, 1994; Gray et al., 1996). The physiological range for CO_2 partial pressure (pCO_2) is 31–54 mm Hg (Altman and Dittmer, 1971). The RQ and CO_2 transport relations (Royce and Thornhill, 1991) can be used to estimate pCO_2 for different aeration strategies. CO_2 accumulation is greatest in cultures oxygenated with a low flow rate and small bubbles of pure O_2 – with pCO_2 predicted to reach 150–200 mm Hg (Kimura and Miller, 1996; Gray et al., 1996; Zupke and Green, 1998). pCO_2 values in this range have been reported for a 200-L culture sparged with small amounts of O_2 (Aunins et al., 1993), 1800–2500-L production bioreactors (Drapeau et al., 1990; Zupke and Green, 1998), and a high-cell-density perfusion bioreactor (Taticek et al., 1998).

pCO_2 values in the range of 120–200 mm Hg have been shown to inhibit growth and recombinant protein production by CHO and other cells. CHO cell growth and cell-specific M-CSF production (q_{M-CSF}) were both ~40% lower at a pCO_2 of 165 mm Hg than at 53 mm Hg (Drapeau et al., 1990). Similarly, Gray et al. (1996) reported dose-dependent decreases in CHO cell density, viability, and specific production rate – such that total productivity of a recombinant viral antigen in a 10-L perfusion bioreactor decreased by 69% at 148 mm Hg. We have previously reported that CO_2 inhibits CHO cell growth in a dose-dependent manner, with a greater decrease when there is no compensation for increased osmolality (Kimura and Miller, 1996). The specific growth rate (μ) decreased by 43% at 250 mm Hg pCO_2 without osmolality compensation. q_{tPA} decreased to 73% of control at 140 mm Hg without osmolality compensation (Kimura and Miller, 1996). Decreases in specific productivity of BHK-21 cells have been observed beginning at 50–80 mm Hg (Taticek et al., 1998). It has also been reported that recombinant protein production by infected *Sf*-9 insect cells is markedly delayed under 115 mm Hg pCO_2 (Garnier et al., 1996). Early cell death has been observed in NS/0 myeloma cell cultures with a final pCO_2 of 120 mm Hg (Aunins and Henzler, 1993). However, there was no change in the specific production rate of these cells.

Hybridomas are used extensively for monoclonal antibody production. Industrial hybridoma cultures may reach very high cell densities in large-scale reactors, so that CO_2 accumulation is likely to be a problem. Here, we present the effects of elevated pCO_2 – with and without osmolality compensation – on hybridoma growth, death, glucose metabolism, and antibody production in well-plate cultures. Continuous culture and batch experiments in a controlled bioreactor were used to further characterize the effects of concurrent increases in pCO_2 and osmolality, including changes in oxygen consumption.

Materials and methods

Cell line and medium

The hybridoma AB2-143.2 cell line (University of California, San Francisco) is an Sp2/0-derived mouse-mouse hybridoma that produces an IgG_{2a} monoclonal antibody against benzene-arsonate (Hornbeck and Lewis, 1985). Unless noted, all reagents were purchased from Sigma (St. Louis, MO). AB2-143.2 cells were grown in a custom low-salt (35.5 mg L^{-1}) DMEM:F12 mixture with 15 mM HEPES (Irvine Scientific; Santa Ana, CA) for well-plate experiments and a standard DMEM:F12 mixture with 25 mM HEPES for bioreactor experiments. Both media were supplemented with MEM nonessential amino acids (at 1.15x), sodium bicarbonate (29 mM), glutamine (to 6 mM; Gibco; Grand Island, NY), HEPES (to 25 mM), sodium selenite (20 nM), ethanolamine (20 μm), bovine insulin (5 mg L^{-1}), bovine holo-transferrin (30 mg L^{-1}), and 2% (v/v) fetal bovine serum (Gibco).

Low-salt medium used for stock cultures, control experiments at 36 mm Hg pCO_2 (5% CO_2), and high pCO_2 experiments not corrected for osmolality was supplemented with 5.71 g L^{-1} of NaCl. Media for experiments at 140, 195, and 250 mm Hg pCO_2 with partial osmolality compensation were supplemented with 4.73, 4.17, and 3.62 g L^{-1} of NaCl, respectively. Media for control pCO_2 experiments with osmolalities of 342, 357, and 376 mOsm kg^{-1} were supplemented with 6.75, 7.24, and 7.86 g L^{-1} of NaCl, respectively. Media for well-plate experiments were supplemented with 100 U mL^{-1} penicillin G and 100 μg mL^{-1} streptomycin sulfate. Stock cultures and reactor experiments were conducted without antibiotics. The AB2-143.2 cells tested negative for mycoplasma (Mycoplasma TC Kit, Genprobe; San Diego, CA).

Well-plate experiments

Three types of experiments were conducted in six-well-plates: (1) elevated pCO_2 (140, 195, and 250 mm Hg)–partial osmolality compensation (337, 361 and 401 mOsm kg^{-1}, respectively); (2) elevated pCO_2 (140, 195, and 250 mm Hg) – uncompensated osmolality (370, 415, and 469 mOsm kg^{-1}, respectively); and (3) elevated osmolality (342, 357, and 376 mOsm kg^{-1}) at 36 mm Hg pCO_2. Cells in well-plates were exposed to high pCO_2 in a modular incubator chamber (Forma Scientific; Marietta, OH). The chamber was fed with controlled amounts of air and CO_2 using flowmeters (Cole-Parmer; Vernon Hills, IL) with needle valves. pCO_2 of the gas mixture was checked using a blood gas analyzer (Instrumentation Laboratories Model 1306; Lexington, MA). The inlet gas line was sparged through a stainless steel frit into a beaker of water to keep the chamber humidified.

Medium for well-plate experiments was transferred into T-150 flasks and pre-equilibrated in a 5% CO_2 incubator or in a modular incubator chamber gassed with the appropriate CO_2/air mixture for 8–12 hr. For experiments at 140, 195, and 250 mm Hg pCO_2 the pH was then adjusted to 7.25±0.05 by adding 7.1, 12.5, and 17.8 μL, respectively, of 5 M NaOH per mL of medium, after which the medium was re-equilibrated for 8–12 hr in the high pCO_2 environment.

Mid-exponential AB2-143.2 cells from stock cultures were inoculated at a density of 5×10^4 cells mL^{-1} into 5 mL of medium pre-equilibrated at the appropriate pCO_2, after which the well-plates were placed either in a 5% CO_2 incubator or in a modular incubator chamber. These hybridomas do not grow well at seeding densities less than 1×10^4 cells mL^{-1}. The higher seeding density assured that any effects observed were due only to the test conditions. Samples were taken at 12-hour intervals. pH was measured by drawing medium from a well directly into a glass syringe, and injecting the medium into the blood gas analyzer. A 2-point calibration for pH, pO_2 and pCO_2 was performed before each sample.

Bioreactor experiments

Continuous culture and batch experiments were performed in a 3-L Applikon (Foster City, CA) bioreactor equipped with pH (Broadley-James; Santa Ana, CA) and dissolved oxygen (Ingold; Wilmington, MA) probes. The working volume was 1.5 L and the impeller speed was 175 rpm. The medium pH was controlled at 7.20±0.02 by addition of 1 M NaOH. A stainless steel mesh filter with a pore opening of 75 μm was attached to the base line so that the NaOH would be diluted with medium inside the filter and then diffuse out. There were no signs of cell damage around the NaOH filter. At elevated pCO_2 more NaOH was required to maintain constant pH. Therefore, increases in pCO_2 were accompanied by increases in medium osmolality.

Medium pCO_2 and pH were determined off-line using the blood gas analyzer. Samples were taken through a sampling system that consisted of a four-way valve (Whitey SS-43YTFS1-049, Swagelok; Solon, OH) and stainless steel HPLC tubing (Waters; Milford, MA). The system was purged with 10–30 mL of medium using a 50-mL syringe. A sample was extracted into a 2.5-mL glass syringe and immediately injected into the blood gas analyzer. Any drift in the pH probe was corrected based on the blood gas analyzer pH measurement.

For continuous culture experiments, the dilution rate was maintained at 0.022 hr^{-1} for one run and at 0.028 hr^{-1} for two others. A spin filter with a mesh size of 75 μm was attached to the impeller to be used when the cells were in danger of washing out.

Assays

The viable cell density and viability were measured with a hemacytometer using the Trypan blue dye exclusion method. Glucose and lactate concentrations were determined using a YSI model 2700 biochemistry analyzer (Yellow Springs Instruments, Yellow Springs, Ohio). The osmolality of culture medium equilibrated at different pCO_2 values, with or without osmolality compensation, was measured using both a Wescor (Logan, UT) Model 5520 Vapro vapor pressure osmometer and an Advanced Instruments (Norwood, MA) 3D3 freezing point osmometer. Samples for osmolality measurement were pipetted into an 0.5-mL Eppendorf (Brinkman, Westbury, NY) microcentrifuge tube (completely filled), capped to minimize degassing, and analyzed. The elapsed time from sampling to analysis was less than 2 minutes. Reported osmolalities are the average of the vapor pressure and freezing point values, since both methods gave comparable results (manuscript in preparation).

Antibody concentrations were determined using a sandwich ELISA in 96-well plates coated with polyclonal goat anti-mouse IgG antibody (Pierce; Rockford, IL). Bound mouse IgG was detected using horse radish peroxidase-conjugated goat anti-mouse IgG antibody (Pierce) and developed with an *o*-phenylenediamine dihydrochloride solution. ELISA standards were prepared from pooled hybridoma supernatant using an Affi-Prep protein A column (Pharmacia; Piscataway, NJ).

Calculation of metabolic parameters

Specific growth (μ) and death (k_d) rates for the well-plate and batch reactor experiments were based on the material balances for total and dead cells:

$$\frac{dX_t}{dt} = \mu X_v \quad (2)$$

$$\frac{dX_d}{dt} = k_d X_v \quad (3)$$

where X_t, X_v, and X_d are the total, viable, and dead cell densities, respectively. Equations (2) and (3) were integrated assuming constant μ and k_d. Thus, the values for μ and k_d were obtained by plotting the value of X_t and X_d, respectively, vs. the integral of (X_v dt) at each time point. Similarly, the specific glucose consumption rate (q_{glc}), lactate production rate (q_{lac}), and antibody production rate (q_{Ab}) were obtained by plotting the glucose, lactate, and antibody concentration, respectively vs. the integral of (X_v dt). The apparent yield of lactate from glucose ($Y_{lac,glc}$) was determined from q_{lac}/q_{glc} and the apparent yield of cells from glucose ($Y_{n,glc}$) from μ/q_{glc}. In several experiments the final data point(s) for X_t or X_d deviated significantly from the straight line for constant μ or k_d, respectively. In these cases, the final time point(s) was not used for determining culture parameters.

The volumetric oxygen uptake rate (Q_{O_2}) for batch reactor cultures was obtained from on-line measurements of the O_2, N_2, and CO_2 gas flowrates (Sierra Instruments model 820 and 840 mass flowmeters; Carmel Valley, CA) into the reactor headspace. The calculated values for pO_2 were consistent with the headspace values measured using the blood gas analyzer. The specific oxygen uptake rate (q_{O_2}) was obtained by dividing $Q_{O_2}(t)$ by a curve-fit of $X_V(t)$ and taking the average from 40–60 hr. The ratio of oxygen consumption to glucose consumption ($Y_{O_2,glc}$) was determined from q_{O_2}/q_{glc}.

The derivative method was employed to calculate the various parameters for continuous culture, as previously described (Miller et al., 1988). Derivative terms were determined after 3-point smoothing.

Statistical analysis

The Kruskal-Wallis test (Glantz, 1981) was used to determine the statistical significance of the effects of elevated pCO_2–partial osmolality compensation, elevated pCO_2–uncompensated osmolality, and constant pCO_2–elevated osmolality for the well-plate experiments. The Kruskal-Wallis test is the non-parametric analog of one-way ANOVA, and does not assume a normal distribution or constant variance.

Results

Well plate experiments

In order to decouple the effects of elevated pCO_2 from those of elevated osmolality, three types of experiments were performed. In the first type, the initial medium osmolality was adjusted so that the final osmolality, after equilibration and pH adjustment at the desired pCO_2, would be about midway between the control level (310 mOsm kg^{-1}) and the osmolality in uncompensated cultures. (Our intention was to achieve full osmolality compensation. However, we recently discovered that the osmolality values that we had reported for these conditions (Kimura and Miller, 1996) are not correct; see Note in Proof). In the second type, the initial medium osmolality was not adjusted. Finally, the third type of experiments evaluated the effects of elevated osmolality at control pCO_2. pH in cultures at 36 mm Hg pCO_2 remained at 7.25–7.3 for the first 36 hr, and then decreased – reaching values of 7.1–7.2 at 48 hr and 6.9–7.1 at 60 hr (Kimura, 1996). In contrast, the pH changed by only ~0.1 unit during culture at elevated pCO_2. More stable pH at elevated pCO_2 has also been observed for CHO cells (Kimura and Miller, 1996), and is probably due to the greater bicarbonate concentration.

Cell growth and death

Representative growth curves for cells exposed to elevated pCO_2 (with and without osmolality adjustment) are shown in Figure 1. The effects of pCO_2 on cell viability are shown in Figure 2 for the same experiments. Higher pCO_2 was associated with greater inhibition of cell growth, as well as more extensive cell death. For both parameters, the effects were more severe when osmolality was not adjusted. There was some variability between the growth curves for replicate experiments, but large differences were only observed

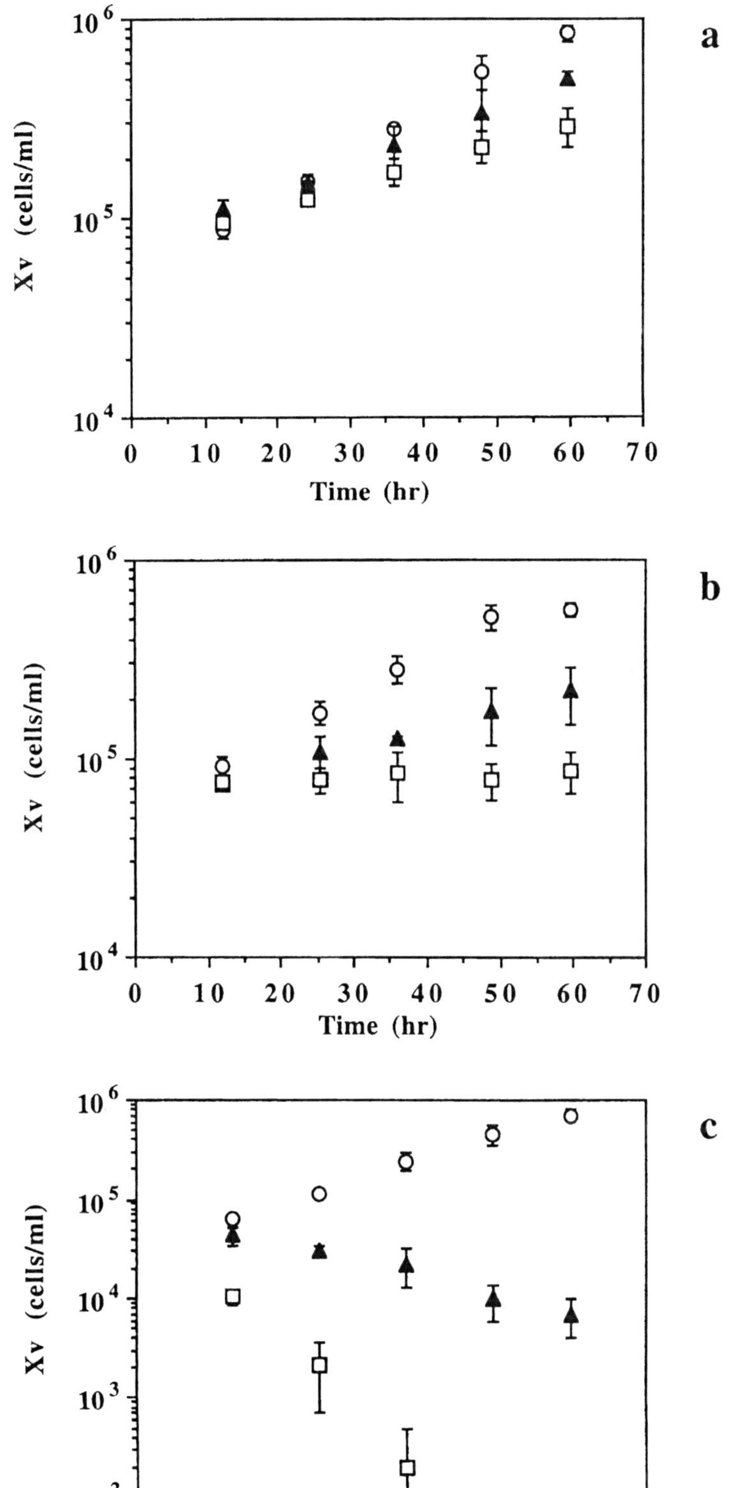

Figure 1. Growth curves for representative well-plate experiments at (a) 140, (b) 195, and (c) 250 mm Hg pCO_2 with (filled triangles) and without (open squares) osmolality adjustment. Control cultures at 36 mm Hg and 310 mOsm kg^{-1} (open circles) are shown for each experiment. Data represent the mean and standard deviation for 3 replicate wells.

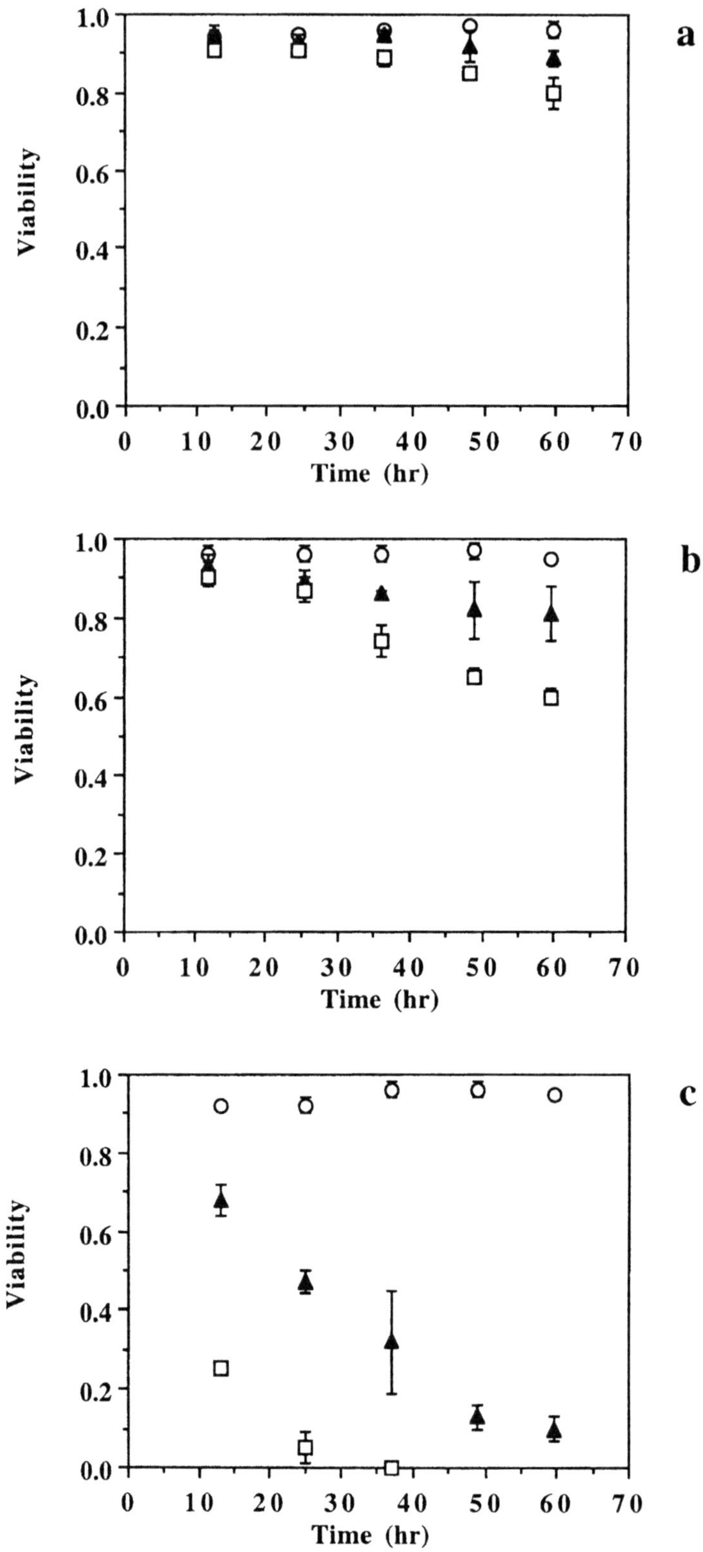

Figure 2. Cell viability for the well plate experiments shown in Figure 1. The symbols are the same as in Figure 1.

Table 1. Effects of elevated pCO_2 and/or osmolality on AB2-143.2 hybridoma cells in well-plate cultures. Based on the Kruskal-Wallis test, both elevated pCO_2-partial osmolality compensation and elevated pCO_2- uncompensated osmolality had a significant effect on specific growth rate ($P < 0.010$), specific death rate ($P < 0.025$), specific glucose consumption rate ($P < 0.025$), and specific lactate production rate ($P < 0.025$). Values are given as mean ± standard deviation, with the number of replicates indicated in parentheses

pCO_2 and osmolality	Cell growth rate	Cell death rate	Specific antibody production rate	Specific glucose consumption rate	Specific lactate production rate	Apparent yield of cells from glucose	Apparent yield of lactate from glucose
	(hr^{-1})	(10^{-3} hr^{-1})	(10^{-13}g $cell^{-1}$ hr^{-1})	(10^{-13} mol $cell^{-1}$ h^{-1})		(10^{10} cells mol^{-1})	(mol/mol)
36 mm Hg 310 mOsm kg^{-1}	0.049±0.005 (7)	2.08±1.33 (7)	7.23±2.02 (7)	5.16±0.59 (7)	10.50±0.12 (7)	9.62±1.50 (7)	2.03±0.08 (7)
140 mm Hg 337 mOsm kg^{-1}	0.038±0.000 (2)	3.70±0.12 (2)	6.29±1.25 (2)	3.17±1.77 (2)	5.14±0.62 (2)	14.2±7.92 (2)	1.79±0.82 (2)
195 mm Hg 361 mOsm kg^{-1}	0.027±0.006 (4)	7.86±0.71 (4)	6.66±1.74 (4)	2.64±0.61 (4)	4.98±0.67 (4)	11.0±4.14 (4)	1.83±0.15 (4)
250 mm Hg 401 mOsm kg^{-1}	ND	64.2 ±0.23 (2)	ND	ND	ND	ND	ND
140 mm Hg 370 mOsm kg^{-1}	0.032±0.001 (2)	4.48±3.97 (2)	9.16±4.86 (2)	3.26±1.16 (2)	5.35±2.01 (2)	10.6±4.19 (2)	1.63±0.04 (2)
195 mm Hg 415 mOsm kg^{-1}	0.018±0.007 (4)	13.0 ±12.7 (4)	9.57±4.27 (4)	2.95±0.62 (4)	5.18±1.33 (4)	6.53±3.10 (4)	1.61±0.20 (4)
250 mm Hg 469 mOsm kg^{-1}	ND	133±50.7 (2)	ND	ND	ND	ND	ND
36 mm Hg 342 mOsm kg^{-1}	0.042±0.003 (2)	2.12±0.69 (2)	10.50±4.77 (2)	6.21±0.31 (2)	12.6 ±1.06 (2)	6.78±0.80 (2)	2.02±0.03 (2)
36 mm Hg 357 mOsm kg^{-1}	0.045±0.001 (3)	2.62±1.60 (3)	8.40±2.42 (3)	5.82±0.35 (3)	12.0 ±1.18 (3)	7.74±0.32 (3)	2.06±0.08 (3)
36 mm Hg 376 mOsm kg^{-1}	0.042±0.008 (3)	3.25±0.2 (2)	9.64±4.07 (3)	5.77±1.27 (3)	11.5 ±2.79 (3)	7.56±2.76 (3)	1.98±0.11(3)

ND indicaties not determined.

for cultures at 195 mm Hg. For example, two sets of experiments at 195 mm Hg exhibited similar inhibition to that shown for 140 mm Hg in Figures 1 and 2, while another set of experiments at 195 mm Hg exhibited inhibition intermediate between those shown for 195 and 250 mm Hg in Figures 1 and 2.

The effects of elevated pCO_2 and/or osmolality on AB2-143.2 hybridoma cells are summarized in Table 1. With partial osmolality compensation, μ decreased to 78 and 55% of control at 140 and 195 mm Hg pCO_2, respectively. Without osmolality compensation, μ decreased to 65 and 37% of control at the same pCO_2 values. Growth rates were not calculated for 250 mm Hg pCO_2 because the small changes in X_t made it difficult to obtain reliable values. Elevated osmolality alone decreased μ to 86% of control at 342 mOsm kg^{-1}, with no further inhibition at 357 or 376 mOsm kg^{-1}. With partial osmolality compensation, k_d was 1.8, 3.8, and 31 times that of control cultures at 140, 195 and 250 mm Hg pCO_2, respectively. Without osmolality compensation, k_d was 2.2, 6.2, and 64 times that of control at the same pCO_2 values. Elevated osmolality alone increased k_d by 1.3 and 1.6 times that of the control cultures at 357 and 376 mOsm kg^{-1}, respectively. Thus, elevated pCO_2 and osmolality appear to have a synergistic effect on μ and k_d.

Glucose metabolism

q_{glc} was 61% of that in control cultures at 140 mm Hg pCO_2 and 337 mOsm kg^{-1} (Table 1). In contrast to what was observed for μ and k_d, there was relatively little further decrease in q_{glc} at 195 mm Hg, and little if any effect of osmolality compensation. A similar trend was observed for q_{lac}. However, the decrease in q_{lac} was 10–20% greater than that in q_{glc}, so that $Y_{lac/glc}$ was slightly lower at elevated pCO_2. Increased osmolality alone did not significantly affect q_{glc} or q_{lac}, although the values tended to be higher at elevated osmolality. Taken together, the above results suggest that elevated pCO_2 decreases glycolysis and that increased osmolality does not contribute to this effect. There was no significant difference in $Y_{n,glc}$ with pCO_2 with or without osmolality compensation. However, the lower value of $Y_{n,glc}$ at 195 mm Hg and 415 mOsm kg^{-1} is consistent with lower values of $Y_{n,glc}$ for elevated osmolality at 36 mm Hg.

Antibody production

q_{Ab} was not significantly affected when cells were exposed to elevated pCO_2 and/or osmolality (Table 1). However, q_{Ab} values tended to be higher without osmolality compensation. This is consistent with a tendency towards higher q_{Ab} values with elevated osmolality at 36 mm Hg.

Continuous flow reactor experiments

The effects of elevated pCO_2 were also evaluated in continuous culture. When pCO_2 was increased, a greater amount of NaOH was needed to maintain pH 7.2. Thus, these experiments are analogous to the well-plate experiments without osmolality compensation.

Results from experiments conducted at dilution rates (D) of 0.022 hr^{-1} and 0.028 hr^{-1} are shown in Figures 3 and 4, respectively. pCO_2 was increased in steps from 50 to 90 and then 140 mm Hg. There was little change in X_v or viability (f_v) when pCO_2 was increased from 50 to 90 mm Hg (Figures 3a and 4a). Although pCO_2 was only maintained at 90 mm Hg for 4 days in these experiments, X_v was essentially constant for 8 days at 90 mm Hg in a third experiment (also at D = 0.028 hr^{-1}; not shown). X_v decreased shortly after pCO_2 was increased to 140 mm Hg, and reached a new steady state at a lower value after 5–8 days. In both cases, k_d was slightly lower at 140 mm Hg, as evidenced by an increase in f_v. The greater decrease in X_v (40% vs. 28%) and the transient decrease in f_v for the experiment shown in Figure 4 may be due to the higher dilution rate (0.028 vs. 0.022 hr^{-1}). In this latter experiment, μ (= D/f_v) was 0.033 hr^{-1} at 90 mm Hg (f_v = 0.85) and 0.031 hr^{-1} for the new steady state at 140 mm Hg (f_v = 0.9). This is close to the value of 0.032 hr^{-1} obtained at 140 mm Hg and 370 mOsm kg^{-1} in well-plate culture. Borderline stability (D near μ_{max} at 140 mm Hg pCO_2) at D = 0.028 hr^{-1} is suggested by a continuing decrease in X_v after the increase in pCO_2 to 140 mm Hg in a second experiment at D = 0.028 hr^{-1} (not shown). However, the lack of recovery in this latter case may have been due in part to a transient increase in pCO_2 to 160 mm Hg (pCO_2 > 140 mm Hg for ~24 hr).

Cell metabolic patterns are shown in Figures 3b and 4b. There was no consistent trend in q_{glc} at 90 mm Hg – with either a slight increase (Figure 3b), a slight decrease (Figure 4b), or no change (not shown). However, there was a consistent increase in q_{glc} at 140 mm Hg. Changes in q_{lac} were generally similar to those for q_{glc} for each experiment. $Y_{n,glc}$ did not change at 90 mm Hg, but decreased by ~20% at 140 mm Hg (not shown). Similarly to q_{glc}, q_{O_2} was not affected by 90 mm Hg pCO_2, and increased by 20–30% at 140 mm

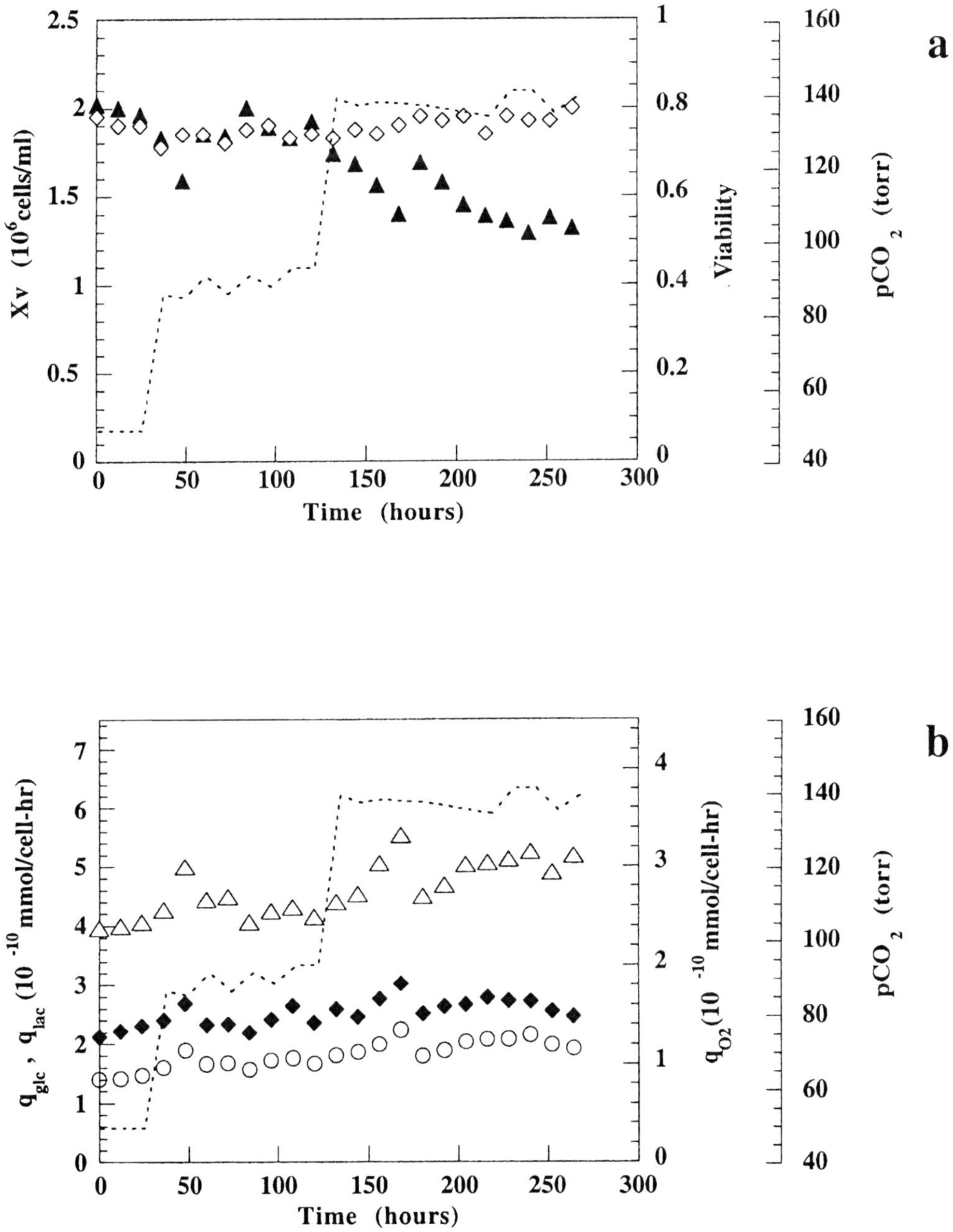

Figure 3. Hybridoma continuous culture at a dilution rate of 0.022 hr^{-1}. (a) Viable cell density (solid triangles) and viability (open diamonds). (b) Specific glucose consumption rate (open circles), lactate production rate (solid diamonds), and oxygen consumption rate (open triangles). The pCO_2 profile is indicated by a dashed line.

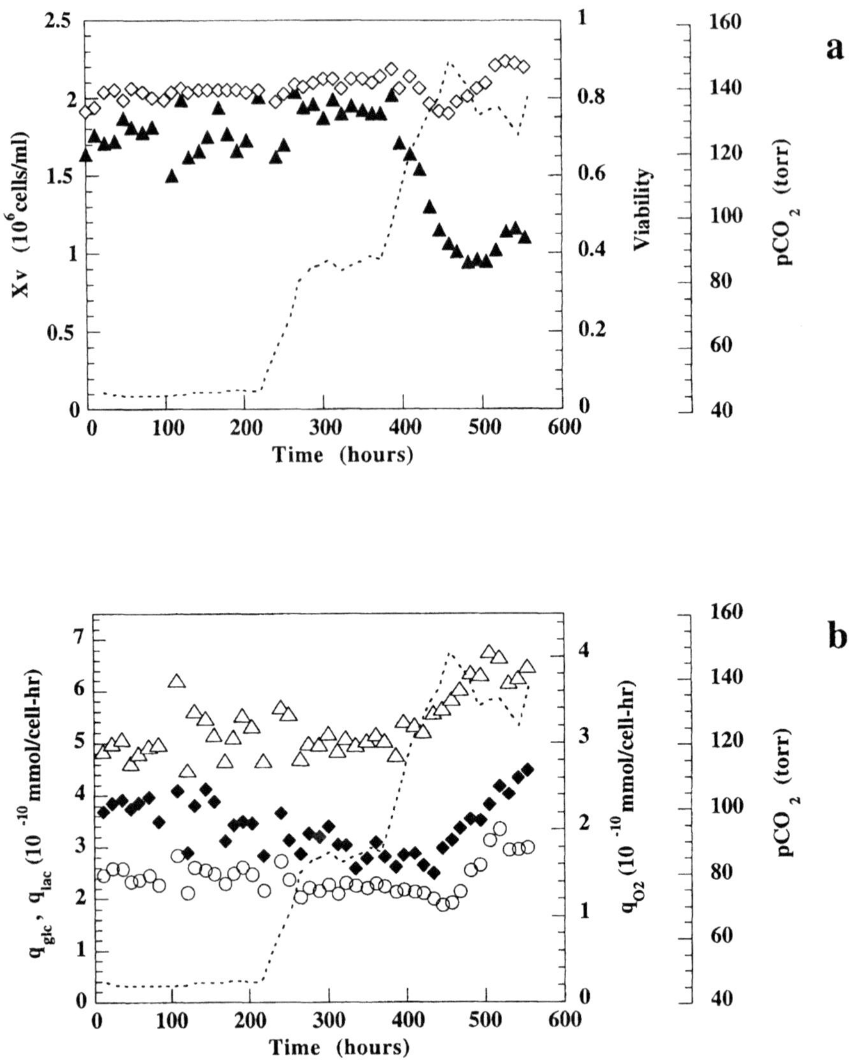

Figure 4. Hybridoma continuous culture at a dilution rate of 0.028 hr^{-1}. The symbols are the same as those shown in Figure 3.

Hg. q_{O_2} was not likely affected by changes in the glucose concentration, which never dropped below 3 mM. $Y_{O_2,glc}$ was very steady at a value of ~1.4 mol/mol, except for a transient 35% increase during the increase in pCO_2 to 140 mm Hg for the experiment shown in Figure 4. There was no consistent effect of 140 mm Hg pCO_2 on q_{Ab} – with a 100% increase (with considerable scatter) for the experiment shown in Figure 3 and a slight decrease for the experiment shown in Figure 4 (not shown).

Batch bioreactor experiments

Several batch reactor experiments were performed at 140 mm Hg pCO_2 and 370 mOsm kg^{-1} to determine whether the differences observed between well-plate and continuous culture experiments (see Discussion) were due in part to the presence of pH and DO control in the continuous culture experiments. Representative growth curves for control and 140 mm Hg pCO_2 cultures are shown in Figure 5. Although the decrease in viability at 140 mm Hg (Figure 5b and results not shown) is not as great as that shown in Figure 2a, it is similar to that observed in a replicate well-plate culture (Kimura, 1996). Apoptosis was identified as the primary mechanism of cell death in the controlled batch cultures, with somewhat more extensive apoptosis in cultures at 140 mm Hg pCO_2 (not shown).

Table 2. Effects of elevated pCO_2 and osmolality on AB2-143.2 hybridoma cells in batch reactor cultures. Values are given as mean ± standard deviation (n = 2 for each condition)

pCO_2 and osmolality	Cell growth rate	Cell death rate	Specific antibody production rate	Specific glucose consumption rate	Specific lactate production rate	Specific oxygen consumption rate	Apparent yield of cells from glucose	Apparent yield of lactate from glucose
	(hr^{-1})	(10^{-3} hr^{-1})	(10^{-13}g $cell^{-1}$ hr^{-1})	(10^{-13} mol $cell^{-1}$ h^{-1})			(10^{10} cells mol^{-1})	(mol/mol)
36 mm Hg 310 mOsm kg^{-1}	0.051±0.005	1.71±0.65	19.8±6.29	4.91±1.17	9.94±2.35	6.8±2.8	10.50±1.48	2.03±0.01
140 mm Hg 370 mOsm kg^{-1}	0.037±0.001	4.17±1.03	33.4±5.09	3.89±0.27	7.11±0.33	3.3±0.2	9.41±0.83	1.83±0.21

The effects of 140 mm Hg pCO_2 and 370 mOsm kg^{-1} in batch reactor culture are summarized in Table 2. The 27% decrease in μ and the 2.4-fold increase in k_d were similar to those in well-plates (Table 1). In contrast, the 21% decrease in q_{glc} and 28% decrease in q_{lac} in the bioreactor were only about 60% as extensive as those in well-plates. As for the well-plates, there was no significant change in $Y_{n,glc}$. q_{O_2} was 51% lower at 140 mm Hg. The larger difference in q_{O_2} relative to q_{glc} resulted in a 38% lower value of $Y_{O_2,glc}$ at 140 mm Hg. Specific antibody production was about 70% greater at elevated pCO_2.

Discussion

AB2-143.2 hybridoma cells exposed to elevated pCO_2 in well-plates experienced both an inhibition of growth and an enhancement of cell death, with greater detrimental effects without osmolality compensation. Cell growth was also inhibited by pCO_2 in continuous culture (CC). The decrease in X_v at 140 mm Hg was due to a transient decrease in μ, and is consistent with the 35% decrease in μ at 140 mm Hg and 370 mOsm kg^{-1} in well-plates and the 27% decrease in μ in batch reactors (Table 3). There was also a trend towards higher q_{Ab} at 140 mm Hg and 370 mOsm kg^{-1} in all 3 culture systems. In contrast, k_d decreased slightly at 140 mm Hg in CC (after a transient increase for the experiment shown in Figure 4), while in well-plate and batch reactor culture k_d was more than twice as great as that for the control at 140 mm Hg. While q_{glc} and q_{lac} were 40–50% lower than control at 140 mm Hg in well-plates and 20–30% lower in batch reactor culture, q_{glc} and q_{lac} increased by 20–30% at 140 mm Hg in CC. Similarly, q_{O_2} was 50% lower at 140 mm Hg in batch culture, while CC q_{O_2} increased by ~30% at 140 mm Hg. Further, $Y_{O_2,glc}$ was ~40% lower at 140 mm Hg in batch culture, while $Y_{O_2,glc}$ remained the same or transiently increased at 140 mm Hg in CC. Finally, there was a consistent decrease in $Y_{n,glc}$ at 140 mm Hg in CC, while $Y_{n,glc}$ was not significantly affected by pCO_2 in batch culture.

Although differences between the responses in batch culture and those in CC may be partially explained by cell adaptation to elevated pCO_2 in CC, they can also be explained by differences in growth limitation for the two systems. Cell growth in CC is limited by the concentrations of nutrients and/or byproducts. Thus, k_d for control cultures was greater in CC (0.006–0.008 hr^{-1}) than in well-plates or batch reactors. The dilution rates of 0.022 and 0.028 hr^{-1} are lower than μ of 0.032 hr^{-1} for cells exposed to 140 mm Hg and 370 mOsm kg^{-1} in well-plates (or 0.037 hr^{-1} for cells in batch reactors). Since D is less than the effective μ_{max} for cells exposed to 140 mm Hg and 370 mOsm kg^{-1}, the cells in CC were able to compensate for inhibition by elevated pCO_2 and reach a new steady state at lower X_v due to the increase in nutrient levels and decrease in byproduct levels that accompanied the decrease in X_v. This explanation is consistent with the decrease in k_d at 140 mm Hg in CC. However, even after the decrease, k_d in CC at 140 mm Hg (0.003–0.006 hr^{-1}) was similar to that in well-plates and batch reactors at the same pCO_2. Also, even after the increase at 140 mm Hg, q_{glc} in CC was less than or equal to that in well-plates or batch reactors. Glucose utilization is generally more efficient at lower glucose consumption rates (Miller and Blanch, 1991). This is consistent with the observation that, even after the decrease, $Y_{n,glc}$ in CC was greater than or equal to that in well-plates or batch reactors (not shown). The lower q_{O_2} at 140 mm Hg in batch culture is consistent with an overall inhibition of cell metabolism. While the increase in q_{O_2} with pCO_2 in CC is consistent with an overall increase in cell metabolic activity, the increase in q_{O_2} cannot be attributed to an increase in DO, as was the case for higher q_{glc} with higher residual glucose concentration. Rather, the similar increases in q_{glc} and q_{O_2}, together with the fact that the glucose concentration was always $\geq$3 mM, suggest that cell metabolism was limited by depletion of another nutrient(s) or by a metabolic byproduct(s) not monitored in these experiments. Glutamine was not limiting for the experiment shown in Figure 4 (0.96 mM before the shift to 140 mm Hg and 3.2 mM after the decrease in X_v at 140 mm Hg), but other amino acids may have been depleted.

Although results for well-plate and batch reactor cultures were qualitatively similar, the decrease in μ and especially the decrease in the extent of glycolysis (q_{glc}, q_{lac}, and $Y_{lac,glc}$) at 140 mm Hg and 370 mOsm kg^{-1} were less extensive in batch reactor culture. These differences cannot be attributed to a lack of pH control in the well-plates because well-plate cultures at 140 mm Hg showed very little change in pH. If anything, one would expect that well-plate control cultures would exhibit lower growth rate and glucose conversion to lactate than in control batch reactor cultures due to the lower pH in the well plates. However, control cultures for the two systems had similar μ, q_{glc} and q_{lac} values (Tables 1 and 2). Sim-

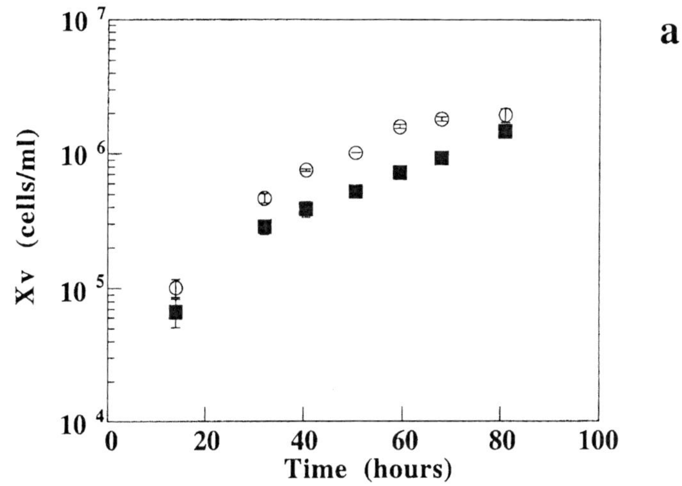

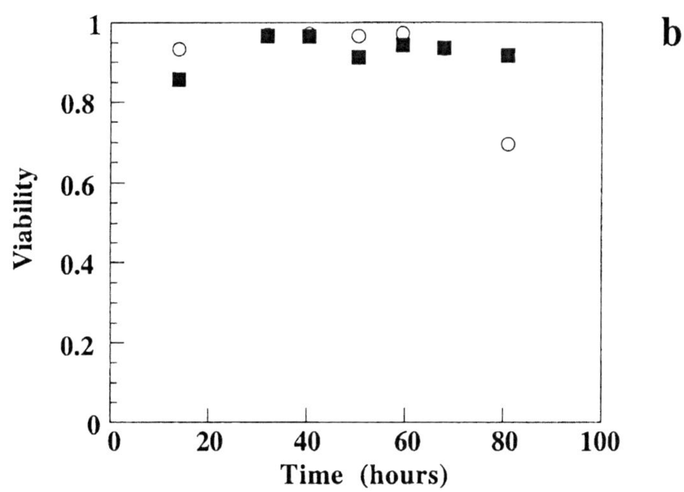

Figure 5. Cell growth (a) and viability (b) for representative batch bioreactor experiments at control conditions (open circles; 36 mm Hg pCO_2, 310 mOsm kg^{-1}) and 140 mm Hg pCO_2 and 370 mOsm kg^{-1} (solid squares). Cultures were seeded with 5×10^4 cells mL^{-1} at zero hr. Data represent the mean and standard deviation for 2 cell counts.

Table 3. Normalized values of various culture parameters (relative to those at 36 mm Hg and 310 mOsm kg^{-1}) at 140 mm Hg and 370 mOsm kg^{-1} in different culture systems. The respective control conditions are used for each culture system

Culture	μ	k_d	q_{Ab}	q_{glc}	q_{lac}	q_{O_2}	$Y_{n,glc}$	$Y_{lac,glc}$	$Y_{O_2,glc}$
Well-plates	0.65	2.2	1.27	0.63	0.51	ND	1.10	0.80	ND
Batch reactor	0.73	2.4	1.69	0.79	0.72	0.49	0.90	0.90	0.62
Continuous culture	Transient decrease; 0.94 @ SS	Slight decrease	1.4	1.3	1.2	1.3	0.8	0.9	Same or transient increase

ilar μ and q_{glc} values in well-plate and batch reactor control cultures also argue against the possibility that oxygen was depleted in the control well-plates.

Hybridoma AB2-143.2 cell growth in well-plates is more sensitive to elevated pCO_2 and osmolality than is that of CHO MT2-1-8 cells. Hybridoma μ decreased by 63% at 195 mm Hg and 415 mOsm kg^{-1}, while that of CHO MT2-1-8 cells decreased by only 24% (Kimura and Miller, 1996). (The osmolality values reported in Kimura and Miller, 1996 are in error; the correct osmolality values are essentially the same as those shown in Table 1 for the same conditions). In addition, the hybridoma k_d increased by 64-fold at 250 mm Hg pCO_2 and 469 mOsm kg^{-1}, while there was no increase in the (low) CHO MT2-1-8 cell k_d. Caution must be used when comparing results obtained in different culture systems. However, since we obtained similar qualitative results in well-plates and batch reactors, we feel that it is appropriate to directly compare results for different cell lines in batch culture. Several reports indicate that some CHO cell lines are more sensitive to pCO_2 than are MT2-1-8 cells. Drapeau et al. (1990) reported that μ for M-CSF-producing CHO cells in batch culture with pH and DO control decreased to 86 and 59% of control at 114 and 167 mm Hg pCO_2, respectively, when osmolality was controlled. Interpolation of the values in Table 1 suggests that this CHO cell line is similar in sensitivity to pCO_2 as are AB2-143.2 hybridomas. Also, another CHO cell line exhibited a dramatic decrease in shake-flask batch growth rate even at 76 mm Hg pCO_2 when osmolality was maintained between 310–340 mOsm kg^{-1} (Gray et al., 1996).

Hybridoma AB2-143.2 cell q_{Ab} did not decrease significantly at elevated pCO_2 with partial osmolality compensation and even increased slightly without osmolality compensation. This increase in q_{Ab} is similar to that observed for increased osmolality at control pCO_2, which is consistent with reports that q_{Ab} for these cells is increased by elevated osmolality in both batch and continuous culture (Reddy and Miller, 1994). As for hybridoma AB2-143.2 cells, recombinant NS/0 cells showed no decrease in q_{Ab} at 120 mm Hg pCO_2 when osmolality was allowed to increase with pCO_2 (Aunins and Henzler 1993). Recombinant protein production by CHO cells appears to be more sensitive to elevated pCO_2. At constant osmolality, CHO cell q_{M-CSF} decreased by 29% at 114 mm Hg and 44% at 167 mm Hg (Drapeau et al., 1990) in batch culture. Without osmolality compensation, CHO cell q_{tPA} decreased by 27% at 140 mm Hg and 21% at 195 mm Hg in batch culture (Kimura and Miller, 1996). Similarly, in high-density perfusion it has been shown that CHO cell q_P for a viral antigen decreased by 26% at 103 mm Hg and 44% at 148 mm Hg (Gray et al., 1996), while BHK-21 cell q_P for Factor VIII decreased by 33% at 130 mm Hg and 50% at 210 mm Hg (Taticek et al., 1998).

Hybridoma AB2-143.2 cell q_{glc} and q_{lac} decreased by 40 and 50%, respectively, at 140 mm Hg in well plates at both 337 and 370 mOsm kg^{-1} – with little further decrease at 195 mm Hg (Table 1). q_{glc} and q_{lac}, as well as q_{O_2}, also decreased at 140 mm Hg in batch reactor culture. Taticek et al. (1998) observed a steady decrease in BHK-21 cell q_{glc} and q_{lac} as pCO_2 was increased in high-density perfusion culture with a constant cell-specific perfusion rate – with 25 and 33% decreases in q_{glc} at 150 and 200 mm Hg, respectively, and 28 and 38% decreases in q_{lac} at 150 and 200 mm Hg, respectively. q_{gln} decreased with increasing pCO_2 in a similar manner (Taticek et al., 1998). Perfusion culture with a constant cell-specific perfusion rate is similar to batch culture – and different from CC – in that nutrient levels will not increase due to inhibition by CO_2.

One possible mechanism through which CO_2 may affect cell growth and metabolism is by changes in the intracellular pH (pH_i), which affects the activity of various enzymes involved in metabolism and protein production (Madshus, 1988). CO_2 is a weak acid that diffuses across cell membranes so rapidly that pCO_2 inside cells is only rarely different from that outside (Thomas, 1995). Once inside, CO_2 reestablishes equilibrium with H^+ and HCO_3^-, so that pH_i initially decreases in response to an increase in pCO_2. However, the decrease in pH_i is counteracted by ion transport systems. Depending on the cell type and conditions, the steady-state pH_i may be less than (Simchowitz and Roos, 1985), the same as (Krapf et al., 1988; Mellergård et al., 1993), or greater than (Putnam and Grubbs, 1990; England et al., 1991) pH_i prior to the increase in pCO_2. Of particular interest for recombinant protein production, Taticek et al. (1998) reported that the steady-state pH_i of BHK-21 cells in perfusion culture at pH 6.8 decreased steadily with increasing pCO_2 from 7.1 at 35 mm Hg to 6.92 at 270 mm Hg. We are currently investigating in more detail the mechanisms through which CO_2 affects the intracellular environment.

Note in Proof

Well-plate experiments at elevated pCO_2 with full osmolality compensation, and at control pCO_2 (36 mm Hg) with osmolality values similar to those in uncompensated cultures at elevated pCO_2, are now in progress. Preliminary results demonstrate a continuing trend of less extensive effects of elevated pCO_2 with greater osmolality compensation. Also, cell growth is significantly inhibited by 403 and 454 mOsm kg^{-1} at 36 mm Hg, although to a slightly lesser extent than by exposure to 195 and 250 mm Hg, respectively, with full osmolality compensation. Thus, elevated pCO_2 and elevated osmolality appear to have a similar and synergistic contribution to inhibition of cell growth by elevated pCO_2 without osmolality compensation.

Acknowledgments

Custom low-salt DMEM:F12 medium was a generous gift of Irvine Scientific. Funding for this work was provided by National Science Foundation grant BES-9402030. V. M. deZengotita was partially supported by a Predoctoral Biotechnology Training Grant (NIH GM08449). We are grateful for the use of osmometers loaned to us by Joanne Bradna (Rochelle Scientific, distributor for Wescor) and Rick Zampa (Advanced Instruments).

References

Altman PL and Dittmer DS (1971) Respiration and circulation. Federation of American Societies of Experimental Biology, MD.

Aunins JG and Henzler H-J (1993) Aeration in Cell Culture Bioreactors. In: Stephanopoulos G, Rehm H-J, Reed G, Puhler A and Stadler PJW (eds.) Biotechnology: A Multi-volume Comprehensive Treatise (Vol. 3), Bioprocessing. VCH Verlag, Weinheim, Germany.

Aunins JG, Glazomitsky K and Buckland B (1993) Animal Cell Culture Reactor Design: Known and Unknown. In: Proceedings of 3rd Intl. Conference on Bioreactor and Bioprocess Fluid Dynamics,Kluwer Academic Publishers, p. 175.

Drapeau D, Luan Y-T, Whiteford JC, Lavin DP and Adamson SR (1990) Cell culture scale-up in stirred tank reactors. Presented at The Annual Meeting of the Society of Industrial Microbiology, Orlando, FL.

England BK, Chastain JL and Mitch WE (1991) Abnormalities in protein synthesis and degradation induced by extracellular pH in BC_3H1 myocytes. Am J Physiol 260: C277–C282.

Garnier A, Voyer R, Tom R, Perret S, Jardin B and Kamen A (1996) Dissolved carbon dioxide accumulation in a large scale and high density production of TGFβ receptor with baculovirus infected Sf-9 cells. Cytotechnology 22: 53–63.

Glantz SA (1981) Primer of biostatistics. McGraw-Hill, New York.

Gray DR, Chen S, Howarth W, Inlow D and Maiorella BL (1996) Carbon dioxide in large-scale and high density CHO cell perfusion culture. Cytotechnology 22: 65–78.

Hornbeck PV and Lewis GK (1985) Idiotype connectance in the immune system. II. A heavy chain variable region idiotype that dominates the antibody response to the p-azobenzenearsonate group is a minor idiotype in the response to trinitrophenyl group. J Exp Med 161: 53–71.

Kimura, R (1996) Effects of elevated carbon dioxide partial pressure on recombinant Chinese hamster ovary and hybridoma cell cultures. PhD Dissertation, Northwestern University.

Kimura R and Miller WM (1996) Effects of elevated pCO_2 and/or osmolality on the growth and recombinant tPA production of CHO Cells. Biotechnol Bioeng 52: 152–160.

Krapf R, Berry CA, Alpern RJ and Rector Jr, FC (1988) Regulation of cell pH by ambient bicarbonate, carbon dioxide tension, and pH in the rabbit proximal convoluted tubule. J Clin Invest 81: 381–389.

Kurano N, Leist C, Messi F, Kurano S and Fiechter A (1990) Growth behavior of Chinese Hamster Ovary cells in a compact loop bioreactor: 1. Effects of physical and chemical environments. J Biotechnol 15: 101–112.

Lovrecz G and Gray P (1994) Use of on-line analysis to monitor recombinant mammalian cell cultures. Cytotechnology 14: 167–175.

Madshus IH (1988) Regulation of intracellular pH in eukaryotic cells (Review Article). Biochem J 250: 1–8.

Mellergård P, Ouyang Y and Siesjö BK (1993) Intracellular pH regulation in cultured rat astrocytes in CO_2/HCO_3^--containing media. Exp Brain Res 95: 371–380.

Miller WM and Blanch HW (1991) Regulation of Animal Cell Metabolism in Bioreactors. In: Animal Cell Reactors. Ho CS and Wang DIC (eds.) Butterworth-Heinemann, Stoneham, MA.

Miller WM, Blanch HW and Wilke CR (1988) A kinetic analysis of hybridoma growth and metabolism in batch and continuous culture: Effect of nutrient concentration, dilution rate, and pH. Biotechnol Bioeng 32: 947–965.

Ozturk SS and Palsson BO (1991) Effect of medium osmolarity on hybridoma growth, metabolism, and antibody production. Biotechnol Bioeng 37: 989–993.

Putnam RW and Grubbs RD (1990) Steady-state pH_i, buffering power, and effect of CO_2 in a smooth muscle-like cell line. Am J Physiol 258: C461–C469.

Reddy S and Miller WM (1994) Effects of abrupt and gradual osmotic stress on antibody production and content in hybridoma cells that differ in production kinetics. Biotechnol Prog 10: 165–173.

Royce PNC and Thornhill NF (1991) Estimation of dissolved carbon dioxide concentrations in aerobic fermentations. AIChE J 37: 1680–1686.

Simchowitz L and Roos A (1985) Regulation of intracellular pH in human neutrophils. J Gen Physiol 85: 443–470.

Taticek R, Petersen S, Konstantinov K and Naveh D (1998) Effect of dissolved carbon dioxide and bicarbonate on mammalian cell metabolism and recombinant protein productivity in high density perfusion culture. Presented at Cell Culture Engineering VI, San Diego, CA.

Thomas RC (1995) Bicarbonate briefly CO_2-free. Nature 374: 597–598.

Zupke C and Green J (1998) Modeling of CO_2 concentration in small and large scale bioreactors. Presented at Cell Culture Engineering VI, San Diego, CA.

Cytotechnology **28:** 229–235, 1998.
© 1998 *Kluwer Academic Publishers. Printed in the Netherlands.*

pQuattro vectors allow one-step multigene metabolic engineering and auto-selection of quattrocistronic artificial mammalian operons

Short Communication

Martin Fussenegger, Samuel Moser & James E. Bailey*
Institute of Biotechnology, Swiss Federal Institute of Technology, ETH Zurich, CH-8093 Zurich, Switzerland

Received 25 August 1998; accepted 25 August 1998

Key words: CITE, EMCV, GFP, *IRES*, picornavirus

Abstract

Based on internal ribosomal entry sites (*IRES*) of picornaviral origin we constructed a novel family of mammalian expression vectors. pQuattro vectors contain quattrocistronic artificial eukaryotic operons which link, in a single transcript, the simultaneous and coordinated as well as adjustable expression of up to three independent genes of interest to a terminal neomycin (*neo*) resistance marker. Due to the strict genetic linkage of the transgenes and the terminal selection marker, this genetic configuration enables, by the selection on neomycin, multigene metabolic engineering of mammalian cells in a single step (one-step metabolic engineering). Furthermore, selection on the terminal cistron of multicistronic expression units enforces cocistronic expression of all upstream encoded genes and maximises genetic integrity of the eukaryotic operon in stable mammalian cell lines, since clones harbouring damaged multicistronic expression units become neomycin-sensitive and are automatically counterselected (auto-selection). The modular set-up and the abundance of restriction sites in pQuattro vectors facilitate the movement of individual genes between multicistronic expression vectors and guarantees high compatibility with genetic elements of a wide variety of existing mammalian expression vectors.

Introduction

Current metabolic engineering technology is in most cases based on the transfection of a single transgene, the expression of which is maintained and selected for by cotransfection of a second vector encoding a selectable marker which often confers resistance to an antibiotic. However, due to unlinked expression of the selection marker and the gene of interest, rarely more than 30% of the stable integrants expressing the antibiotic resistance also express the desired transgene (Fussenegger et al., 1998a). This phenomenon either results from a deletion or inactivation of the gene of interest during or following plasmid integration or from the exclusive integration of the vector encoding the resistance marker. Even if stable cell lines with concomitant expression of the selection marker and the gene of interest could be generated, the expression level of the transgene is unpredictable and often decreases during extended cultivations since there is no positive selection for the maintenance and the transcriptional activity of the transgene.

Furthermore, as contemporary basic research continues to reveal more and more details on the molecular complexity of metabolic pathways and networks, classical one-gene metabolic engineering is becoming increasingly limited in its potential to realised desired global phenotypes. In order for metabolic engineering to keep up with the current detailed knowledge of physiological pathways and metabolic networks simultaneous expression of multiple genes is a prerequisite to achieve difficult-to-attain metabolic characteristics or cell culture states (Papoutsakis, 1998; Fussenegger et al., 1998b). Although current expression technology may in principle enable such multigene metabolic engineering by subsequent rounds of transfections with a single gene, such methodology is tedious, time consuming, and increases the risk of generating secondary effects unrelated to the transgenes

* Author for all correspondence.

at every round of transfection due to random integrations of the different vectors into the host chromosome (Fussenegger et al., 1998a).

Results and discussion

Based on specific viral genetic elements which enable cap-independent translation initiation (internal ribosomal entry sites, *IRES*; Fussenegger et al., this issue for a review), multicistronic expression technology was recently established and successfully used for multigene metabolic engineering of Chinese hamster ovary cells (CHO) (Papoutsakis, 1998; Fussenegger et al., 1998b; Fussenegger et al., 1998c). The concept of multicistronic eukaryotic expression leading to the quattrocistronic family of mammalian expression vectors is based on the previously reported pTRIDENT system (Fussenegger et al., 1998d). The tricistronic expression unit contained in plasmid pTRIDENT1 consists of a tetracycline-repressible promoter ($P_{hCMV*-1}$) and three multiple cloning sites which are separated by two identical *IRES* elements of polioviral origin (Fussenegger et al., 1998d). In order to extend this tricistronic expression unit to the quattrocistronic level by the addition of a selectable marker (pQuattro1), the *Ssp*I/*Bgl*II fragment of pTRIDENT1 ($P_{hCMV*-1}$-MCSI-*IRESI*-MCSII-*IRESII*-MCSIII) was fused to the corresponding fragment of pIRES*neo* (*Ssp*I/*Bgl*II; Clontech, Palo Alto, CA; Rees et al., 1996) containing a cap-independent translation enhancer (*CITE*) an *IRES* element of the encephalomyocarditis virus (EMCV) fused to the neomycin resistance gene as well as the pUC-derived origin of replication (*ori*) for high copy number amplification in *E. coli* (Figure 1; Table 1). Additional quattrocistronic vectors pQuattro2, pQuattro3 and pQuattro4 containing different regulatable as well as constitutive promoters such as the ecdysone-responsive promoter (P_{EC}; pQuattro2), the SV40 promoter (pQuattro3) and the MPSV promoter (pQuattro4) were constructed by replacing $P_{hCMV*-1}$ of pQuattro1 by the respective promoter elements (*Ssp*I/*Eco*RI) of pIND (Invitrogen), pSBC1 (Dirks et al., 1993), and pMPSVEH (Dirks et al., 1994), respectively (Figure 1).

In order to test the quattrocistronic expression unit, pQuattro3 and pQuattro4 derivatives pMF143 and pMF144 were constructed which harbour the secreted alkaline phosphatase (*seap*, Berger et al., 1988), the tumor suppressor gene p21 (El-Deiry et al., 1993), the green fluorescent protein (GFP; Crameri et al., 1996) and the neomycin resistance gene (Rees et al., 1996; Table 1). In addition, a pQuattro1 derivative was constructed to assess tetracycline-dependent regulation of the multicistronic expression unit (data not shown; Fussenegger et al., 1998d). pMF143 and pMF144 resulted from an *Eco*RI/*Bgl*II fusion of the pTRIDENT1 derivative pMF132 (Fussenegger et al., 1998d) and the respective pQuattro vectors (Table 1). pMF143 and pMF144 were transfected into wildtype CHO-K1 cells (ATCC: CCL61) as described before (Fussenegger et al., 1997a). Transfected populations were grown in FMX-8 medium (Messi Cell Culture Technology, Switzerland) supplemented with 10% fetal calf serum (FCS, Boehringer Mannheim, Lot Nr. 14713602) and for the selection of stable transfectants also with 400 μL mL^{-1} G418. The neomycin-resistant cell population was qualitatively tested for expression of SEAP, p21, and GFP using an enzymatic assay or (immuno)fluorescence microscopy (Berger et al., 1988; Fussenegger et al., 1997a, Figure 2). Besides the SEAP activity measured in the cell culture supernatant the transfected cells show a coordinated expression of p21 and GFP (Figure 2). Owing to the cytostatic effect of the tumor suppressor gene p21, transfected cells stop their cell cycle and cell division following transfection of the quattrocistronic expression vector. Consequently, cells harbouring the pQuattro derivatives pMF143 and pMF144 are not able to proliferate and remain single for extended time periods of at least 4 weeks and probably beyond that period (Figure 2). Because of the instantaneous proliferation arrest of CHO cells transfected with pMF143 and pMF144, SEAP, p21 and GFP expression levels were assessed in transient transfection experiments. The SEAP production derived from pMF143 and pMF144 was identical to levels expressed from pTRIDENT derivatives or from isogenic mono- or dicistronic expression vectors (Fussenegger et al., 1997a; Fussenegger et al., 1998d). Similarly, (immuno)fluorescence intensities of p21 and GFP were comparable to the ones derived from isogenic di- or tricistronic configurations (Fussenegger et al., 1997a; Fussenegger et al., 1998d).

Despite this complete proliferation block the cells display no morphologies suggestive of apoptosis and seem to remain viable since their cell membranes retain intracellular GFP during the entire time period (Figure 2; Fussenegger et al., 1997a). Certainly, p21 has been previously shown to arrest CHO cells at G1-phase in transient transfection experiments at high intracellular concentrations (Fussenegger et al., 1997a)

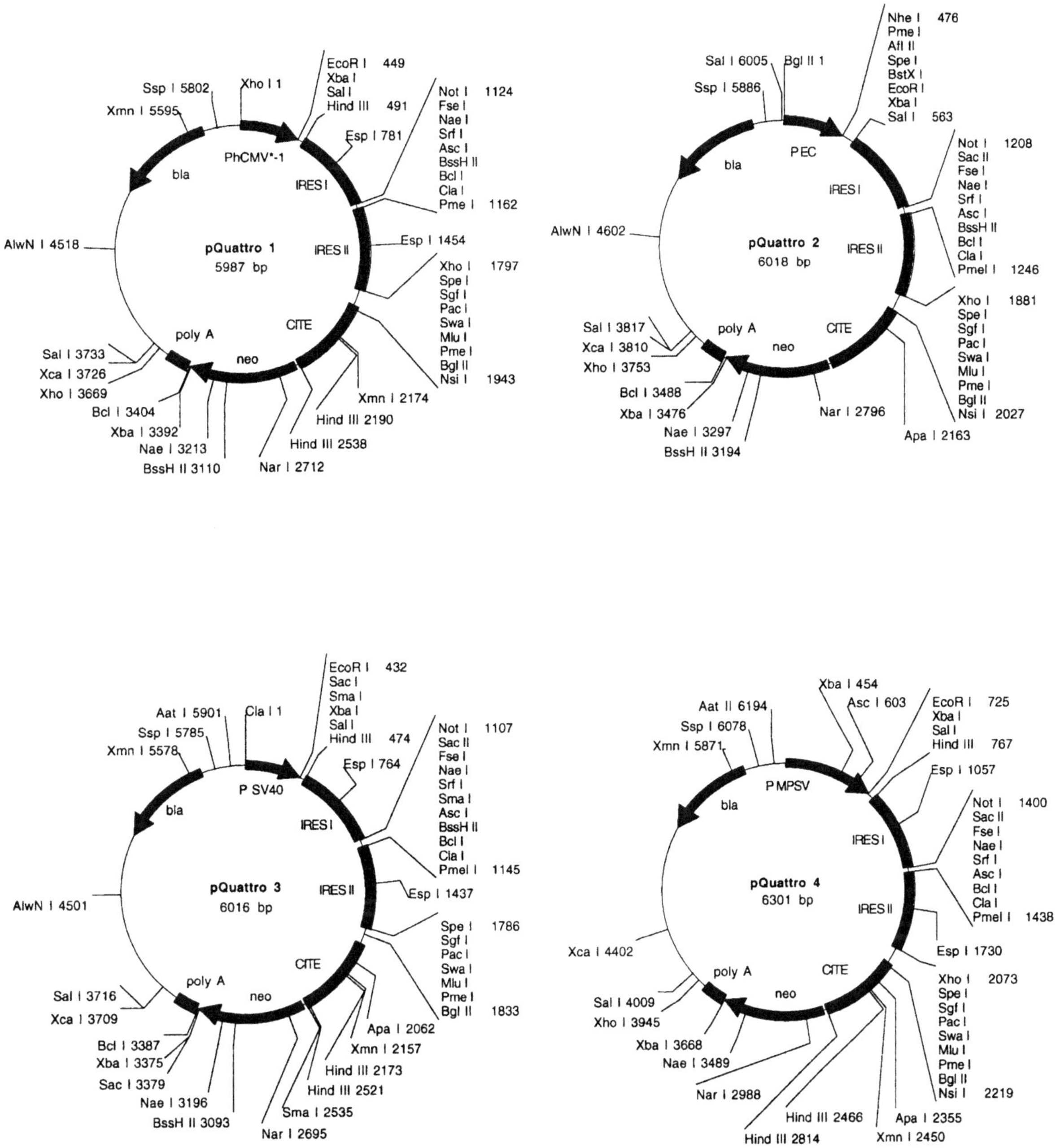

Figure 1. Maps of pQuattro1-4. All mammalian expression vectors contain the high copy number pUC-derived *ori* as well as the ampicillin resistance gene (*bla*) for high level amplification and selection in *E. coli*. The quattrocistronic expression units consist of three multiple cloning sites, eukaryotic promoter elements $P_{hCMV*-1}$ (tetracycline-responsive), P_{EC} (ecdysone-responsive) or the constitutive viral promoters $P_{SV40}$and P_{MPSV}, two internal ribosomal entry sites (*IRES*), a cap-independent translation enhancer (*CITE*) fused to the G418 resistance gene (*neo*), and a terminal polyadenylation signal (*pA*) derived from bovine growth hormone.

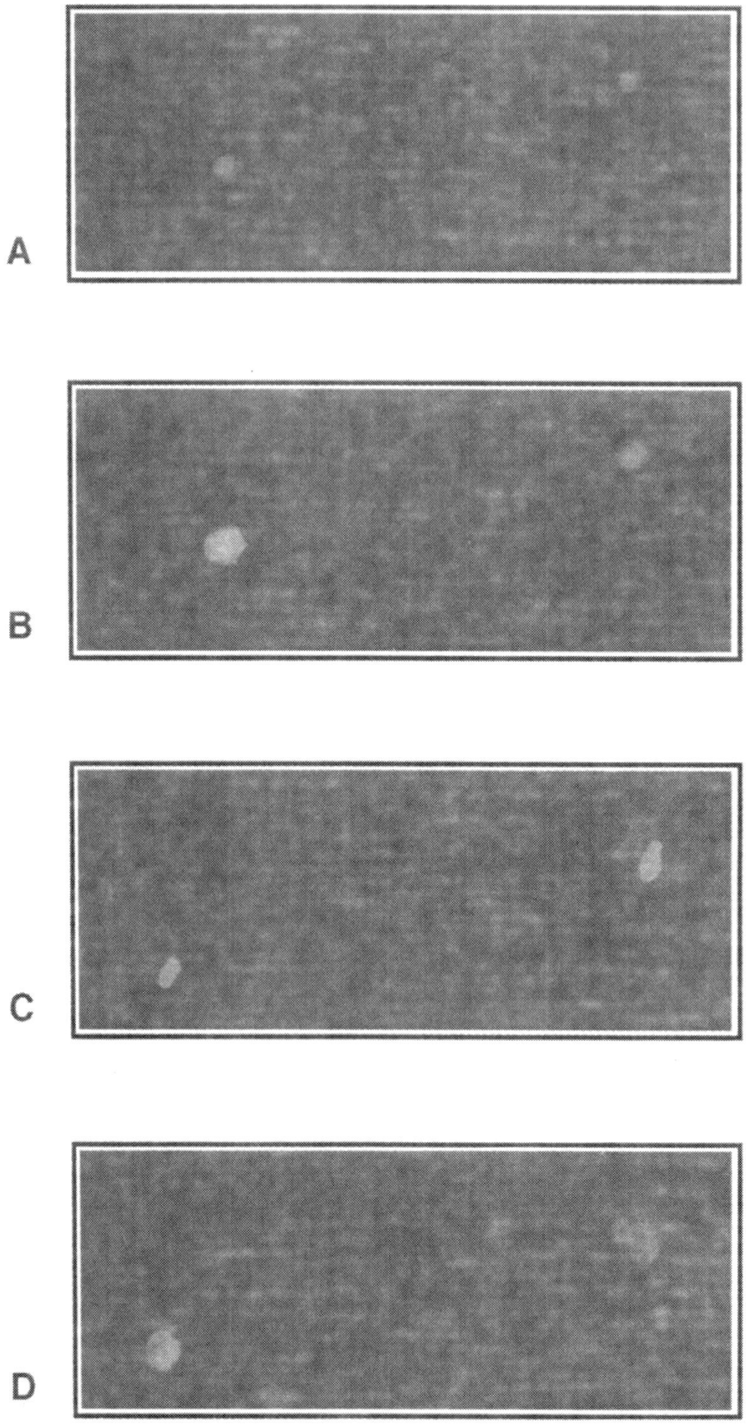

Figure 2. Immunofluorescence of CHO-K1 cells stably transfected with pMF143 (A and B) and pMF144 (C and D). A and C show immunofluorescence of p21 encoded by the second cistron whereas B and D show GFP-mediated green fluorescence. The expression of both genes is strictly correlated due to their expression from a single expression unit.

Table 1. Plasmids used in this study

Plasmid	Description	Reference or source
pIND	Cloning vector containing the ecdysone-regulatable promoter P_{EC}	Invitrogen
pIRES*neo*	Dicistronic expression vector encoding neomycin resistance in the second cistron	Clontech/Rees et al., 1996
pMPSVEH	Cloning vector containing the strong, constitutive viral promoter P_{MPSV}	Dirks et al., 1994
pSBC1	Cloning vector containing the strong, constitutive viral promoter P_{SV40}	Dirks et al., 1993
pTRIDENT1	Tricistronic mammalian expression vector	Fussenegger et al., 1997b
pMF132	pTRIDENT1-derivative harboring *seap-p21-gfp* in a tricistronic expression unit	Fussenegger et al., 1997b
pMF143	pQuattro3 containing the tricistronic *seap-p21-gfp*-containing expression unit of pMF132 (*Eco*RI/*Bgl*II)	This work
pMF144	pQuattro4 containing the trisictronic *seap-p21-gfp*-containing expression unit of pMF132 (*Eco*RI/*Bgl*II)	This work
pQuattro1	*Bgl*II/*Ssp*I fusion of pTRIDENT1 and pIRES*neo*	This work
pQuattro2	pQuattro1 containing the Ecdysone-responsive promoter of pIND (*Eco*RI/*Ssp*I)	This work
pQuattro3	pQuattro1 containing the SV40 promoter of pSBC1 (*Eco*RI/*Ssp*I)	This work
pQuattro4	pQuattro1 containing the MPSV promoter of pMPSVEH (*Eco*RI/*Ssp*I)	This work

but stable transfection of regulated, $P_{hCMV*-1}$-based p21-expression vectors such as recently developed for the cytostatic production process were not able to arrest the cell cycle, most probably due to insufficient expression levels (Fussenegger et al., 1997a; Fussenegger et al., 1998c).

The pQuattro family of mammalian expression vectors is a highly modular multicistronic expression system which allows the simultaneous, coordinated and, in some cases, adjustable expression of up to three genes of interest which are transcriptionally linked to the neomycin resistance marker in the terminal cistron. The concept of the quattrocistronic expression vectors presented here has several advantages over existing mammalian expression systems and has been designed for maximum compatibility with existing and future expression technologies:

(1) pQuattro vectors contain three multiple cloning sites harbouring up to 19 unique restriction sites (pQuattro2), 7 of which represent 8 bp target sequences for restriction endonucleases. The abundance of unique restriction sites present combined with the modular configuration of the functional genetic elements (promoter cassettes, *IRES* and *CITE* elements) avoids conflicting cloning situations arising from consecutive gene integrations, guarantees optimal compatibility with existing di- and tricistronic mammalian expression vectors (Dirks et al., 1993; Dirks et al., 1994; Rees et al., 1996; Fussenegger et al., 1998d) and allows the downsizing of pQuattro to tri- and dicistronic vectors by deleting individual *IRES* elements using either restriction sites contained in the MCS (*Pme*I, *Bcl*I/*Bgl*II; *Srf*I/*Swa*I; *BssH*II/*Mlu*I) or in the *IRES* elements (*Esp*I). The vector backbone as well as the neomycin resistance cassette are most easily replaced using a restriction site of the various MCS and the *Ssp*I (or *Xmn*I) sites contained in the promoter region of the ampicillin resistance gene (*bla*). Also, the pQuattro vectors are compatible to the pIRES*puro* (Clontech) vector for *Ssp*I/*Bgl*II replacement of the neomycin cassette with the puromycin resistance gene and to the dual selection marker vectors pNeoEGFP and pHygEGFP (Clontech) via *Ssp*I/*Mlu*I sites. Furthermore, the promoter cassette of pQuattro vectors can be replaced by corresponding *Ssp*I/*Eco*RI promoter elements contained in a great variety of existing mammalian expression vectors (Dirks et al., 1993 and 1994; Fussenegger et al., 1998). The modular structure of the pQuattro vector family allows straightforward adaptation for special applications and future expression technologies. Additionally, useful promoter-*IRES*I-*IRES*II, *IRES*I-*IRES*II, *IRES*I-*IRES*II-*neo* cartridges can be generated from the pQuattro vectors.

A single *Aat*II site at the 5′ end of the multicistronic expression unit which is present in all pQuattro vectors and allows further modifications and integration of additional genetic elements.

(2) The pQuattro family of multicistronic expression vectors allows the simultaneous and coordinated expression of up to three genes of interest which is most useful for (i) multigene engineering of com-

plex metabolic pathways (e.g. cell cycle and apoptosis or glycosylation; Fussenegger et al., submitted), (ii) evaluation of multiple gene functions (functional genomics), (iii) tissue engineering as well as cell and gene therapy, (iv) auto-regulated expression systems (Fussenegger et al., 1997b), (v) production of multiprotein complexes (e.g. antibodies), (vi) production of viral packaging cell lines (expression of *gag*, *env*, *pol*), (vii) multi-antigen/epitope (DNA) vaccination or (viii) the combination of antigens with immunomodulatory or -stimulatory components (adjuvants). Furthermore, the use of developmentally regulated *IRES* elements (Bernstein et al., 1997) in multicistronic expression units could superimpose additional translational regulation on multigene metabolic engineering approaches which would be especially valuable for technologies such as tissue engineering and gene therapy.

(3) pQuattro allows novel polycistronic ‘one-step auto-select’ expression technology. Since the selection marker is encoded by the terminal cistron and transcriptionally and translationally linked to the multicistronic expression unit (one-step multigene metabolic engineering), selection on neomycin ensures the presence of all 5′ cistrons (‘auto-select’ engineering) and exerts a positive selection pressure for the maintenance of all genes encoded on the multicistronic expression unit in an intact and transcriptional active chromosomal context. Furthermore, after selection in G418-containing cell culture medium, nearly all surviving colonies will stably express the gene of interest, thus decreasing the need to screen large numbers of colonies to find functional clones (Rees et al., 1996). Consequently, the expression of proteins which impair cell viability or genetic configurations imposing high counterselective constraints (such as controlled proliferation engineering, Fussenegger et al., 1997a) are expected to be selectable (Kaufman et al., 1991; Rees et al., 1996; Müller et al., 1997). Auto-selection also decreases the risk of genetic drift in bioprocesses (Rees et al., 1996).

Additionally, the *CITE* element contained in pQuattro harbours a reduction-in-function mutation which has been previously reported for the dicistronic expression vector pIRES*neo* (Rees et al., 1996; Clontech). The lower translation efficiency of the resistance cistron has been repeatedly shown to be compensated by the selection of chromosomal integration sites with overall high transcription activity (Kaufman, et al., 1991; Rees et al., 1996; Krömer et al., 1997).

Despite repetitive *IRES* elements, problems with genetic stability of the pQuattro vectors has never been observed neither during cloning or amplification in *E.coli* nor following transfection into mammalian cells (Fussenegger et al., 1997b and 1998). Taken together, the family of quattrocistronic mammalian expression vectors harbours the potential for coordinated multicistronic, auto-selective as well as regulatable expression of several transgenes in mammalian cells following a single engineering step. Owing to its modular set-up, pQuattro can be easily adapted to any state-of-the-art expression technology currently evaluated for use in tissue engineering and gene therapy.

Acknowledgement

This work was supported by the Bundesamt für Bildung und Wissenschaft (BBW) within the Framework IV Biotechnology Program of the European Commission and the Swiss Priority Program in Biotechnology (SPP BioTech).

References

Berger J, Hauber J, Hauber G, Geiger R and Cullen BR (1988) Secreted alkaline phosphatase: A powerful new quantitative indicator of geneexpression in eukaryotic cells. Gene 66: 1–10.

Bernstein J, Sella O, Le SY and Elroy-Stein O (1997) PDGF2/C SIS mRNAleader contains a differentiation linked internal ribosomal entry site (DIRES). J Biol Chem 272: 9356–9362.

Crameri A, Whitehorn EA, Tate E and Stemmer WPC (1996) Improved green fluorescent protein by molecular evolution using DNA shuffling. Nature Biotechnology 14: 315–319.

Dirks W, Wirth M and Hauser H (1993) Dicistronic transcription units for gene expression in mammalian cells. Gene 128: 247–249.

Dirks W, Schaper F, Kirchhoff S, Morelle C and Hauser H (1994) A multifunctional vector family for gene expression in mammlian cells. Gene 149: 387–388.

El-Deiry WW, Tokino T, Velculescu VE, Levy DB, Parsons R, Trent JM, Lin D, Mercer WE, Kinzler KW and Vogelstein B (1993) WAF1, a potential mediator of p53 tumor suppression. Cell 75: 817–825.

Fussenegger M, Mazur X and Bailey JE (1997a) A novel cytostatic process enhances the productivity of Chinese hamster ovary cells. Biotechnol Bioeng 55: 927–939.

Fussenegger M, Moser S, Mazur X, Bailey JE (1997b) Autoregulated multicistronic expression vectors provide one-step cloning of regulatedproduct gene expression in mammalian cells. Biotech Prog 13: 733–740.

Fussenegger M, Bailey JE, Hauser H and Müller PP (1998a) Genetic optimizationof recombinant protein production by mammalian cells. TIBTECH (in press).

Fussenegger M, Schlatter S, Dätwyler D, Mazur X and Bailey JE (1998b) Controlled proliferation by multigene metabolic engineering enhancesthe productivity of Chinese hamster ovary cells. Nature Biotechnology 16: 468–472.

Fussenegger M, Mazur X, Renner WA and Bailey JE (1998c) Higher productivity of growth-arrested Chinese hamster ovary (CHO) cells expressing the cyclin-dependent kinase inhibitor p27. Biotechnol Prog (in press).

Fussenegger M, Mazur X and Bailey JE (1998d) pTRIDENT, a novel vector family for tricistronic gene expression in mammalian cells. Biotech Bioeng 57: 1–10.

Kaufman RJ, Davis MV, Wasley LC and Micknick D 1991. Improved vectors for the stable expression of foreign genes in mammalian cells by use of the untranslated leader sequence from DMC virus. Nucl Acids Res 19: 4485–4490.

Krömer WJ, Carafoli E and Bailey JB (1997) Purification of the cardiacsarcoplasmic reticulum membrane protein phospholamban from recombinant *E. coli.* Eur J Biochem 248: 814–819.

Müller P, Kirchhoff S and Hauser, H (1998) Sustained expression in proliferation controlled BHK-21 cells. in New developments andapplications in animal cell technology, Merten OW, Perrin P and Griffiths JB (eds.), Kluwer Academic Publishers, Dordrecht, The Netherlands (in press).

Papoutsakis ET (1998) Express together and conquer. Nature Biotechnology 16: 416–417.

Rees S, Coote J, Stables J, Goodson S, Harris S and Lee MG (1996) Bicistronic vector for the creation of stable mammalian cell lines that predisposes all antibiotic-resistant cells to express recombinant protein. BioTechniques 20: 48–56.

Printed in the United States
212885BV00004B/2/P